INFLATION, INSTITUTIONS AND INFORMATION

Axel Leijonhufvud

Inflation, Institutions and Information

Essays in Honour of Axel Leijonhufvud

Edited by

Daniel Vaz
Professor of Economics
University of Uruguay

and

Kumaraswamy Velupillai
Professor of Economics
The Queen's University of Belfast, and
Senior Research Fellow
Statistics Norway

First published 1996 by
MACMILLAN PRESS LTD
Houndmills, Basingstoke, Hampshire RG21 6XS
and London
Companies and representatives
throughout the world

ISBN 0–333–61772–X

A catalogue record for this book is available
from the British Library

10 9 8 7 6 5 4 3 2 1
05 04 03 02 01 00 99 98 97 96

Printed and bound in Great Britain by
Ipswich Book Co Ltd
Ipswich, Suffolk

Contents

List of Tables

List of Figures

Notes on the Contributors

Michael Bordo is Professor of Economics at Rutgers University and an Associate at the National Bureau of Economic Research. He has held a Professorship at the University of South Carolina; he has also been a Consultant to the IMF at various intervals in recent years.

Shu-Heng Chen is an Associate Professor of Economics at the National Chengchi University in Taipei and a member of the editorial board of the *Taiwan Journal of Political Economy*, a journal managed from the department of economics at his University.

Robert W. Clower is the Hugh C. Lane Professor of Economic Theory at the University of South Carolina. He is also an Honorary Fellow of Brasenose College, Oxford and Emeritus Professor of Economics at UCLA. He was a Rhodes Scholar and has been the Managing Editor of *Economic Inquiry* (1973–9) and the *American Economic Review* (1980–5).

Arthur DeVany is Professor of Economics at the University of California at Irvine. He has taught at Texas A&M University. Ever since his days of doctoral studies at UCLA, where he was taught by Axel Leijonhufvud, he has considered himself a student of this gentle *Maestro*.

Luis Hernandez-Aramburo is a Research Fellow in the Economic Division of the Landeszentralbank in Hessen.

Daniel Heymann is Professor of Economics at the University of Buenos Aires and the Istituto Torcuato Di Tella. He is, concurrently, a Senior Economist at the Buenos Aires office of the UN Economic Commission for Latin America and the Caribbean (ECLAC). He has been a Visiting Professor at UCLA several times. He gave the second Arne Ryde Lecture (jointly with Axel Leijonhufvud) at the University of Lund.

Peter Howitt is Professor of Economics and Bank of Montreal Professor of Money and Finance at the University of Western Ontario. He is also an Associate of the Canadian Institute for Advance Research. He is a Past President of the Canadian Economic Association, a Fellow of the Royal Society of Canada and a Member of the Board of the *Journal of Economic Literature*.

Lars Jonung is Professor of Economics at the Stockholm School of Economics and was the Chief Economic Adviser to the Prime Minister of Sweden in the period 1992–4. He has been a Professor at Lund University in Sweden and has taught at the University of Virginia (Charlottesville).

Jenn-Shyong Kuo is a doctoral student in the department of accounting at the National Taiwan University.

Spiro J. Latsis is Chairman of Euromerchant Bank and President of SETE SA (Société d'Etudes Techniques et Economiques SA). He has taught at the London School of Economics, Harvard University and Boston University. He is a member of the Court of Governors of the London School of Economics and Political Science.

Chu Chia Lin is the Chairman of the Department of economics at the National Chengchi University. His publications include *An Empirical Study of Taiwan's Investment on Mainland China* and *Taiwan's Investment Experience in Mainland China.*

Francesco Luna is an Assistant Professor of Economics at the University of Venice and a Research Fellow at LATAPSES in the University of Nice at Sophia-Antipolis. He has been a Research Fellow of OFCE in Paris.

John McCall is Professor of Economics at UCLA and at the Queen's University of Belfast in Northern Ireland. He has been a Fellow of the Hoover Institution in Stanford and a Resident Consultant at the Rand Corporation in Santa Monica. He has taught at several European and Asian Universities.

Lionello Punzo is Professor of Economics at the University of Siena and the Director of the Centre for Research on Central–Eastern European Countries and the Former Soviet Union at the University. He is also the Co-ordinator of the Research Network for Computable Economics financed by the EU under its Human Capital Mobility programme. He has taught in Bielefeld, Vienna and UCLA.

Christof Rühl is an Assistant Professor of Economics at UCLA and Professor of Economics at the Central European University in Prague.

Richard Sweeney is the Sullivan/Dean Professor of International Finance at Georgetown University in Washington, DC. He is a co-editor of

Economic Inquiry and is an associate editor of several journals. He has taught at UCLA, Texas A&M and the Claremont Colleges, and worked in International Affairs at the US, Treasury.

Daniel Vaz is the Director of the Research Department at the Central Bank of Uruguay and Professor of Economics at the University of Uruguay in Montevideo. He is also the Head of the Delegation of the Central Bank of Uruguay at the MERCOSUR plenary sessions.

Kumaraswamy Velupillai is a Senior Research Fellow at Statistics Norway and Professor of Economics at the Queen's University in Belfast. He was the founding Associate Director of the Center for Computable Economics at UCLA and has been a Professor of Economics in Fiesole and Aalborg. He gave the fourth Arne Ryde Lecture at the University of Lund.

1 Introduction

Kumaraswamy Velupillai and Daniel Vaz

Macroeconomics emerged from IS–LM complacency as a result of the Clower-Leijonhufvud revolution in the 1960s. Three decades have passed since the decisive contributions by these two macroeconomic masters and turbulence has reigned in the subject during this period, not least because of the fundamental reorientations they pioneered.

It was fitting, then, that Robert W. Clower was present, on the occasion of a splendid celebration in honour of Axel Leijonhufvud's (henceforth AL) 60th birthday on 6 of September 1993, to give the lead talk in the grand aula of the Central Bank of Uruguay in Montevideo.

We are greatly indebted to the Central Bank of Uruguay, to its distinguished President, Dr Ramon Diaz, and to the honourable Minister of Finance of the Government of Uruguay for making this memorable event possible; and for making it a success in the grand traditions of Latin-American hospitality. All participants were met with graciousness, guided with immaculate consideration and treated with immense courtsey. Ordinary words of gratitude are woefully inadequate to express our sense of wonder, happiness and obligation, but these must suffice for the moment.

As joint co-ordinators of a truly satisfactory academic conference, we are also very grateful to all the distinguished participants who responded to the invitations with a promptness and willingness that befits the man we gathered to honour. All the speakers were close friends, former colleagues or students of AL at UCLA and other prestigious universities in Europe and Latin America. Indeed, it is a fitting tribute to a man who has given almost the whole of his professional life to the great academic institution that is UCLA, that every single contributor to this volume has been associated with it as AL's colleague or student at some point in time since the 1960s. It must be a unique distinction for a Festschrift.

Kumaraswamy Velupillai (KV) has known AL as a mentor, colleague and friend for more than twenty years; Daniel Vaz (DV) has known AL for a decade and more, as a student and friend. Legions of students and colleagues will testify to AL's legendary generosity, magnanimity and open-mindedness (see also Daniel Heymann's ruminations on such themes in chapter 14 of this volume). In a UCLA[1] that was itself legendary for fostering libertarian economics and espousing the virtues of untrammelled competition in a free market,[2] there were enlightened liberals such as Clower, Leijonhufvud and John McCall, among others, who kept asking the deeper questions about the foundations of the institutions that could support such philosophies. They posed awkward and serious queries about the limits of rational behaviour in environments that were dysfunctional and where information had to be husbanded with a finesse that required historical visions, not just expected-utility maximisation.

They argued, almost passionately, for a role for history, evolution and institutions in the fleshing of the analytical framework of the economist. They made their case with historical erudition, with a deep understanding of contemporary events and institutions, and with penetrating studies of the classics: from Vico to Marshall; from Oresme to Wiener; and from Laplace to Kolmogorov.

All those of us who were present at the conference had been schooled in the open-minded and eclectic tradition that AL fostered at UCLA, with sterling support from Clower, McCall and others. Many of us took these messages to other continents, but primarily to South and Central America: Argentina, Brazil, Chile, Mexico, Peru and Uruguay – to name but a few of the countries to which AL's students returned, after sitting at the feet of the master for their doctoral studies.

Many also came from the Old World Latin countries: Italy, France and Spain. Indeed it may not be an exaggeration to say that there is even a Leijonhufvud school in Italy, from the shores of which so many students and colleagues came to UCLA to be taught by, and to work with, AL on the shifting frontiers of macroeconomics. It is fitting, therefore, that this volume is produced on the eve of his probable semi-permanent move to the University of Trento. The news of new frontiers in economics will, in the forseeable future, come from the Dolomites, not the Sierras.

There is also the Far East. Students came from Taiwan, Korea and Japan in droves, and went back after working with AL to create a second generation of traditions and macroeconomic culture in their nascent (in the case of Taiwan and the Republic of Korea) but remarkable economies. Indeed, the *Taiwan Journal of Political Economy*, a new journal from the department of economics at National Chengchi University is, literally, a 'house' journal of the department of economics at UCLA. The editorial board is dominated by former UCLA students – most of them, directly or indirectly, AL's students – and present members of the economics faculty at UCLA!

All these continents, and countries were represented at the conference, in the papers collected in this volume and in the celebrations. It is a measure of AL's outstanding international reputation that there is such a flowering of second-generation AL economists all over the world.

ORGANISATION OF THE TEXT

The papers speak for themselves. In this concluding section of the introduction, we summarise briefly the structure and contents of the volume. We must, of course, begin this summary with a word of deep gratitude to the Hon. Dr Ignacio de Posadas, Minister of Finance in the Government of Uruguay, for having taken time off from his busy schedule to inaugurate the conference with a warm and erudite speech. This is reproduced in this volume as Chapter 2.

The volume is organised into five main parts and a subsidiary, somewhat methodologically oriented, opening part.

Chapter 3, by Kumaraswamy Velupillai and Lionello Punzo, is a reflection on the mathematical, philosophical and epistemological foundations of AL's economics, and his way of doing economics.

Spiro J. Latsis's concise, but sharp, chapter (Chapter 4) is a reflection on issues that have been at the centre of the *growth of knowledge* literature. Latsis worked closely with that mercurial philosopher of science, Imre Lakatos, who, in turn, deeply influenced AL's visions on methodology. It is apt, therefore, that we have the Latsis contribution in this volume. It was, after all, Latsis's intellectual and organisational skills that gave rise to the

> 1975 Conference on Growth of Knowledge theories that the Latsis Foundation arranged in Nafplion, Greece, with participation of philosophers, physicists and economists [which] was the *intellectually most rewarding conference I have ever attended.*
>
> (Leijonhufvud, 1981, pp. vii–viii; italics added)

Latsis's chapter is also a very personal piece; it puts in perspective, in relief so to speak, an almost lost aspect of the growth of knowledge wing of the philosophy of science: the nature of the impulses of Lakatos, the personal dimensions between the giants who fashioned this subject after Karl Popper–Thomas Kuhn, Imre Lakatos and Paul Feyerabend, and Paul Feyerabend, and the place of economics, its epistemology and methodology from these perspectives. We are privileged to have these reflections by Latsis for this volume.

In Part I, Robert W. Clower (Chapter 5) and Peter Howitt (Chapter 6) raise fundamental queries about the nature of macroeconomics. Clower, with characteristic wit and panache, challenges the textbook interpretation of the Keynesian Cross and, by way of a Marshallesque Cross persuasively argues for the Keynesian Crux to replace the orthodox cross. The argument's stated purpose, however, is not only a revamping of the teaching of elementary macroeconomics; it is also to reduce the tension that exists between what AL has called different *cosmologies* in economics (Leijonhufvud, 1985, esp. pp. 187–8) by describing the genesis of these differences in terms of the familiar tools of aggregative analysis. To put it in the language of AL (*ibid*., pp. 188 ff.) Clower's alternative crosses model the theories underlying different cosmologies in such a way that the tensions between the latter are reduced while the conflict of visions implicit in them are brought into relief, even graphically. In the process, we meet Clowerian vignettes of classic authenticity,[3] and even learn the source of the phrase 'fumbling fingers' that humanises the 'invisible hand'.

Howitt's theme is equally fundamental and, in a certain sense, tackles a related issue. 'Whatever happened to the Keynesian revolution' is a query that has often been posed by AL, in different contexts (and with fascinating answers each time). Clower went back to the basics to answer one set of aspects of such a question. Howitt trudges a different path with the same kind of aim.

Various forms of the orthodox code – generically termed the *neowalrasian code* by Howitt – transformed the economics of Keynes and the economics of information; themes lying at the heart of AL's early investigations. Howitt traces the intellectual metamorphosis of yet another of the conceptual underpinnings of AL's attempts to formulate the theory of a dynamic macroeconomy, peopled by agents with 'fumbling fingers': namely, Clower's theory of budget constraints. It is a masterly analysis, with a dose of methodological speculation too. But with even more fruitful alternative conceptual suggestions to reposition Clower's theory of budget constraints where it was intended to belong: the theory of effective demand in a monetary macroeconomic framework, where observationally meaningful trade takes place.

Since the early 1980s, AL has been broaching a theme in a way Borges is known to have speculated in other contexts: about roads not taken and imaginative thought experiments; about possible worlds and alternative outcomes in imagined cosmologies. Not just as an exercise in intellectual curiosity, but as a way of extracting possible worlds when the actual world of theories and experiments reach an impasse. Such issues are discussed by John McCall (Chapter 7) and Francesco Luna (Chapter 8), in Part II.

McCall, a lifelong colleague and friend at UCLA, has been investigating these worlds – of roads not taken – in precise ways by going back to the great Vico. McCall has done more than almost any other economist to study the role of information in evolutionary institutions as it pertains to the ordinary decision processes of economic agents. He traces the origins of *his* thoughts, and the framework *he* has built, in the roads that were pointed out by Vico, but were not taken by most social scientists. In doing this, he links these ideas to AL's evolutionary and interdisciplinary concerns, implicit in the main body of AL's *magnum opus*, but explicit in the first of two postscripts in that famous book. This is vintage McCall: vast erudition, massive capacity to link seemingly unrelated contributions to the study of information and institutions in an evolutionary way, and, above all, the uncanny ability to pinpoint the roads not taken.

In yet another of his many incarnations, AL has been trying to formalise, combinatorially, the idea behind Adam Smith's notion of increasing returns in manufacturing production. These interests have been the focus of some of AL's work for a number of years, and began long before the so-called *new growth theory* legitimised such concerns (but emphatically not in a combinatorial way). However, this is only one aspect of his deep interest in Smithian themes. The fumbling fingers alternative to the invisible hand is another.

Luna's elegant analysis traces a path back along the decision tree of conceptual developments in economic theory, to where Smith was a node (indeed almost the 'root' of the tree, in graph theoretic terminology), with many paths emanating from it. Luna then evaluates the different contexts in which Smith developed, and repeatedly used, the fertile concept of the invisible hand, to give us new insights in the uses of the past (Leijonhufvud, 1991). In particular, a refreshing look at Smith the supreme *inductor*, who was subverted by David Ricardo, the master *deductor*, who mesmerised the whole profession into employing that method for the purposes of economic analysis.

The question of monetary regimes has a distinguished Swedish pedigree. Erik Lindahl's framework for the theory of economic policy was against an explicit backdrop of monetary regimes[4] as *the* macroeconomic setting in which monetary authorities acted. AL has followed this tradition in the footsteps of his distinguished Swedish predecessor; in particular, for discussing the 'costs and consequences of inflation' and, eventually, to reduce such costs and consequences to the real economy.

In the first contribution of Part III, Chapter 9, Michael Bordo and Lars Jonung subject AL's postulates on monetary regimes, and much else that these are predicated upon, to a searching empirical analysis. In this empirical testing of theoretical visions, it is refreshing – and certainly not surprising, at least to us – to note that AL's vision, his cosmology, is amply vindicated.

Knut Wicksell was, of course, Lindahl's *maestro*, and AL treads the path carved by his two great Swedish predecessors in almost every question of monetary macroeconomics. The so-called new monetary economists are the newest school to claim Wicksell as a forefather. This is on the basis of an inadequate understanding of his famous thought experiment of a pure credit economy. In the other chapter in Part III, Chapter 10, Luis Hernandez-Aramburo tries to set the record straight on the issue of price stability in a cashless economy by sorting out the theoretical bases of the question of whether a cashless economy can be said to replicate the behaviour of a barter economy. Hernandez-Aramburo's discussion, on this issue, highlights yet another of AL's conceptual bases: co-ordination of disparate decisions. Moreover, in his more recent work on high and hyperinflationary economies, AL's analytical concerns have impinged upon the role of the medium of exchange and standard of value functions of money and index number problems in general. It is refreshing to see Hernandez-Aramburo's explication of Wicksell's views on this in the context of a rational reconstruction of a debate, as a thought experiment, with Eugene Fama and his followers. We concur, for what it is worth, with Hernandez-Aramburo's penetrating critique of Don Patinkin's interpretation of Wicksell. It is, to some extent, similar to lines of argument pursued by Professor Guglielmo Chiodi of the University of Perugia in a series of Wicksellian writings stretching over a period of fifteen or twenty years.

Many of us, under the UCLA umbrella, have long awaited the possible publication of a novel intermediate macroeconomics text by AL and Richard Sweeny. KV has used some of the chapters with great success at UCLA and in Prague. In the absence of that long-awaited text, the best we can offer is a solid piece by Sweeny, Chapter 11 in Part IV, on a key issue in AL's work in monetary macroeconomics: the interaction between expectation and speculation. We should point out and that, in our opinion, the agents in AL's economies are non-rational expectationists; along the same lines, we would contend that the market economies in which non-rational expectational agents act do not satisfy the celebrated *efficient market hypothesis* (EMH). How could they, if the agents, with fumbling fingers, are non-rational expectationists? In measured tones, Sweeny surveys some of the empirical literature, in financial economics and open-economy macroeconomics, to investigate AL's visions on the destabilising role of speculation in the macroeconomy. AL has long maintained that persistence of far-from-equilibrium configurations characterise dysfunctional economies. Sweeny looks at the evidence on the persistence of destabilising speculation and finds that there may be support for AL's conjectures, at least in dysfunctional economies.

AL had – *has* – very high personal and professional regard for the economics and philosophy of Friedrich von Hayek. He counted Hayek and Sir John Hicks among his friends and *maestros*. His own philosophy was, perhaps, more akin to that espoused by the latter rather than the former. AL, like Hicks, probed the outer limits of the normal; somewhat like studying counter-examples to theorems as mathematicians do (rather than investigating the *para*normal as psychics do).[5] It brings into sharp relief the boundaries of the hypotheses; of the axioms; and of the concepts. It is not simply an observational belief that imperfections are the rule – in rational behaviour, in markets and in policy, but it is as if the study of pathologies guides an understanding of the healthy.

Hayek felt that local imperfections of various sorts are smoothed out by the impersonal forces of the market in the aggregate. AL and Hicks, on the other hand, are not sanguine about the smoothing properties of market forces. In very recent years, AL, now in his incarnation as the director of UCLA's Center for Computable Economics has been studying the possibility of simulating actual economics, peopled by a large number of interacting agents with 'fumbling fingers', as parallel distributed processors, rather than the central processing unit metaphor conveyed by the *Walrasian Auctioneer*.

In the second chapter in Part IV, Chapter 12, Shu-Heng Chen and his co-authors model a quintessential AL system in the setting of experimental double auction markets. They suggest, tentatively, that perhaps Hayek was wrong after all. Hayek, however, has interesting ways of being wrong, and this is clearly one of them. But we will not dwell upon that theme in this introduction.

Finally, in Part V, Chapter 13, Art DeVany, in a characteristically imaginative piece, investigates co-ordination in a non-general equilibrium world; and Daniel Heymann (Chapter 14) and Christoph Rühl (Chapter 15) 'descend' into the messy and complex world of experiences in pathology in the Southern Cone and the Eastern Chill respectively.

DeVany takes up the challenge of substantiating the title of this volume within the framework of an economy characterised by Axelian elements: co-ordination failures, increasing returns, information transmission, inflation, the structure of the market as an institution, and the corridor hyothesis; and all this against the backdrop of explicit dynamic considerations. It is very much in the philosophical spirit that Velupillai and Punzo identify as being characteristically Axelian: the search for the outer limits of standard formalisms and frameworks. In this sense, DeVany sets the scene for the ensuing pieces by Heymann and Rühl in this part.

Heymann's lucid cataloguing of possible pathologies, the inadequacy of standard theoretical technologies to characterise them, and hopeful alternatives all follow the classic lines of argument AL has pursued in all his professional writing, teaching and conversation. It is written from the perspective of a first-rate analyst who has soiled his hands repeatedly in the world of policy in massively dysfunctional economies.

Rühl, on the other hand, sets the emerging market economics east of Sweden, and their analysis, in frameworks that have had AL's blessings in more recent years. In particular, the emergence of complex economic systems viewed from the point of view of institutions. Rühl emphasises the important fact that to understand and implement transitions it is necessary to have a firm historical grasp. The legacy, particularly the institutional legacy, of decades of central planning can thwart all well-intentioned schemes of reform and revision. He, interestingly, wonders about the sub-optimal nature of vertical integration implied by the activities of a rational central planner. This may be contrasted with DeVany's conclusion about the need for markets to be hierarchical for overcoming co-ordination failures in systems with interrelated, complex, production structures.

It is not always an easy task to edit a book. In this case, however, it was not only a privilege, it was also a tremendous pleasure. We are paying homage to a princely macroeconomist who continues to explore new frontiers: analytically, conceptually and methodologically.

In paying homage to AL we have not forgotten the role played by close members of his family in sustaining the cheerfulness, optimism and generosity that has been the hallmark of a life that has itself been, at times, difficult. Foremost, of course, his wife, Earlene Craver-Leijonhufvud; then there are his children: Karl, Gabriella and Christina. At the roots, his mother and sisters. To them, too, we are grateful and, in a way, the book is a celebration of that part of his life that they filled with happiness of a different kind.

One final indebtedness must be acknowledged: to the respective research and academic institutions that have housed us during the period that this book was in preparation. DV wishes to acknowledge the sterling support – material, financial and human capital – provided by the Central Bank of Uruguay and, in particular, its research division. KV, on the other hand finishes these lines with the majestic Lake Como beneath him and the imperial Italian alps all around him to remind him of the exciting years he spent at UCLA as Associate Director to AL at the Center for Computable Economics in the department of economics. There he had the gentle, but toiling, hands of Caroline Shin to take so much of the administrative and other chores off his rough hands. But much of the 'penultimate' logistics were completed as Senior Research Fellow at Statistics Norway. The enlightened freedom made available to him by Dr Olav Bjerkholt, Asst. Director General of Statistics Norway, was crucial at a decisive moment in the preparation of the manuscript. To him, too, he is grateful.

Notes

1. Sometimes fondly deciphered as the *U*niversity of *C*hicago at *L*os *A*ngeles!
2. As AL noted with characteristic wit, in the preface to Leijonhufvud (1981, p. viii): '*In the UCLA spirit, responsibility for remaining errors will be allocated to the highest bidder*'!
3. Example: fn 10, where we find a priceless comment, a part of which we cannot resist quoting here: 'before this footnote becomes pointless, let me make its point'.
4. A monetary regime is defined as a system of expectations that governs the behaviour of the public and that is sustained by the consistent behaviour of the policy-making authorities (Leijonhufvud, 1983, p. 3), a definition that could also have come from any of Lindahl's classics from the 1920s.
5. It may be apposite to quote Clower's closing lines in his contribution to this volume: The more difficult question is whether self-adjustment capabilities of modern economies are strong enough to withstand prolonged periods of economic adversity. That question can be answered only by a gifted *psychic*. (italics added)

References

Leijonhufvud, A. (1981) *Information and Coordination: Essays in Macroeconomic Theory* (Oxford University Press).

Leijonhufvud, A. (1983) *Inflation and Economic Performance* (Tübingen: J. C. B. Mohr (Paul Siebeck)).

Leijonhufvud, A. (1985) 'Ideology and Analysis in Macroeconomics', in Peter Koslowski (ed.) *Economics and Philosophy* (J. C. B. Mohr (Paul Siebeck)). Tübingen.

Leijonhufvud, A. (1991) 'The Uses of the Past'. Paper written for the History of Economics Society Conference, Harvard University, Cambridge, Mass., June 1987 (revised November 1991).

2 Opening Speech

Hon. Dr Ignacio de Posades

The principal difficulty in introducing the economist in whose honour this conference is being held is not in his work, which is well known and, in spite of its depth and originality, quite clear. The real difficulty lies in pronouncing his name. One way of getting around this was hit on by Professor Robert Solow of MIT who, when asked how to say the economist's name said, 'That's simple, it's pronounced Axel.' Others have resorted to the English translation of his Swedish family name, calling him Professor Lionhead, which in any case does not sound so different from the original.

Professor Lionhead's prominence dates back to his doctoral thesis, entitled 'On Keynesian Economics and on the Economics of Keynes'. This work, carried out by means of a shrewd and patient effort, sought to separate out the lasting content of the work of the great twentieth-century economist from current interpretations which have become common knowledge. Axel's thesis seems to be full of paradoxes. One after another he shows us that the most typical Keynesian propositions cannot be found in Keynes at all. Much of Keynes' original thought was never picked up by the conventional version of his doctrine.

The dichotomy posed by Professor Lionhead is supported by Keynes' own words. In the final stages of the Second World War, when Keynes was working with economists of the Treasury Secretary to create the monetary system later consecrated in Bretton Woods, he wrote to his wife in London that, 'Last night I went out to dinner with four treasury economists; the only non-Keynesian was myself.'

Shortly thereafter, Keynes passed away. In the following years his thought, which had been influential, though controversial, for more than a quarter of a century, began to shape a new orthodoxy. In the process, those features of his doctrine which directly opposed old neoclassic orthodoxy grew, assuming rather absurd proportions. 'Currency does not matter' became the motto of typical Keynesians in the 1960s. Of course, interest rates were not important either, neither to determine investments, nor to determine consumption. One of Professor Lionhead's greatest contributions lies in his demonstrating that this triviality of interest rates can be found among Keynesians, but not in Keynes.

This was part of the word carried out by Axel and a handful of authors who were able to extract Keynes' thought from what had become an oversimplified doctrine containing a strong element of voluntarism. Axel redeemed Keynes' economic thought, which had been lost in the wanderings taken by the Keynesians who created this oversimplified and voluntaristic version.

While this work alone assures our illustrious guest a place in the history of economic thought, it is obviously by no means his only contribution. However, it would be presumptuous on my part to tell the members of this audience of Axel's academic background and endeavours; of the works he has authored and of his past and current activities in UCLA; of his often controversial relationships with economists of varied orientations, Keynesians as well as non-Keynesians. I have attempted to point out those aspects which in my limited view particularly justify the pride I feel upon inaugurating this conference, and being able to welcome among us such an outstanding figure of international economic theory.

3 Co-ordinating the Not-Too-Rational: Leijonhufvud's Macroeconomic Methodologies

Kumaraswamy Velupillai and
Lionello Punzo

Axel Leijonhufvud (AL) has almost always worked at the frontiers of macroeconomics; in his early professional career the frontiers of monetary macroeconomics were defined and delineated by his work. However, almost immediately after his *magnum opus*, he began introducing a series of conceptual innovations as desiderata in the formalisms of market dynamics, monetary regimes, macrodynamics, rational behaviour, consumer theory, production theory and also doctrine and monetary history.

To take the last mentioned item, just as an example, his *magnum opus* and some of the essays in *Information & Co-ordination* (I&C) – in particular, chs 7 and 8 – are considered to be classics, not only in modern monetary macroeconomics, but also as fundamental contributions to doctrine history in that field. But, in a certain sense, this is orthodox doctrine history.

On the other hand, his work on 'Capitalism and the Factory System' (AL, 1986) and 'The Uses of the Past' (AL, 1991) break new ground also in the methodology of doctrine history. The former, while going back to Adam Smith's discussion for a setting, suggests new conceptual and mathematical formalisms and implies novel ways of handling *increasing returns to scale* in a model of a multiagent market economy. The latter, on the other hand, gives the methodological underpinnings for doing what he did in going back to Smith and coming forward to the frontiers of analytical economics.[1]

This chapter is therefore structured as follows. In the next section we dabble in methodological ruminations. We try to find out where we can place AL, as an economic scientist,[2] in a methodological stratification.

Having discussed, in the next section, how and why AL introduced novel new concepts, we then ask why they have not become part of *normal science* (*pace* Kuhn) or the *positive heuristic* (double *pace* Lakatos) of a core[3] and suggest that the reason is that AL's intuition ran ahead of the theoretical technology available to the economist, even at the frontiers of economic analysis, when he introduced the revolutionary concepts.[4] We also think that the time is yet to come for these conceptual innovations to be absorbed into the normal scientist's repertoire of tools, concepts and so on. This comment leads to our third section, where we try to discuss the nature of the code (see fn 3) that can absorb AL's individual conceptual innovations (and relate it to his work as the founding director of the newly-established Center for Computable Economics in the department of economics at UCLA).

The various threads are brought together in the concluding section of the chapter.

AL's METHODOLOGICAL FOUNDATIONS

We want to suggest that AL, were he a mathematician, would be a George Polya in the way he *does* economics; we have already hinted at this link in the Introduction to this volume. A Polya not only as a practising methodologist (if he tried to reconstruct – rationally or not – his way of doing economics in the sense of suggesting postulates, combining them into testable propositions and so on) – as Daniel Heymann also suggests in his insightful comments on the way AL even *looks* at the economic world around him; but also a Polya in the way he writes and teaches economics. He seeks suggestive metaphors to characterise the concepts of economic analysis when they have to be stretched to their definitional limits (for example, the *Walrasian demon* invoking images of the *Maxwellian demon* and what it can and cannot do); he searches for imaginative analogies to tame the complexities of economic activity.

As we have suggested, in passing, also a Polya in the sense of working at counter-examples, in its strict mathematical usage. Mathematicians use counter-examples to learn the limits, the outer boundaries, of their theorems and the building blocks of their mathematical edifices. But only mature mathematicians, those who have grasped, not just the formal façade of their mathematics, but also the intuitive content of their symbolic structure and methods of proof, can work at counter-examples. It is only when the intuitive content has been grasped that the new concepts can be introduced to go beyond the limits of the old; only when one has intuitively seen the centralised processing nature of the *Walrasian auctioneer* can one *think* of the limits inherent in the *Walrasian demon* and then seek the asynchronous parallel processor as a new metaphor.[5] Working at counter-examples, therefore, forces one to introduce new concepts, new tools to go beyond the official boundaries. At that point the craftsman and the technician takes over.

Sir John Hicks, Nicholas Georgescu-Roegen and Axel Leijonhufvud are the classic examples of the Polyas of economic analysis. We cannot easily think of any others.

We have felt, ever since one of us (KV) had the pleasure and privilege of listening to AL delivering his remarkable Marshall Lectures at Cambridge in 1974, that he introduced fundamental concepts such as the *corridor hypothesis*, the *Marshallian consumer*, the Walrasian demon, the fumbling fingers, the *complexity of contracts*, markets as *coupled oscillators* and much else into the repertoire of the analytical economist through his counter-example activities.

Simply put, if asked to advise the transition economies of Eastern Europe, or the hyperinflationary regimes of Latin America, the Hicks's and the Leijonhufvud's would not suggest standard shock therapies; to clear a ravine, one could be asked to leap across, as with Sachs's advice; or one could use the age-old device of building a bridge – which takes time. AL would suggest the building of a bridge because of his deep knowledge and interest in the *limits* of the institutions that characterise market economies, the evolutionary origins of the institutions, and the nature of the behavioural basis of agents who have adapted to such institutions. The building of the bridge gives the breathing space for the adaptation, for the learning that is inherent in the processes that underline market dynamics.

On the basis of these general observations on the philosophy underlying AL's way of doing economics, interpreting the world around him and then acting on it, is it possible to pigeonhole him as a methodologist in one of the standard ways suggested by the growth of knowledge wing of the philosophy of science literature? AL has, himself, reflected upon Kuhn's visions and Lakatos's worlds (as we have also mentioned in the Introduction to this volume).

But we are not satisfied with the sociological bases of Kuhn's normal scientists and cannot fit the way AL has worked into that framework of paradigms and other much maligned concepts. Normal scientists do not work at counter-examples; they work *with* examples. That, after all, is the *raison d'être* of the paradigmatic world of normal science.

What, then, about the Lakatosian worlds of *MSRP* and *Proofs and Refutations*? Can one of these worlds encapsulate the way AL has tried to interpret and theorise about economic activity? For this to be possible a case must be made that AL accepts the standard core: that is, the neowalrasian code. For example, an element in the core of the neowalrasian code, is preferences; but AL's Marshallian consumers with fumbling fingers do not maximise preferences; indeed, they may not even *have* preferences in the traditional sense. In other words, the Lakatosian world of MSRP cannot be harnessed for the purposes of AL's economics.[6] Is there, then, a sense in which the world of Lakatos's Methods of Proofs and Refutations can be harnessed for aspects of the way AL does his economics? We reject this alternative too, but obviously for different reasons. A prime reason is the following. Herbert Simon would like us to be problem solvers,[7] with a repertoire of heuristics at our disposal. In this sense, *his* methodology would fit squarely into the Lakatosian scheme of Proofs and Refutations where mathematicians try to find rules for solving problems. But that is emphatically *not* what AL (or, for that matter, Hicks) does. He works at counter-examples using induction and analogies. There is more to this rejection of the Lakatosian worlds than we can say here, and we must reserve it for a different exercise.

Finally, we return to Polya and his world of plausible reasoning, induction and analogies. Polya contrasted plausible reasoning with demonstrative reasoning. Clearly, the latter uses deductive logic and formal methods, not induction and analogies. Polya wanted to stress that 'There are two kinds of reasoning ... *demonstrative reasoning* and *plausible reasoning*' (Polya, 1954, p. vi; *italics added*). He had led the reader to this distinction after prefacing his discussion with the following observations:

> Strictly speaking, all our knowledge outside mathematics and demonstrative logic ... consists of conjectures ... We secure our mathematical knowledge by *demonstrative reasoning*, but we support our conjectures by *plausible reasoning*. A mathematical proof is demonstrative reasoning, but the inductive evidence of the physicist, the circumstantial evidence of the lawyer, the documentary evidence of the historian, ***and the statistical evidence of the economist belong to plausible reasoning***.
>
> The difference between the two kinds of reasoning is great and manifold. Demonstrative reasoning is safe, beyond controversy, and final. Plausible reasoning is hazardous, controversial, and provisional. Demonstrative reasoning penetrates the sciences just as far as mathematics does, but it is in itself (as mathematics is in itself) incapable of yielding essentially new knowledge about the world around us. Anything new that we learn about the world involves plausible reason-

> ing, which is the only kind of reasoning for which we care in everyday affairs. (Ibid., p. v; bold italics added)

But he also pointed out that:

> [demonstrative reasoning and plausible reasoning] do not contradict each other; on the contrary, they complete each other. (Ibid., p. vi)

It is in this almost precise sense that we believe AL reasons plausibly towards his tentative postulates. At the next stage he employs analogies to construct (literally) toy models, at which point he employs principles of induction, as formalised in the modern theory of induction (see Velupillai (1994), Appendices), to suggest testable implications *and* the possible constructive mechanisms that can encapsulate the testable implications.

But, to go back to Polya, and the Polya in AL, we claim that *one* of the senses in which demonstrative reasoning and plausible reasoning complement each other is via the role of counter-examples in deriving or stating propositions of a general nature:

> A *counter-example* explodes *a* statement purporting to apply to all objects of a certain category: The counter-example is an object of the proper category to which the allegedly general statement does *not* apply … A problem to prove is concerned with a clearly stated assertion *A* of which we do not know whether it is true or false: we are in a state of doubt. The aim of the problem is to remove this doubt, to prove *A* ***or to disprove it*** … we face a decision: should we try to prove the assertion *A* or try to disprove it? We have here a choice between two different directions. To prove *A* we should look for some propositions from which, or for some strategy by which, we could derive *A*. ***To disprove A we should look for a counterexample***. (Polya, 1981, pp. 49–51; bold italics added)

It is in this sense that we find the Polya in AL; that is, using plausible reasoning in the search for counter-examples which 'explodes a statement purporting to apply to all objects of a certain category'; using induction and analogies to construct[8] 'an object of the proper category to which the allegedly general statement does *not* apply'.

Where, then, are the counter-examples in AL? They can be found in almost every single piece of writing by him. More concretely, for example, the way he suggests the postulate of the Marshallian consumer and then, from that step, towards the type of observable behaviour such a construct depicts and then back, by induction, towards a behavioural postulate to supplant maximisation which is based on this 'object [the Marshallian consumer] of the proper category to which the allegedly general statement [of maximization] does *not* apply'.

Another counter-example is the way coupled oscillators are used to discuss the simultaneous adjustment of prices and quantities[9] and the nature of the stability properties of interacting markets, and then back, by way of induction, to the sort of behavioural postulates that would sustain an aggregate economy with all these elemental ingredients.

A third counter-example is the way AL conceived a formalism for modelling increasing returns to scale, by going back to Adam Smith, but reading it again in the light of the deficiencies of standard production theory to be consistent with some of the broader facts of economic history and Smith's predictions. Here, plausible reasoning suggested an entirely new type of mathematical representation of production techniques, and the analogy with parallel processing architecture reinforced the plausible reasoning to which a further dose of conventional historical induction and the modern theory of induction were then applied. Finally, AL suggests how this formalism can be used to interpret some of the stylised facts of contemporary economic events.

Then there is the brilliant suggestion of the *corridor hypothesis*. Here is how AL proposed the postulate:

> The system is *likely* to behave differently for large than for moderate displacements from the 'full co-ordination' time-path. Within some range from the path (referred to as *'the corridor'* for brevity), the system's *homeostatic mechanism* works *well*, and deviation-counteracting tendencies increase in strength. Outside that range these tendencies become weaker as the system becomes increasingly subject to 'effective demand failures'. (Leijonhufvud (1981), pp. 109–10; italics added)

We have emphasised the key words and phrases that show the Polya in AL. First of all, a plausible statement is suggested: a reasonable guess – a guess that is more reasonable than not. And the reasonableness is gauged by the facts of history. Next note the imaginative usage of analogies. Where is the place of induction in the strict sense of the *modern theory of induction*? That lies in the implicit parsimonious definition of effective demand and its failures. And the whole scheme is set against the backdrop of constructing, as usual, the 'object of the proper category to which the allegedly general statement does *not* apply'.

An additional feature of the Polya in AL can be discerned from the above quote. The suggested analogies must take into account the fact that behaviour must bifurcate: inside the corridor there is one type of behaviour; outside it, another, qualitatively different type of behaviour. This rules out certain types of formalism (linear, nonstochastic dynamics, for example) and suggests other types (nonlinear dynamics), but the basis of the modern theory of induction requires, via the theory of effective demand, parsimonious selection of models.

The explication of these examples, in the way Polya dissected his examples (or, as Lakatos did with *his* examples), is left as a suggestion for others to do a Polya on AL, in, perhaps, the commemorative volume that will be issued on the occasion of his 75th birthday in the year 2008.

ON CODIFYING AL's ECONOMICS

In working at counter-examples, AL felt compelled to introduce the fertile new concepts that we have referred to above. Information, co-ordination and institutions are the basic cementing media in which these new and fertile concepts were set in their paces. Computational, algorithmic and combinatorial complexity are the theoretical technologies for the media to encapsulate the concepts. For example, the Marshallian consumer would be defined algorithmically to be able to process information emanation from institutions, and co-ordinate activities with other algorithmic agents.

Why, then, is there no normal science in terms of such a scheme? A scheme where Marshallian consumers exist in a market economy where co-ordination is by fumbling fingers; where the Walrasian demon meets all the computational impossibilities that his cousin, the Maxwellian demon, is known to possess when information has to be processed algorithmically; where the economic system is capable of behaving classically only within a corridor of stability, outside of which dysfunctional behaviour is manifested widely. In a much more concrete way, and using concepts that have been transformed and incorporated into the neowalrasian code, Peter Howitt asks a related question in Chapter 6 of this volume. On page 72 there is an illuminating discussion of the way the neowalrasian code acts as 'black hole'. In fact, there is an orthodox way of describing Howitt's examples via the Lakatosian scheme of MSRP. In this case, the neowalrasian code is described by the standard neoclassical core (preferences, endowments and technology), and the *positive heuristic, the protective belt* and so on. AL's economics of information, his reinterpretation of Keynes's economics and his use of Clower's theory of budget constraints were, according to Howitt's persuasive analysis, adopted by macroeconomists with regard to the analytical apparatus they contain, but were rejected as far as the economic ideas that were at their roots. In Lakatosian terms of the MSRP, these become classic elements of the positive heuristic.

Howitt is, of course, aware of what has to be produced if the theories AL was advocating are to succeed: a new conceptual framework. He proceeds to outline some of the possible elements of such a new conceptual framework. Perhaps he has not gone far enough? We suggest a more general series of observations to achieve the same purpose, outlined more concisely and pungently by Howitt. We do this because of our fear that Howitt's own innovations will become part of the positive heuristic, of the neowalrasian code, almost before this volume is published. We feel that the only way out is to devise a whole new code; let us call it the MALE code (after MArshall + LEijonhufvud), provisionally or, alternatively, in the interests of neutrality, the fumbling finger (FF) code. We shall, in this essay, use the former, in view of the simplicity of the acronym and its mnemonic features.

What would the conceptual basis, formalism and theoretical technology of such a code be? We ask this question, keeping in mind the nature of these three for the neowalrasian code, and for its neoclassical core. Instead of given preferences (or their usual functional representations) we would have the Marshallian consumer represented algorithmically; for example, as a set of *decision rules*. They may or may not be given exogenously. It may well be that the task of the applied economist is to conduct sample surveys at Gelson's, say to extract a reasonable collection of sets of decision rules profiling different types of consumers. This is neither formally nor computationally a difficult task; but it makes a fundamental shift in the nature of the subject: it becomes explicitly experimental!

Instead of technologies being represented by production functions, in the classical sense – both economically and mathematically – we would represent them combinatorially and by partial functions (in the recursion theoretic sense) with consistent use of the imperative form in all the formalisms. This is, partly, the way Scarf (1981) has proceeded to formalise production. Optimisation problems become decision problems as in combinatorial optimisation and the economy will be pervasively one of fumbling fingers.

What of endowments, the third of the three elements of the neoclassical core and the neowalrasian code? We suggest that the MALE code would dispense with this in its core and, instead, characterise ownership, and the interdependence implied by it, more directly. The Marshallian consumers, as algorithmically rational agents (ARAs) and production processes represented combinatorially, say on a lattice, where nodes or lattice points are connected locally, as on a network, with evolutionary and other rules specified on the basis of historical observations, experimental results and intuitive arguments.

This code would process *and* represent information also algorithmically; it would co-ordinate, synchronously or asynchronously, as the case may be. The whole economy, as a lattice, and as a combinatorial object, would exhibit perfectly rational, albeit algorithmically rational, behaviour by ARAs that would, in turn, depict global dynamics that can persist in an out-of-equilibrium trajectory for considerable periods of time (*soliton* representations, for example). It could be wired, so to speak, such that minor perturbations, in the form of disturbances to a collection of local rules, do not lead to significant changes in the global dynamics (the corridor hypothesis).

It could be used to model competitive, co-operative, evolutionary and even altruistic behaviour (depending on the qualitative and quantitative bases for the local rules) at different levels of cohesion and coherence; that is, for different collections of lattice points (each point a production process or ARA, and they, in turn, lattices themselves, stacks upon stacks of them, or algorithms with collections of subroutines). Then information processing could be modelled laterally and transversally; synchronically and asynchronically. The limits of the computational and information processing capabilities of the Walrasian auctioneer, and his alter ego, the Walrasian demon, can be demonstrated both experimentally and theoretically. The latter, in terms of classic undecidability and uncomputability results of recursion theory.

But there are several details to be added, to fill in. The believer in the neowalrasian code may ask the cosmologist with faith in the MALE code for an explanation of the behavioural rules that ARAs use. In other words, what is the analogue, for ARAs, of maximisation (under constraints)? The MALE code cosmologist has a plausible answer: to your maximisation postulate, in a deductive system, the analogue in my cosmology is *simplicity* in an *inductive system*, as formalised in the modern theory of induction.

To this the neowalrasian code believer would retort: I can use Kuhn-Tucker, Lagrangians, Hamiltonians, the Principle of Optimality, the Maximum Principle and such like in maximising under different frameworks. What are the theoretical technologies that can implement simplification and induction systems such that all the behaviour we can explain can also be explained by your MALE code?

To this query, the MALE code cosmologist has an even more considered reply. First of all, says s/he, let me remind you that the Copernican system did not overthrow the Ptolemic cosmology simply because the former explained or predicted better than the latter; we chose, eventually, the former cosmology, in view of the simplicity of its postulates. I put it to you, says the MALE code cosmologist, that my code is simpler than your epicycle system of *ad hoc*eries, and, moreover, I can give you a quantitative measure of the degree to which my system is simpler than yours.

Secondly, continues the MCC (MALE code cosmologist), I not only have the theoretical analogues of your Kuhn-Tucker *et al.*, I also have an empirically implementable framework of approximations to my theoretical constructs. But to explain that to you, I must first give you a small lesson on the Modern Theory of Induction. For this I refer you to the Ryde Lectures given by one of us (KV). Once you have grasped that, I will explain Rissanen's MDL principle. So, instead of doing my theory on Mondays, Wednesdays and Fridays and then the applied, econometric, part of the theory by *ad hoc* assumptions on Tuesdays and Thursdays, I have a consistent, unified, framework for the whole code; so that both the theoretical and the empirical progress of the whole research programme can be evaluated with the customary rigour. I have, therefore, one unified framework, for the decision-maker as a consumer, producer, policy-maker, financier or whatever, for the theory and its empirical implementation.

The neowalrasian code believer is, naturally, reluctant to give up his core, the protective belt and so on. But even the Vatican finally absolved Galileo – although it took a matter of five centuries for them to see the light, so to speak. Our hope, and we think we echo AL's hopes and visions, is this: that the neowalrasian code cosmologists behave more like the Ptolemic physicists who accepted the Copernican system on aesthetic principles and work within the MALE code sooner than it took the Vatican to accept Galileo.

CONCLUDING NOTES

In earlier work we tended to formalise AL's conceptual innovations in isolation. Thus, when we tried to formalise the corridor hypothesis, it was as a bifurcation phenomenon in a suitable dynamical system. It was only recently that we were able to formulate his whole vision in one coherent, computable, framework. This is fitting, since AL is the founding Director of the first Center for Computable Economics in any academic establishment in the world – at least so far as we know.

None of the theoretical technologies are unduly demanding, compared to the *ad hoc* paraphernalia the average graduate student needs to master in any of the better graduate schools in the Western world. Such a student, when satisfied that s/he has mastered the Maximum Principle, has to suddenly change gear and master the Girsanov transformation, for which, first the Radon–Nikodym theorem must be understood, and so on *ad infinitum*.

On the other hand, to practice a research programme along MALE code lines requires a mastery of a wholly different way of doing economics: both theoretically and empirically. In this sense, therefore, we return, with sympathy, to a Lakatosian characterisation. We are, indeed, talking about research programmes in the sense that Lakatos did in MSRP. It is not a question of falsifying or replacing this or that postulate in the core or the protective belt of the neowalrasian code; it is the wholesale supplanting of one way of doing economics for another.

AL has given ample guidelines for the implementation of such a programme: the MALE code. It is, perhaps, going to be the research programme of UCLA's Center for Computable Economics under AL's direction.

Notes

1. Just for the record, we would also like to point out that AL's bracketing of his production formalism, in 'Capitalism and the Factory System', with that of Georgescu-Roegen's, is a brilliant insight for a very important *mathematical* reason too: both of these formalisms eschew the use of the traditional Dirichlet definition of a function, which economists blindly use in their work. But this is a topic of an entirely different type of work from the one we are trying to write here.
2. After all, the Nobel Memorial Prize is given for economic *science*!
3. Say, of the *neowalrasian code*, as defined and discussed by Howitt in his brilliant contribution to this volume.
4. This does not mean that they – the theoretical technologies – were not available in the general scientific literature of the time in mathematics, physics, computer science or cognitive psychology.
5. And also requires a vast intellectual culture and a basic feel for interdisciplinary richness, all of which characterise AL in his daily professional life of writing, teaching and conversing.
6. But it must not be forgotten that MSRP is to be used for supplanting whole research programmes, not piecemeal substitutions of isolated elements of the core.
7. Not problem solvers in the sense in which Kuhn defined normal science and its activities.
8. We have always felt that Polya was a natural *constructivist* in the strict mathematical sense; that AL, too, is a natural constructivist not only follows from our 'Polya in AL' story – but also, even more clearly, from the way AL derives propositions in economics: there is never a demonstration that

relies on proof by contradiction, nor any reliance on the law of the excluded middle in the derivation of general propositions (such as indexing over a continuum of agents).

9. Conflating Walras and Marshall!

References

Leijonhufvud, A. (1981) *Information and Co-ordination* (Oxford University Press).

Leijonhufvud, A. (1986) 'Capitalism and the Factory System', in *Economics as a Process: Essays in the New Institutional Economics*, ed. Richard Langlois, (Cambridge University Press).

Leijonhufvud, A. (1991) 'The Uses of the Past'. Paper written for the History of Economics Society Conference, Harvard University, Cambridge, Mass., June 1987. (Revised November 1991).

Polya, G. (1954) *Mathematics and Plausible Reasoning*, vol. 1 (Princeton, NJ: Princeton University Press).

Polya, G. (1981) *Mathematical Discovery* (New York: John Wiley).

Scarf, H. (1981) 'Production Sets with Indivisibilities, Parts I & II', *Econometrica*, vol. 49, no. 1, pp. 1–32; no. 2, pp. 395–423.

Velupillai, K. (1994) 'Rationality and Learning in a Computable Setting: Part II: Computable Learning', UCLA Center for Computable Economics, WP no. 16.

4 Nafplion Revisited

Spiro J. Latsis

In September 1974, a Colloquium on Research Programmes in Physics and Economics was held in Nafplion, Greece. Revisiting Nafplion after twenty years is the task I set myself as I sit down to write a few words in honour of my friend Axel Leijonhufvud.

A year before the Colloquium – in 1973 – Imre Lakatos and I went to the US on a double-barrelled promotion trip. It was a whirlwind affair visiting seven cities in nine days, with Lakatos lecturing and promoting his book on the methodology of scientific research programmes,[1] and me helping him to recruit participants to the Colloquium. We ended up in Los Angeles towards the end of our trip and met Leijonhufvud and Alchian in the bar of the Century Plaza Hotel. None of us particularly frequented bars, but Lakatos had decided that Axel would need a stiff drink before he heard what type of paper he would be asked to deliver. As it turned out, Lakatos had been overcautious, and Axel quickly agreed to see if he could tell the story of the Keynesian revolution through 'research programme' tinted spectacles. Lakatos and I boarded the plane back to London via San Francisco with most of our US recruits for the Colloquium in the bag!

The intellectual mood of the London School of Economics (LSE) at the time was one of turmoil internally – because of Lakatos's heresy – and one of defensiveness externally towards Thomas Kuhn's paradigms and Paul Feyerabend's anarchism. Although not sufficiently recognised at the time, the advent of Kuhn's *Structure of Scientific Revolutions*[2] was a major sociological and intellectual event for the rationalist camp. By 'rationalist camp' I mean the camp of Plato, René Descartes, Immanuel Kant, Friedrich Frege and, eventually, Karl Popper. According to the rationalists, we seek true knowledge about the world and ourselves. This 'expressionistic' knowledge, imposed 'top-down' on the world, must have a certain objectivity – an existence outside ourselves; it must be discovered and expressed by propositions and formal manipulations; it must finally have some empirical justification through a certain correspondence to real events, which attributes truth, falsehood or 'truthlikeness.' Depending on which rationalist we are talking about, the criteria for the virtue of knowledge could be quite stringent.

Kuhn's philosophy and history of science can best be likened to a non – invasive medical examination, one that leaves the patient-the evolution of science in this case – as little interfered with as possible, while telling us a little bit about him. We sometimes trade off clarity, specificity and excitement in telling the story of science, but may avoid the risk of catastrophically affecting the object of our investigation.

Although Kuhn tells us the story of the evolution of natural science in a certain way, he does not impose the same rigour as a rationalist or an empiricist would. Kuhn's approach seems to combine Thucidedean detachment on the one hand with an acceptance of the evolution of science as an inescapably historical, social and human event and not one that solely inhabits the privileged world of Platonic knowledge.

The only time I met Kuhn was with Lakatos at Peter Hempel's house in Princeton. We were eager to get Kuhn to give a talk in Nafplion, but he declined. It was a curious evening, with Lakatos – as usual – highly excited about his latest paper on the History of Science and its Rational Reconstructions, and Kuhn polite, but rather distant. Lakatos's legendary charm had failed to move him. On our way back to The Nassau Inn, Lakatos was rather depressed. He confided to me that he was looking forward – after publishing his research programmes work in a more integrated form – to returning to his first love, the psychology of discovery.

In terms of rationalist philosophy of science, the stakes were high. Not only were the internalist rational reconstructions of scientific progress severely questioned, but the empirical standards set for the evaluation of scientific theories were at best relegated to a secondary role. So empiricism as well as rationalism was affected by Kuhn's views. The demarcation between positive and normative 'internalist' and 'externalist' questions disappears with paradigms, and rationality is no longer an absolute but a relative matter. The associated philosophical baggage of dualism, interactionism, indeterminism and free will might as well be left at home. A Heideggerian nightmare lurked around the corner.

The Nafplion Colloquium was in a sense a response to what was seen as the threat of Kuhnian relativism and the mood was such. Lakatos's methodology of scientific research programmes – despite scuffles within the rationalist camp – was seen by many as a welcome attempt to accommodate Kuhn's insights without abandoning the basic tenets of rationalism.

For me personally and for Greece, 1974 was an eventful year. Lakatos, whose Methodology of Scientific Research Programmes was to be the subject of the Colloquium, died suddenly in February of that year, at the height of his powers and in the midst of preparations for the September event. The Colloquium had been his pet project for the last few months of his life, and I, along with a supportive organising committee and some solid friends such as Adolf Grünbaum and Bob Cohen, decided to 'soldier on.'

In June of the same year, Turkey invaded Cyprus, setting off a seemingly inevitable chain of events which led to the collapse of Greece's seven-year-old military government and set the scene for the emergence of a democratic Greece out of the troubled post-war era.

Greece would never be the same again, and my own life changed, but for different reasons. I had assumed that I would remain Lakatos's research assistant for another couple of years before undertaking the perilous exercise of finding a more permanent academic job. His death, combined with other developments in my family's business, contributed to my decision to give up academic life.

The participants of the Nafplion Colloquium – among them Leijonhufvud – flew in to Hellenikon airport almost at the same time as Constantine Karamanlis, who was arriving from Paris to head Greece's first democratic government of national unity. Considering the circumstances, the Colloquium went well as a social and as an intellectual occasion. I, bearing the burden of the organiser, came through it in a sort of hyperactive trance and apparently did not enjoy it as much as did some of the others. I was certainly pleased that there were no major disasters. After all, war between Greece and Turkey or some other type of military adventure was at the time not totally out of the question: as we gathered at Nafplion, Greek and Turkish troops were massing along our common border in Thrace. We remained concerned up to the last minute that some of the confirmed participants might not appear. We need not have worried, however. Even Paul Feyerabend, who had threatened repeatedly to send us a tape of his views, finally turned up!

Looking back at the economics proceedings of Nafplion, I cannot help but feel that the contributions of Leijonhufvud and Hicks[3] were the most damaging to the rationalist programme. Both Hicks and Leijonhufvud were critical of rationalist methodology and historiography of science, but Leijonhufvud also attacked the neoclassical research programme in his usual, deceptively innocuous, style.

In economics, Adam Smith laid the foundations for what must be an exemplary case of a theoretical construct in the rationalist tradition. The main idea of the Smithian approach – which has permeated respectable economics from Adam Smith to the Arrow–Debreu general equilibrium model – is that economic phenomena can be treated as being relatively isolated; that causation comes from individual, rational economic behaviour; and that the resulting structure is an anthropocentric negative feedback system. Modelling this system and discovering its properties was to be the business of theoretical economics.

Leijonhufvud started, challenging the dominant rationalist tradition in the historiography of science, drawing an analogy between the rational reconstructions of individual behaviour in economics and the rational reconstruction of scientific achievement in the history of science.

Both types of reconstruction are anthropocentric, and both use the rationality principle axiomatically – in the sense that actions are appropriate to circumstances. Both consist of hypothetico-deductive schemata which lead to 'single-exit' outcomes.[4] Both fail to give a satisfactory account over a wide range of phenomena and problem areas. And both profess to be criticisable, but turn out to be very tenacious in the face of criticism.

In neoclassical economics we build static or intertemporal models with internalist restrictions and procedures which lead to a single-exit result; for example, a determinate price–output combination, or a set of values at which the economic system is at rest or in dynamic equilibrium. The models we use are purely economic or 'internalist.' Technology, institutions and systemic considerations that could affect economic phenomena and processes remain unchanged by assumption. The economic agent's inner environment is similarly assumed to be 'given' and thereby deprived of a causal role. Whatever the non-economic inner or outer environment, the 'single exit' will be reached from purely economic (or internalist) considerations. Externalist considerations may serve to explain the particular timing of events, but are irrelevant for understanding the working of the economy or for assessing the epistemological virtue of our theories.

In rationalist history of science, something analogous is the case. The reconstruction of scientific achievement consists of internalist premises concerning scientific theories, initial conditions, observational statements, standards of empirical justification and the inevitable 'rationality principle' which allows the deductive step that leads to the victorious theory. Externalist considerations can be added to the account, but, by and large, are an irrelevant side story. As in economics, so in the history of science: as we move away from single-exit to multiple-exit situations,[5] we open the way to externalist factors in the inner and outer environment which play a determining role in the achievement and validation of the eventual outcome. Few would doubt the importance of what rationalists call 'internalist' elements in the history of science, but Leijonhufvud criticises the spurious demarcation and signposts the restrictiveness of a philosophy of science which rationalises and legitimates purely internalist accounts.

In Pittsburgh we visited Adolf Grünbaum and Herbert Simon. Adolf – though clearly no Popperian rationalist – was an early supporter of the Nafplion venture and besides, he cured a bad cold I had managed to acquire during the trip. For me, it was the start of a long friendship with Adolf and his wife Thelma.

Herb Simon was also supportive but cautious: He offered to contrast what he called 'substantive' with 'procedural' rationality, but could not promise a 'research programmes' approach.

The difficulty lies not only in terms of how we explain, but more particularly in how we *conceive* the phenomena and/or problem areas we seek to explain. Sterilised in their internalist/rationalist disguise, these explanations are often uninteresting, though 'explanation-friendly.' More in the social sciences than anywhere else, the problems or phenomena to be explained cannot be divorced from their context. Their outer environment is historical, political, cultural, moral, technological, and so on. Problem areas are not and cannot be adequately described as purely economic, purely ethical or purely technological. The dominance of the neoclassical programme itself cannot be explained by using either purely rationalist or purely empiricist criteria of epistemological virtue.[6]

But it can be made intelligible by its historical, political and social context, by the processes which allowed it to emerge, and by the tools available when it was developed. The tools seem to come first, the grand theories later. The 'toolmakers' are often searching for a way without knowing where they are going. This is why they open the occasional Pandora's box! The tools of medicine were once herbal remedies; then they were the tools of chemistry. Today they are the tools of molecular biology. The tools of economics were initially arithmetic and statistics, and later they were infinitesimal calculus and set theory. Tomorrow they could be the tools of chaos theory and brain science. Leijonhufvud concludes that the Keynesian 'revolution' did not realise its potential because it occurred within the neoclassical toolbox. Keynesianism lacked the technology to become a fully-fledged research programme, while the neoclassical programme, having been constructed by borrowing technology (infinitesimal calculus and axiomatic set theory) from the hard sciences, remained dominant notwithstanding its explanatory and empirical poverty. But, as Leijonhufvud shows, once questions concerning instability, miscoordination and system breakdown are asked, 'old' technology appears to be insufficient. To sum up, neither empirical data nor iterations of conjectures and refutations explain satisfactorily the dominance of research programmes in economics and the other social sciences. What a rationalist would call 'externalist' considerations appear to have a causal role.

Kuhn's merit has been to free historiography of science from its rationalist and empiricist straightjackets. He writes characteristically: 'The historians' problem is not simply that the facts do not speak for themselves, but that, unlike the scientists' data, they speak exceedingly softly. Quiet is required if they are to be heard at all.[7]

Once scientific evolution is seen as a complex sociohistorical phenomenon, autonomous rationalist or empiricist accounts are, to say the least, imperilled. For they only tell us part of the story – and not necessarily the most important part.

In the social sciences, phenomena and problem areas are not only complex and interdisciplinary, but they also actively interact with whatever knowledge we have of them much more directly than they seem to in the physical sciences. They evolve more quickly, too. Intervening in a living system is not the same as intervening in a non-living system, and practically any attempt at 'knowing' is, or becomes, an intervention. The interaction between history, technology, institutions, knowledge of economic and social phenomena, and the phenomena themselves creates new types of phenomena at great speed. As Leijonhufvud and Hicks point out, by the time we have attempted to grasp and explain the nature and causes of an economic problem or event, it has already altered or evolved out of our grasp.

It is often not rational economic actions which cause these changes, but subtle transformations in the architecture of complex systems, in the quality and intensity of the links between elements, and their interaction with other complex systems. Both rationality and knowledge become relative notions, and so does complexity. In certain situations what is unknown, uncertain and highly complex for an economic agent may be familiar and unproblematical for an organised assembly of agents. It may be demonstrated easily that collectivities of interconnected artificial agents facing an unknown environment fare much better than collectivities of identical, unconnected artificial agents.[8] Under conditions of perfectly competitive equilibrium, where there is a 'single exit', it does not much matter whether the agents are unconnected or interrelated. But in cases where the environment is unknown and we have 'multiple exits', it matters a great deal. As information technology places in our hands the capacity to create systems of increasing complexity and to make them interact with others, it would follow that often solutions are found to problems 'without a knowing subject'. In the world of finance, examples abound where important investment decisions are taken on the basis of software incorporating learning and adaptation to complex circumstances. As the software learns and evolves, it adopts new internal architectures in order to cope with the environment, engages in decision-making and becomes more and more of a 'black box' for its designers and users.

A living, or even an artificial, system may therefore evolve successfully in its environment despite severe limitations of its constituent elements – for example, limitations in sensory apparatus, computational capacity and so on. It is not necessary to have knowledge in the anthropocentric sense for a system to adapt to a changing and largely unknown environment. The proof of the pudding is in the eating – not in the ingredients.

Lakatos and I only spent eighteen hours in Chicago. It was memorable for me because I had my first Chinese meal there and later met Friedman and Stigler in Friedman's home. There was a brief discussion of methodological issues, which suggested that Friedman's well-known methodological views had further evolved towards instrumentalism. His main concern, however, was to demonstrate to us his new colour Polaroid camera which – after some persuasion – yielded the only photograph I have of myself and Lakatos together! All in all, I found Friedman and Hicks to be fairly close in their appreciation of what a methodology and history of economic thought could achieve, and fundamentally pessimistic about the attainment of absolute, durable knowledge of human affairs.

On another level, real economic agents are highly complex systems themselves and do not behave as the Pavlovian homeostats of neoclassical economic theory. Between the economic stimulus and the rational economic response, a great deal happens which is relevant. And the responses themselves may belong to any part of a vast spectrum, from reflex and involuntary behaviour to highly deliberate planned behaviour. In neoclassical theory the inevitable operation of market forces ensures a behaviourist 'pleasure–pain' reinforcement, where deviants are either eliminated or brought to the right path. Once the rational agent is in a particular economic situation, all he has to do is to throw a digital switch.

According to Donald Hebb,[9] learning, memory and even cognition reside to a large extent in neural networks that are evolving continuously and in the quality, strength and architecture of their connections. Human action is not the result of digital switches being thrown, nor is it necessarily analysable into the usual anthropocentric categories of rationalism. It is rather the result of subtle changes in complex patterns. The 'yes' or 'no' of digital logic gives way to the 'often', 'sometimes' and 'rarely' of fuzzy logic: the rationality principle is unpacked into countless algorithms.

Rationalism and its fellow travellers, anthropocentrism and negative feedback, are doubly restrictive: on the one hand by imposing on us a simplistic internalist account of an individual economic agent engaging in

Pavlovian behaviour; and on the other by discouraging questions about the evolution and properties of complex living systems.

Hicks suggests that economics consists of temporary artifices we construct to resolve pressing problems in society. These artifices depend on circumstances, available knowledge, and technology. There are no guarantees that the 'right' technology will be there when we need it, nor that we will use it correctly once we have it. Out of their broader context, our theoretical efforts in the social sciences are difficult to comprehend. Building large theoretical systems and attempting to anchor them in even more elaborate philosophical frameworks seems to be a task out of proportion to the inherent limitations of the enterprise. So, rationalist, objectivist history and philosophy of science seems a futile undertaking. Normative philosophy of science, however, is open to all.

As a postscript to Nafplion, I would add that some of the technology that seemed to be missing in the 1940's has started to become available through advances in information technology, mathematics, molecular biology and the neurophysiology of the brain.[10] But this would take me far beyond revisiting Nafplion.

Since Kuhn was not coming to Nafplion, we had to get Feyerabend! On our way back London, we visited him at his earthquake-proof house in Berkeley, California. He was watching professional wrestling on TV, and Lakatos and I sat and watched with him for an hour without talking much. My family's shipping company had just purchased a couple of Second World War tugs from the 'graveyard' in San Francisco and our general manager from New York – Father Basil Gregory – was there to take delivery. All four of us went for dinner in Ghirardelli Square and Feyerabend, Lakatos and I quickly engaged in an animated discussion on anarchism. It ended up with Lakatos getting rather excited and loudly accusing Feyerabend of 'copping out' and failing to solve any of the interesting problems in the philosophy of science. Feyerabend, who could see that Father Gregory was eating busily and not paying attention to the conversation, suddenly asked him: 'And what do you think about all this, Father?' Gregory was unfazed. He looked up briefly from his plate and said: 'You only hurt the ones you love!'

Notes

1. Lakatos, I. (1970).
2. Kuhn (1962).
3. Leijonhufvud (1976); Hicks (1976).

4. Latsis (1976a).
5. Latsis (1976*a*).
6. See Latsis (1972), (1976a), (1976b).
7. Kuhn (1980).
8. di Primio and Muller (1994).
9. Hebb (1949); Hayek (1952).
10. Leijonhufvud (1993).

References

Hayek, F. A. (1952) *The Sensory Order: An Inquiry into the Foundations of Theoretical Psychology* (Chicago: University of Chicago Press).

Hebb, D. O. (1949) *Organization of Behavior* (New York: John Wiley).

Hicks, J. R. (1976) "Revolutions" in Economics' in S. J. Latsis (ed.), *Method and Appraisal in Economics* (Cambridge University Press), pp. 207–18.

Kuhn, T. S. (1962) *The Structure of Scientific Revolutions* (Chicago University Press).

Kuhn, T. S. (1980) 'The Halt and the Blind: Philosophy and History of Science', *The British Journal for the Philosophy of Science*, 31, pp. 181–92.

Lakatos, I. (1970) 'Falsification and the Methodology of Scientific Research Programmes', in I. Lakatos and A. Musgrave (eds), *Criticism and the Growth of Knowledge*, pp. 91–195.

Latsis, S. J. (1972) : 'Situational Determinism in Economics', *The British Journal for the Philosophy of Science*, 23, pp. 207–45.

Latsis, S. J. (1976a) 'A Research Programme in Economics', in S. J. Latsis (ed.), *Method and Appraisal in Economics* (Cambridge University Press) pp. 1–41.

Latsis, S. J. (1976b) 'The Limitations of Single-Exit Models: Reply to Machlup', *The British Journal for the Philosophy of Science*, 27, pp. 51–60.

Leijonhufvud, A. (1976) 'Schools, "Revolutions" and Research Programmes in Economic Theory', in S. J. Latsis (ed.), *Method and Appraisal in Economics* (Cambridge University Press), pp. 65–108.

Leijonhufvud, A. (1993): 'Towards a Not-Too-Rational Macroeconomics', *Southern Economic Journal*, Vol. 60, pp. 1–13.

Primio, F. di and Muller, B. S. (1994), 'Problem Solving Abilities of Systems of Artificial Agents', *From Perception to Action*, pp. 322–5.

Part I

5 On Truth in Teaching Macroeconomics[1]

Robert W. Clower

It seems paradoxical beyond endurance to rule that a manufacturer of shampoos may not endanger a student's scalp but a premier education institution is free to stuff his skull with nonsense. (Judge Howlson, reported in Stigler (1975), p. 191)

INTRODUCTION

My title is a misnomer. Considering Judge Howlson's finding, I start by asserting that teachers are concerned not so much with truth as with intellectual coherence. In other capacities, teachers may strive for truth as they conceive it; but in lectures and textbooks, their aim is to communicate ideas with sufficient clarity to enable students to form an accurate impression of what has been said. Whether the ideas themselves are true or false, useful or useless, is neither here nor there; the essential requirement is that teaching should be unambiguous, or else it will produce confusion and misunderstanding instead of knowledge. Properly speaking, therefore, my subject is not truth but coherence in teaching macroeconomics.

Recent years have witnessed mounting dissatisfaction with undergraduate macroeconomics. The reasons are many;[2] it seems, however, that contemporary discontents mostly derive from a common source: the 'conflict of visions' to which modern macroeconomics owes its existence. I refer to the distinction drawn by J. M. Keynes[3] between the 'orthodox' belief that market economies have an inherent tendency towards self-adjustment and the 'heretical' view that the orthodox faith is mistaken. This conflict of visions has never been resolved, nor is it conceivable that it should be: the conflict concerns not differently perceived facts but different cosmologies.[4]

Before Keynes the conflict was invisible, because the heretics 'could find no place within the citadel of established theory to mount a challenge. After the *General Theory* everything changed. The heretics took control of the citadel and Keynes' conflict of visions became the agenda for doctrinal squabbles that continue to this day. But time marches on. The Keynesian Revolution is history, the heretics have fallen on hard times, and the self-adjusting school again dominates the citadel, united by adherence to an apparatus of formal analysis that derives from John Hicks and Arrow–Debreu but accommodates ideas that earlier would have seemed incompatible with any but 'Keynesian' views[5]. The non-selfadjusting school, now commonly labelled Keynesian and occasionally pronounced disapprovingly, remains within the citadel, unable to ignore the intellectual force of new orthodox analysis but still convinced that there must be something more to the economics of J. M. Keynes.[6]

The conflict of visions is presently in abeyance, but the tension of opposed ideologies continues unabated and so has become a source of chronic intellectual unease. My purpose in this chapter is to initiate expression of the conflict in a non-threatening context: macroeconomic models that have explicitly-stated antecedents in the literature. My object is not to resolve the conflict but to end the tension which, if I am correct, mainly requires dispassionate communication between contending schools.

PRELIMINARY IDEAS

Keynes was not unfamiliar with Walras's *Elements* (see Hicks CEETII, p. 296 fn.: Edgeworth, 1925), but his background, education and life as a Fellow of King's College, all encouraged 'Marshallian' habits of thought. On doctrine and method, there was in any case little reason for Keynes to regard Marshall and Leon Walras differently; in the period before 1936 it was natural to treat both as leading interpreters of the classical tradition (see Phillips (1924), pp. 236–7; and, of special relevance, Schumpeter (1955), pp. 837–40). In *Elements* (Jaffe (1956) pp. 83–4), for example, Walras says:

> Value in exchange ... 'arises spontaneously in the market as the result of competition. As buyers, traders make their *demands* by *outbidding* each other. As sellers, traders make their *offers* by *underbidding* each other ... The more perfectly competition functions, the more rigorous is the manner of arriving at value in exchange. The markets which are best organized from the competitive standpoint are those in which purchases and sales are made by auction, through ... agents who centralize transactions [so] that the terms of every exchange are openly announced and

> an opportunity is given to sellers to lower their prices and and to buyers to raise their bids ... Besides these markets, there are ... fruit, vegetable and poultry markets, where competition ... functions fairly effectively ... City streets with their stores and shops of all kinds – baker's, butcher's, grocer's, tailor's, shoemaker's, etc. – are markets where competition, though poorly organized, nevertheless operates quite adequately. Unquestionably competition is also the primary force in setting the value of the doctor's and lawyer's consultations, of the musician's and the singer's recitals, etc.

In much the same vein, we read in Marshall (1920, p. 341):

> The forces of demand and supply have free play ... there is no ... combination among dealers ... and there is much free competition; that is, buyers generally compete freely with buyers, and sellers compete freely with sellers. But though everyone acts for himself, his knowledge of what others are doing is ... generally sufficient to prevent him from taking a lower or paying a higher price ... This is assumed ... to be true both of finished goods and of their factors of production, of the hire of labor and of the hiring of capital.

In modern writing, the contrast between so-called Marshallian and Walrasian perspectives mainly entails the treatment by 'Marshallians' of quantities as being independent variables, and demand and supply prices as being dependent, whereas 'Walrasians' regard prices as being independent variables and treat quantities demanded and supplied as being dependent.[7] In this respect, the US literature has a distinctly Walrasian look. But there can be no doubt that, at least in reference to goods markets, Keynes was a quantity-into-price Marshallian.

By 'classical economist' Keynes meant the *followers*, and not the predecessors, of David Ricardo (Keynes (1971–), JMKVII, p. 3 fn). In truth, the followers he had in mind – Mill, Marshall, Edgeworth, Pigou – did not deviate much from Adam Smith. But neither did Walras or, for that matter, Keynes before 1936. So it is not surprising to find a hint of Keynes' *General Theory* in *The Wealth of Nations* (bk. I, Sec. vii):

> The quantity of every commodity brought to market naturally suits itself to the effectual demand. It is in the interest of all those who employ their land, labour or stock in bringing any commodity to market, that the quantity never should exceed the effectual demand; and it is in the interest of all other people that it never should fall short of that demand.

Of course, Smith, like other writers before Keynes, was referring to a particular 'market' or 'industry'; but there is no reason why the same line of argument should not be applied to the economy as a whole by supposing that the economy consists of a multitude of firms, all of which produce and sell a single type of commodity output 'bundle' (Q).[8] So let us imagine a world with a large number of independent market-making producers of (Q), here called *marketers*,[9] each of whom fixes his/her output q_i ($i = 1, 2, \ldots, N$) and asking price p_i^s, and offers units of (Q) for sale to a horde of prospective buyers called *marketees*, in exchange for units of a commodity called 'money' (M). Imagine also that each marketer is economically contiguous ('has a common geographic border' or 'is electronically connected with, by telephone, direct computer link, etc.') to at least one other marketer, so the set of marketers and marketees is 'economically connected'.[10]

On these assumptions, it is plausible to imagine that each marketer acts competitively in the sense that each plans its output in the expectation that its probable sale price, p_i^e, is independent of its output:

$$\partial p_i^e / \partial q_i \equiv 0$$

Similarly, imagine that each marketer operates a market for inputs of factor services (N) and in fixing the planned input of factor services n_i and the associated bid price w_i^d proceeds on the competitive presumption that the probable wage rate required to attract and hold factors is independent of the planned quantity demanded:

$$\partial w_i^e / \partial n_i \equiv 0$$

Notice that we do not suppose planned sales equal realised sales, or that planned purchases equal realised purchases; on the contrary, we suppose generally that neither planned sales nor planned purchases are realised. But, to say within the classical tradition, we suppose that trading in every market is sufficiently brisk, hence markets sufficiently thick,[11] that the subjective price expectation assumptions defining competitive marketer behaviour, namely,

$$\partial p_i^e / \partial q_i \equiv 0 \text{ and } \partial w_i^e / \partial n_i \equiv 0 \qquad [C]$$

are not disconfirmed by experience. In short, marketers not only believe, but also 'feel' their markets to be 'thick' or, more colourfully, 'experientially continuous'.

A MARSHALLESQUE AGGREGATIVE MODEL[12]

Continuing now to state a 'traditionalist' macromodel, I. suppose that short-period output, denoted by $Q \equiv \Sigma q_i \equiv Q^s$, is initially[13] at the level Q_0, so that the (unit) supply price of output, P^s, is indicated by P^s_0, as shown by the Aggregate Supply [Price] [14] Curve labelled $S(W_0)$ in Figure 4.1. I have drawn $S(W)$ as an upward-sloping curve with a discontinuity at 'capacity' output (Q_{MAX}), on the assumption that short-run marginal cost rises[15] up to capacity.[16] In principle, the level of $S(W)$ will depend on the stock of money as well as the given money wage rate, W_0 (see Keynes (1971–), JMKVII, p. 27).

Suppose next that when aggregate output is Q_i the [unit] demand price is P^d_i as shown by the Aggregate (output-adjusted)[17] Demand [Price] Curve $P^d \equiv \bar{D}\ \{Q, \dot{A}\}$[18] in Figure 4.1. Given the supply and demand relations, we define *equilibrium* price, P^*, and *equilibrium* quantity, Q^*, by the coordinates of the point of intersection of S with $\bar{D}$ (the point X in Figure 4.1). As Marshall argued (1920, p. 345):

> When ... the amount produced ... is such that the demand price is greater than the supply price, then sellers receive more than is sufficient to make it worth their while to bring goods to market to that amount; and there is at work an active force tending to increase the amount brought forward for sale. On the other hand, when the amount produced is such that the demand price is less than the supply price, sellers receive less than is sufficient ... so that those who were just on the margin of doubt as to whether to go on producing are decided not to do so, and there is an active force ... tending to diminish the amount brought forward for sale. When the demand price is equal to the supply price, the amount produced has no tendency either to be increased or decreased; it is in equilibrium.

Analogously, in Walras (Jaffe (1956), p. 224, sec. 187) we have:

> in the real world ... purchases and sales take place according to the mechanism of competitive bidding. When you go to a shoe manufacturer to buy a pair of shoes he acts as an entrepreneur delivering the product and receiving the money ... If more products are demanded than supplied, another consumer will outbid you; if more products are supplied than demanded another producer will underbid your shoe manufacturer.

Figure 4.1 Marshallesque cross

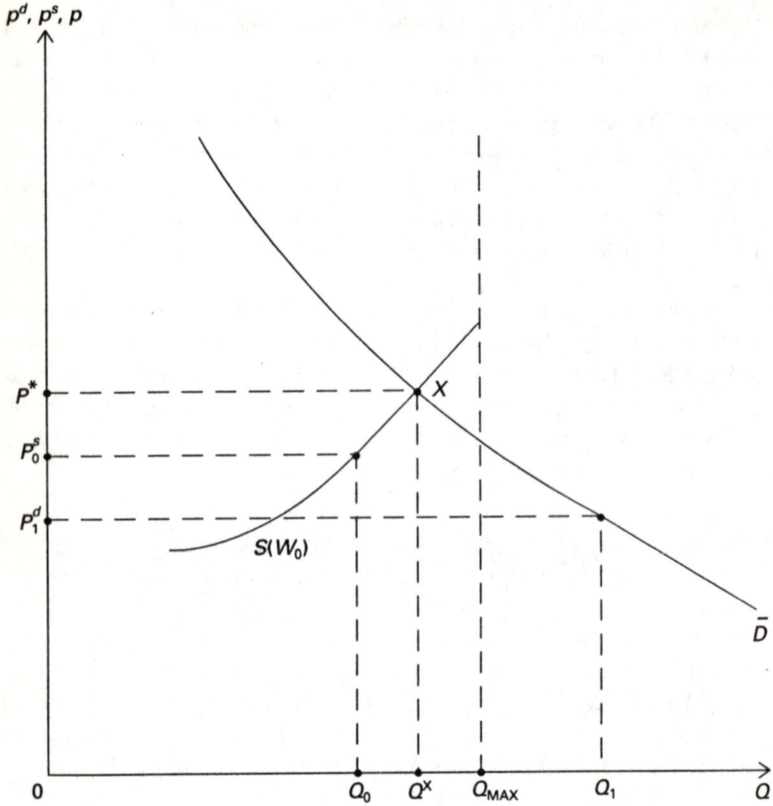

These different views, one quantity-into-price, the other price-into-quantity, are so similar that a hasty reader might find it hard to tell which is Marshall and which Walras.

In both Walras and Marshall we find some hint of adherence to what I call 'the classical stability postulate', misguidedly identified by Keynes with 'Say's Law' (Keynes (1971–), JMKVII, p. 26). The classical stability postulate presumes that the economic system is inherently self-adjusting. It does not derive from J. B. Say's original 'Law of Markets', from pre-Keynesian statements about the impossibility of a 'general glut', or from any notion involving explicit demand and supply functions.[19] In truth the hypothesis has no analytical or evidentiary support; it rests on nothing but anecdote and casual historical knowledge.[20] When Marshall (1889), p. 16) wrote:

> The position of the point of intersection[of S and $\bar{D}$] ... [represents] approximately the average amount which would be produced and the average price about which the mean price would oscillate ...

he is presuming short-run 'clearance' of commodity markets, which (Keynes' several *General Theory* insinuations to the contrary notwithstanding) is a far cry from the classical stability postulate. Thus in Marshall (1920, pp. 346–7), we read the qualified statement:

> When demand and supply are in stable equilibrium ... there will be ... forces tending to push it back ... just as, if a stone hanging by a string is displaced from its equilibrium position, the force of gravity will at once tend to bring it back ... But in real life ... the comparison would be more exact if the string were supposed to hang in the troubled waters of a mill race, whose stream was at one time allowed to flow freely, and at another partially cut of.[21]

Similarly, early in Walras (1874, sec. 188) we read:

> It never happens in the real world that the selling price ... is absolutely equal to the cost ... Yet equilibrium is the normal state ... towards which things spontaneously tend under a régime of free competition.

But we find later (sec. 322):

> the market is like a lake agitated by the wind ... But whereas there are days when the surface of the lake is almost smooth, there never is a day when the effective demand for products and services equals their effective supply ... and just as a lake is, at times, stirred to its very depths by a storm, so also the market is sometimes thrown into violent confusion.

Again, we see the similarity between Walras and Marshall, this time with reference to the theoretical versus the practical problem of market 'clearance'. Belief in the more-or-less continuous clearance of commodity markets seems to be common also among contemporary economists (see Tobin (1980), pp. 788–92). Still, one may easily be led into serious confusion if one treats propositions about hypothetically-defined equilibrium states as propositions about observable states of actual economies.[22]

So much for the Marshallesque partial equilibrium macromodel. One could state a comparable general equilibrium model (the comparative statics properties of such a system are outlined on pp. 74–5 of Clower (1992); but what is most needed at present is a more ambitious treatment that would permit unified analysis of continuous time theoretical motions of both Marshallian quantity-into-price and Walrasian price-into-quantity stock-flow systems for a monetary economy. For the discussion of contemporary macromodels that follows, however, our present model will do nicely.

KEYNES ON THE CROSS

Though the macroeconomic literature contains numerous accounts of the so-called 'Keynesian cross', no textbook writer appears to have constructed a 'cross' according to Keynes' description in ch. 3 of *The General Theory*.[23] I propose to construct such a 'cross'. As we shall see later, this construction is related only distantly to 'Keynesian' crosses found in textbook accounts of national income analysis.

Focusing on Keynes's account of aggregate demand and supply, (Keynes (1971–), JMKVII, pp. 24–32), we start by identifying *aggregate supply*, the function $Z(Q) \equiv P^s Q\{N\}$ shown in the two-quadrant diagram Figure 4.2, as a curve that shows for each given level of employment and output 'the expectation of proceeds [which will] just make it worth the while of entrepreneurs' to maintain the given level of output.[24] This definition mimics Marshall's description of normal supply price for the representative firm, namely, 'the price the expectation of which will just suffice to maintain the existing aggregate amount of production' (Marshall (1920), pp. 342–3). More succinctly, $Z(Q)$ is aptly characterised as 'a reconcoction of our old friend the supply function' (Keynes (1971–) JMKXIII, p. 513)

Analogously, we identify *aggregate demand*[25] as the function $E(Q\{N\}, \dot{A}) \equiv P^d(Q\{N\}, \dot{A})Q\{N\}$ illustrated by the curve $E(Q, \dot{A})$ in Figure 4.2, which shows for any given level of output (and employment) 'the proceeds which entrepreneurs expect to receive from the employment of N men' (Keynes (1971–), JMKVII, p. 25). This differs from Keynes's later definition (p. 30) where he writes 'which the entrepreneurs *can expect* to get back' [my italics]'. The latter phrase would make E another reconcoction, this time of the Marshallian demand curve; the former phrase is puzzling (see Patinkin, and Robertson's 3 February 1935 letter to Keynes (Keynes (1971–), JMKXIII, pp. 497–8), but it may be 'just one of those things' that Keynes omitted to alter in the final draft of *The General Theory*.[26]

Suppose, next, that employment is at N_0 and output at Q_0 so that $E(Q_0) > Z(Q_0)$: aggregate demand exceeds aggregate supply. On

Figure 4.2 The Keynesian crux

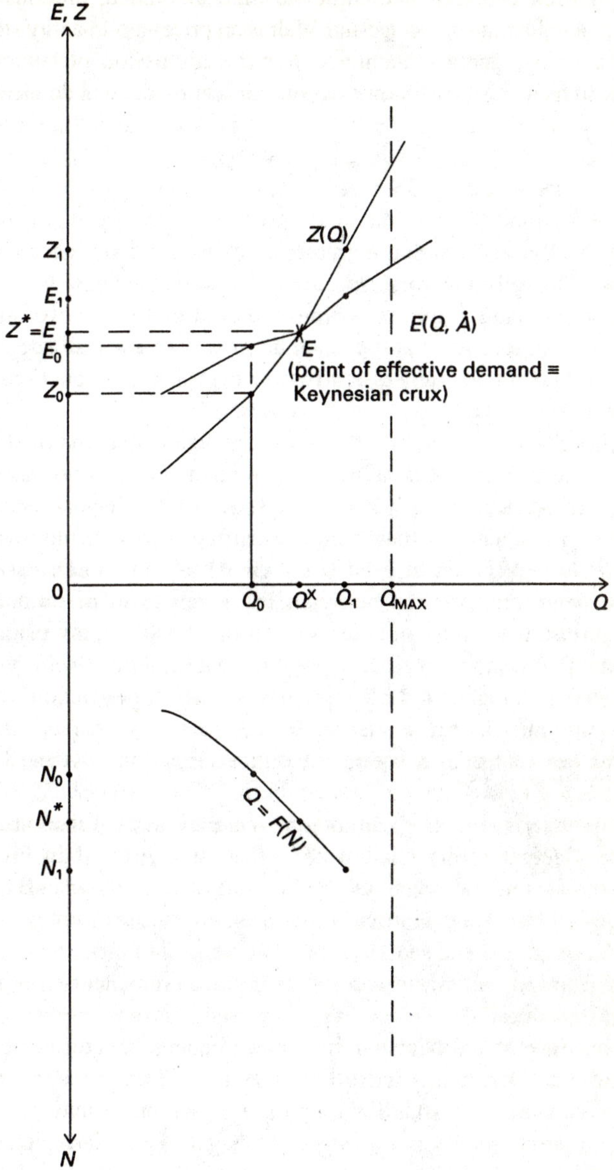

standard maximisation assumptions,[27] some producers will increase employment and output. Dropping further reference to employment, we argue conversely that for Q such that $E(Q) < Z(Q)$ – where aggregate supply exceeds aggregate demand – output will decline. We conclude that the equilibrium level of output (to be faithful to Keynesian terminology we should call this the *effective* or, going back to Adam Smith, the *effectual* level of output) is Q^*, where $E(Q) = P^dQ = P^sQ = Z(Q)$, corresponding to the abscissa of the Marshallesque Cross X in Figure 4.1. It is concordant with the thought of both Marshall and Keynes to presume that market price typically lies between demand price and supply price, so we conclude that equilibrium price in Keynes' system is defined jointly with equilibrium quantity by the point of intersection between aggregate demand and aggregate supply (the point E in Figure 4.2). For convenient reference, I shall refer subsequently to the cross illustrated in Figure 4.2 as 'the Keynesian crux'.

On this showing, Keynes' theory of effective demand is a straightforward reconcoction of Marshallian demand and supply analysis.[28] Taking the proceeding account as a valid representation, Keynes' remarks at the bottom of p. 25: namely, that 'Supply creates its own demand' must mean that $f(N)$ and $\phi(N)$ are equal for *all* values of N', is nonsense; but if Keynes meant instead what he writes in the next sentence, namely 'The classical theory assumes, in other words, that the aggregate demand price ... always *accommodates* itself to the aggregate supply price' [my italics], then both his argument and my restatement make sense. In all that follows I shall adopt this charitable interpretation.[29]

Let us next consider a special case that derives from Alvin Hansen's 1937 review of *The General Theory*. In that review (reprinted in Hansen, 1938), he asserts (p. 321): 'So long as there are unused resources, every increase in demand is matched by an increase in supply.' If the proposition 'supply creates its own demand' is to be called 'Say's Law', we may with equal justice translate Hansen's assertion as 'demand creates its own supply' and call it 'Hansen's Law'[30], but we should interpret the translation charitably, not as a truism but rather as a statement valid only for equilibrium states.

To convert Hansen's Law into a proposition about aggregate supply, let us suppose that marketers set (full cost)[31] reservation prices, p_{i0}, at levels at or above which they stand ready to sell whatever quantities buyers are willing to purchase. Then writing the aggregate asking price index as P_0 we define the corresponding aggregate supply function a $Z^H \equiv P_0Q$, and graph it as shown in Figure 4.3. Combining this aggregate supply function with the Keynes aggregate demand function $E = P^dQ$, defined just as before, we supply the earlier argument to define equilibrium output, Q^*, by the intersection of Z^H with E (the point (Q^*, P_0Q^*) in Figure 4.3). Note that shifts in $\dot{A}$ shift only the equilibrium level of *output* in Figure 4.3: the 'equilibrium' asking price is fixed at a pre-assigned level 'by the theorist outside' the model (the equilibrium value of demand price is, of course, affected by shifts in $\dot{A}$, but demand price is not 'observable' in the Hansen special case of Keynes' model). To all intents and purposes, the Hansen special case consists in the *elimination* of business price-making decisions from the Keynesian model.

THE FUNDAMENTAL EQUATION OF KEYNESIAN ECONOMICS

I next restate Keynes' theory of effective demand as a fixed-point (45°-line) proposition. Because aggregate demand and aggregate supply are both functions of Q, we may for any given value of $\dot{A}$ write E as a parametric function, F, of Z:

Figure 4.3 Hansen special case, Keynesian crux

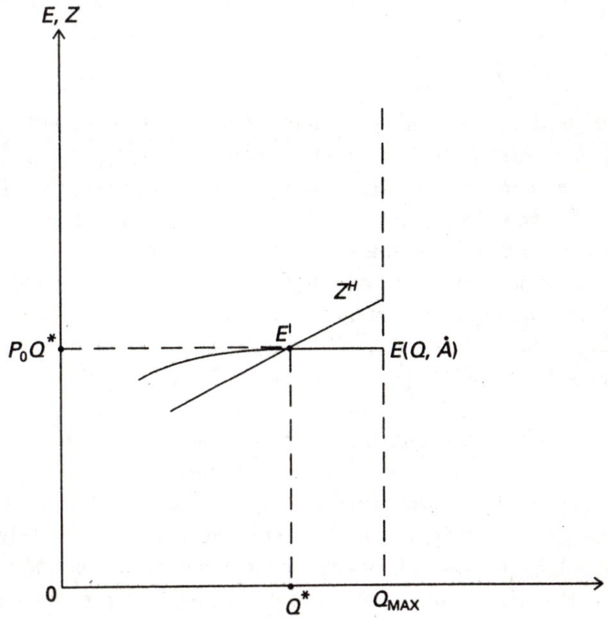

Figure 4.4 Keynesian *F*/*X* diagram

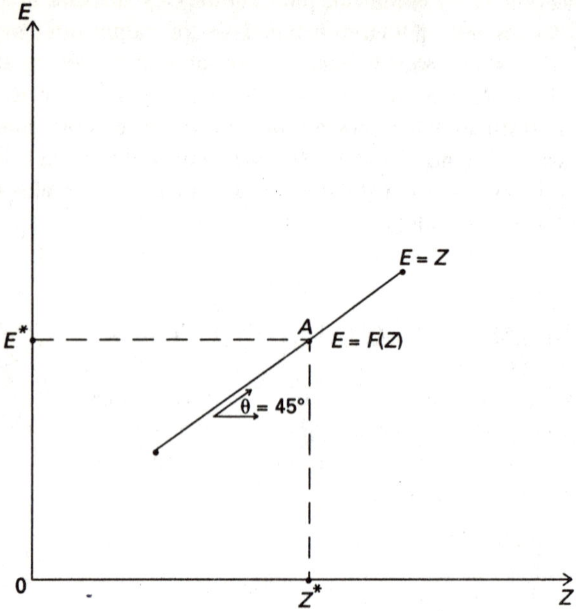

$$E(Q, \mathring{A}) = F(Z\{Q\}). \qquad [H]$$

Then the point of 'effective demand' (Keynes (1971–), JMKVII, p. 25), that is, the point E in Figure 4.4, corresponds exactly to the point $A = (Z^*, E^*)$ in Figure 4.3, where the curve $F(Z)$ intersects the identity relation.[33] Thus the point (Z^*, E^*) in Figure 4.4 is a *fixed point* of the mapping F[34], i.e., Z^* is a solution to the equation $Z - F(Z) = 0$.

If we confine attention to values Q^* of Q that *solve* [H], then for alternative given levels of $\mathring{A}$, we obtain the *identity*

$$E\{(Q^*), \mathring{A}\} = F\{Z(Q^*)\}; \qquad [K]$$

hence, for each alternative value of the 'animal sprits' vector $\mathring{A}$, the expenditure–income *identity* $E^* \equiv Z^*$ is seen to hold.

Because of the central role that Keynes assigns to this hypothesis (Keynes (1971–), JMKVII, p. 25) I shall henceforth refer to [K] as 'the fundamental equation of Keynesian economics'.[35] Note carefully that 'the point of effective demand' corresponds always to the *intersection* of two curves: thus *it holds only for equilibrium states*.[36] This is important because it is only by supposing that commodity markets are always in equilibrium, so that 'effective demand' is the same thing as 'realised' income and expenditure, that we can make sense of the comparative statics exercises that one finds in macroeconomics texts. So-called 'dynamic' multipliers of course make no sense in comparative statics; in that context, a 'multiplier' is simply the change in the equilibrium value of some variable divided by the change in value of the parameter upon which the equilibrium value of the variable is assumed to depend. Keynes himself seems occasionally to forgets this and so refers to 'states' and 'equilibrium states' as if they were equivalent (see, for example, Keynes (1971–), JMKVII, p. 30).

Of course, many contemporary texts are based not on Keynes' theory of effective demand but on the Hansen special case from which Paul Samuelson presumably conceived the so-called 'Keynesian cross' that dominated elementary expositions of national income analysis for at least thirty years after it appeared in Samuelson's *Economics: An Introductory Analysis* (1948). In Figure 4.5 I have drawn the expenditure curve $E(Q) = H(P_0Q, \dot{A}) \equiv H(Y)$ as a function of income (Y or GNP) as the sole independent variable on the assumption that aggregate output in the Hansen model is measured in units such that, at the predetermined reservation price P_0, one unit of money will buy just one unit of output; hence $P_0Q \equiv Q \equiv Y$ (or GNP). Then the intersection of $H(Y)$ with the 45° line (the identity relation $E = Y$) defines a fixed point (shown as point A^1 in Figure 4.5) of the mapping H that corresponds exactly to the point E^1 in the Hansen special case of Keynes' crux drawn in Figure 4.3.

For reasons that are unclear, Paul Samuelson, in his presentation of 'the simple mathematics of income determination' in the Hansen Festschrift volume (Metzler, 1946), pp. 134–5), identified the *conditional* equilibrium condition $E = F(Z)$, equation [H] with the national income accounting *identity* $Y \equiv C + I$, thereby converting an ordinary equation into a definitional identity that holds for *every* admissible value of Y (the identity asserts that the phrase 'market value of output' is literally a synonym for 'national income'). Perhaps Samuelson had in mind the fundamental equation [K], for then his Keynesian cross would be a special case of the fixed-point diagram in Figure 4.4. As usually described, the Keynesian cross diagram defines only a particular solution (assuming output units are chosen so that $P_0 = 1$) to the household expenditure equation $E(Q)–Q = 0$.

Because the Hansen special case effectively dispenses with the business sector of the economy (reducing it to a 'placeholder' for Y, I, $\dot{A}$ and other

Figure 4.5 Hansen special case, *F/X* diagram

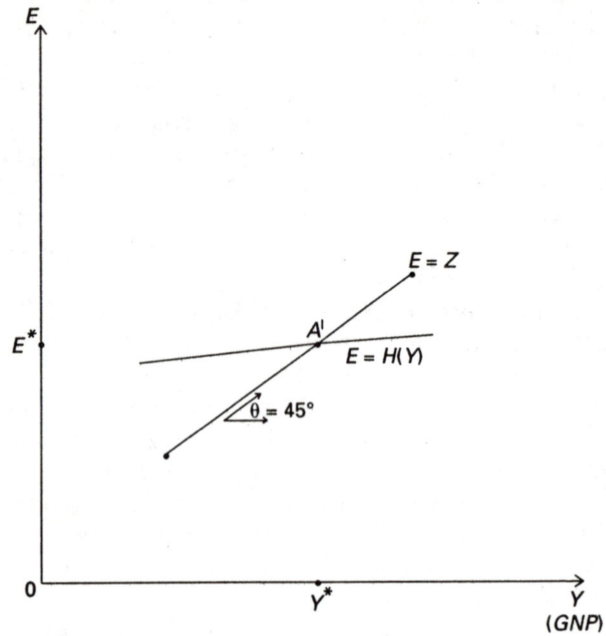

independent variables and/or parameters), the Keynesian cross cannot logically be used to determine 'equilibrium' business decisions respecting output.[37] Given a solution to the household expenditure problem, aggregate output is trivially determined by Hansen's Law ('demand creates its own supply'). Samuelson's three proofs (graphical, tabular, algebraic)[38] to the contrary notwithstanding, the Keynesian cross does not permit us to define a 'Keynesian' equilibrium.

THE TRUE CROSS

The 'genuine' Keynesian cross seems first to have appeared in a 1939 paper by Paul Samuelson[39], but even then it seems to have been an oral tradition among Alvin Hansen's students (on this, see Bishop (1948)). As for its initial source, that appears to be the Hansen assertion quoted earlier (Hansen (1938), p. 321). The diagram seems to have been 'labelled' by Samuelson; but I imagine the label was suggested by Joseph Schumpeter's 1936 review of Keynes' *General Theory*, in which Schumpeter coined the term 'Marshallian cross' and pointed out what he called the 'unwarranted' analogy between the intersecting 'aggregate' demand and supply curves described by Keynes and the intersecting 'ordinary' demand and supply curves drawn by Marshall (Schumpeter (1936), p. 792). However that may be, the diagram with its now familiar title entered the lexicon of economics when Paul Samuelson made it a centerpiece of the first edition of his bestselling text.

I have already expressed doubts about the connection between the *General Theory* and analyses with the Keynesian cross. But that is largely beside the point, because the Hansen–Samuelson model has long since developed an independent existence as various later authors have expounded their version of Samuelson's 1948 classic. By now, any resemblance between Keynesian economics and the economics of Keynes is coincidental (as Leijonhufvud (1968) has argued). So when Joan Robinson dubbed the Hansen–Samuelson *et al.* income analysis 'bastard Keynesianism'[40] (Robinson, 1962) she was not using foul language: she was merely voicing her literal opinion of the doctrinal illegitimacy of one 'Keynesian' school.

The original textbook Keynesian cross is reproduced in Figure 4.[41] Here the variable *Y*(*NI*, sometimes GNP) represents the value of the aggregate quantity (flow) of output produced per unit time, defined more formally as $Y \equiv PQ$, where Q represents 'bundles' of output produced and P represents the (money) price of a single bundle.[42]

Analogously, the curve labelled $C + I$ represents the value of the flow of goods and services demanded per unit time, here defined as $AE \equiv C + I = PQ^d$, where Q^d is the inverse of the demand price function $P^d = D(Q, \dot{A})$ and so represents the number of commodity bundles demanded. The size of the unit bundle is chosen so that at some arbitrary but fixed base price P_0 of P; the *normalised* current price, defined as $\beta \equiv P/P_0$, will be unity and one unit of money will buy just one bundle of goods and services.

Now consider the point E in Samuelson's figure, where the $AE(C + I)$ locus cuts the 45° line. At that point, aggregate household expenditure is equal to the money value of aggregate output ($E = P_0Q_0$): and because the normalised price is $\beta = 1$ (this equation defines Samuelson's 45° line), we may state the central proposition of conventional income analysis:

Figure 4.6 The 1948 Samuelson Keynesian cross

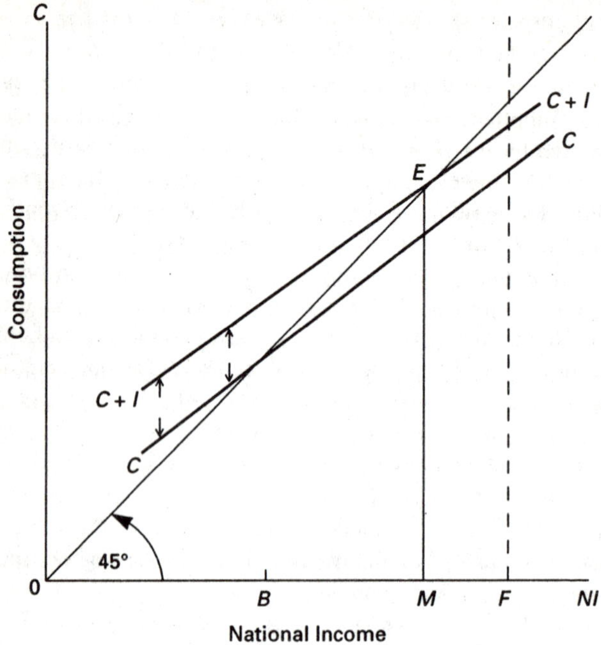

> *Where the expenditure locus cuts the (45°) line, we have not only E = Y, but also Qd = Q^0: the real quantity of goods demanded is equal to the real quantity of goods produced.*

Then, according to Parkin (1990, p. 680): 'This position is the equilibrium. There are no unplanned inventory changes and output remains constant.' Alternatively, this time quoting the matter (p. 260), 'along a 45° line, the level of national income as recorded on the horizontal axis is exactly equal to the sum of the vertical distances representing the amount of consumption plus investment expenditure that people are willing to make. This shows that businesses are getting back just enough to justify their continuing that level of national product.

Most readers will probably find this argument persuasive; millions of students and thousands of instructors have done so over the past fifty years, or else exactly similar statements would not have appeared in virtually every principles text published since 1948. Most textbook authors, however, appear to harbour lurking doubts, or the usual text would not follow Samuelson in offering three or more 'proofs'. If the argument is valid, one proof should suffice. But consider the following:

Case 1

By hypothesis, $\beta = 1$; so equilibrium at *E* entails equality between bundles 'supplied' and 'bundles' demanded. Hence, if *NI* were *lower* than shown by the point *M*, demand would *exceed* supply (*C* + *I* lies above the 45° line everywhere to the left of *E*), there would be 'excess' demand, and, applying elementary microtheory, we would conclude: (i) *either* firms would increase output back towards *M*; *or* (ii) *households would bid up the market price* until quantity demanded was reduced to match output at the level below *M*. But why not both (i) *and* (ii)? Doesn't point *E* define a particular solution to the aggregate expenditure function obtained when $\beta = 1$? If so, then *E* defines just one of an infinity of possible solutions to the household sector equilibrium problem one for each $\beta > 0$); hence the standard 'cross' argument is specious as applied to decisions by business firms.

Case 2

Suppose that when *NI* = *M* and current market price is $P = P_0$ (so that $\beta = P/P_0 = 1$), all firms are covering average variable costs but no firm is covering either its average total cost or its marginal cost. Then some firms surely would reduce output[43]. Suppose output declines to a point where, with price still at P_0, *all* firms just cover marginal cost. Then *NI* less than *M* would be an equilibrium for firms, but not for households (that is, as in Case 1, there would be excess demand with *NI* less than *M*). Suppose households then bid up the market price. At some stage (through higher output induced by higher price), firms would reach a new equilibrium at an output level less than *M* but higher than initially posited; and, at the higher price, households would attain a 'demand equals' supply solution to the aggregate expenditure function at a level of *NI* less than *M*. As in Case 1, therefore, point *E* does not define an equilibrium.

Counter-examples could be multiplied endlessly, each constructed by postulating price–cost conditions consistent with textbook presentations of the Keynesian cross but about which the textbooks are silent. The usual presentation is not logically erroneous; it is meaningless, unless one believes in Snarkhunter 'proofs'. In truth, the argument leading to the 'three assertions proof' turns on the presumption that market prices are 'rigid' (see Tucker, 1958).[44] On this presumption, provided that firms are willing to sell any amount demanded at the rigid price (invariably assumed to be equal to the base price so the normalised price, β, is unity), one cannot quarrel with the logic of the standard analysis (see, for example, the guarded presentation of Peterson (1984), pp. 107–10).[45] Even so, one must recognise – and protest – that the standard analysis: (i) reflects a dubious theory of price and output behaviour;[46] (ii) fails to state crucial background assumptions; and (iii) confronts students with a notion of equilibrium that is technically misleading and incompatible with standard accounts of profit-maximising business behaviour. Note especially that the fundamental equation of Keynesian economics [K] is obtained by imposing as equilibrium conditions the conventional maximising assumptions of both households and firms (see Davidson and Smolenski (1964); and Amadeo (1989)). So the fixed-point diagram of Figure 4.4, although it is geometrically indistinguishable from the Keynesian cross, is grounded on different assumptions.

CONCLUSION

My aim has been to establish a coherent basis for communication among contending schools in macroeconomics. Towards that end, I have presented a Marshallesque macromodel to clarify the 'classical' foundations and establish a basis for reconsideration of contemporary views of Keynes' theory of effective demand. I then restated Keynes' theory of effective demand (the Keynesian crux of Figure 4.2) and derived a related fixed-point (45° line) version of Keynes' theory to replace the 'Keynesian cross' analysis of Samuelson and later expositors (notably Don Patinkin – see Clower (1991), pp. 251–2). Finally, I provided a restatement of Samuelson's cross and constructed cases to demonstrate its factitiousness. I now conclude with general remarks intended to promote fresh discussion of the 'conflict of visions' described at the beginning of the chapter.

The aim of Hicks and other precursors of the Arrow–Debreu theory (especially Marshall and Walras) was to construct a formal model to interpret aspects of experience that would reveal the 'fumbling fingers',[47] as it were, of Adam Smith's invisible hand. By contrast, the aim of modern contributors to neowalrasian theory has been to invent worlds in which the invisible hand works perfectly. In the first case, the aim is to correlate theory with experience; in the second, to correlate theory with preconceived ideas. The distinction would hardly amount to a difference if the implications of neowalrasian theory corresponded to a first approximation with experience. Were that the case, the neowalrasian analysis of unreal worlds might be regarded as a tidying-up operation such as that which heralded the beginning of the end of Newtonian conceptions of 'reality' at the end of the nineteenths century. But the analogy is false. Hordes of nineteenth-century physicists would have staked their lives on the truth of Newtonian mechanics; no neowalrasian theorist would stake even the ninth life of a stray cat on the empirical verisimilitude of general equilibrium theory.

In truth, few economists take neowalrasian theory seriously, except as an intellectual game; it is commonly regarded, indeed, as a fanciful description of a non-existent world – useful as an instrument for academic abuse of PhD students but otherwise the academic economists' equivalent of the stage costumes of Elvis Presley. However that may be, neowalrasian theory is 'a doubtful tale from faery land, Hard for the non-elect to understand'.

Neowalrasian theory is perhaps most obviously worthless for discussing exchange processes, because its logic restricts its purview to strictly *virtual* phenomena (elsewhere I have called such data 'notional'). No trades are possible because no one, not even the 'auctioneer', has any information as to who wants to exchange what specific goods for which other goods, and no mechanism for discovering this is provided by the theory.[48] Given its inanity as a theory about real-time activities, it is ironic that an offshoot of neowalrasian theory, commonly called 'the neoclassical synthesis' or ISLM analysis, largely dominates the thinking of economists who fancy themselves to be 'Keynesian' in economic policy (I have in mind, specifically, Solow, Tobin, Mankiw, Blinder and Stiglitz). The central message of the neoclassical synthesis is that we have learned nothing from Keynes that we did not already know (see Leijonhuvud, 1983). But this finding merely emphasises the intellectual sterility of attempts to pour Keynesian wine into neowalrasian bottles (see Mankiw, 1990).

Personally, I view the Keynesian revolution as being helpful despite its dismissal by 'serious' theorists who contributed to the Keynesian debate. Apart from its catalytic effect on a discipline grown stale with subjectivist platitudes, Keynes' *General Theory* encouraged economists to re-examine critically the foundations of their discipline for the first time since Ricardo. So far that seems to have produced only one definitive analytical result: the formal structure of contemporary theory precludes meaningful analysis of monetary economies of the sort to which Keynes supposed his argument was applicable. The Keynesian debate showed that Emperor Keynes had no trousers, but it also showed that Keynes' critics were themselves stark naked. To be sure, some economists think the Keynesian debate merely confirmed the futility of regarding economics as anything more than a convenient platform for soapbox oratory. But economists who take Keynes seriously – those who believe the purpose of economics is to enhance intellectual understanding of observed phenomena – see matters differently. The bankruptcy of traditional theory as a vehicle for realizing Keynes' 'vision' is viewed as a challenge to provide something better. This might involve little more than a restatement and elaboration of accepted doctrine. If recent developments in monetary theory are any guide, however, what will be entailed is a radical reconstruction of economic theory along lines that are only dimly foreshadowed in the existing literature.[49] In our present situation we are able to provide improved answers to the comparative statics questions posed by classical writers, but unable to offer anything more than *ad hoc* answers to the finite-time dynamical questions that were effectively raised by Keynes. Recorded history afford ample evidence that, if individuals enjoy minimal property rights and enforcement of exchange contracts, private ownership economies are self-organising. The more difficult question is whether the self-adjustment capabilities of modern economies are strong enough to withstand prolonged periods of economic adversity. That question can be answered only by a gifted psychic.[50]

Notes

1. This paper was written to commemorate more than three decades of delightful intellectual companionship with Alex Leijonhufvud and is dedicated to him on the occasion of his sixtieth birthday.
2. Some are described in Leijonhufvud (1983), pp. 64–90; also Hicks (1974 and 1983).
3. See Keynes (1971–), JMKXIII, pp. 486–9. A revised version of the same paper appeared in the 20 February 1935 issue of *The New Republic*, pp. 35–7, under the title 'A Self-Adjusting Economic System?' [The abbreviation JMK[Vol #] refers to the specified volumes of Keynes's *Collected Writings* (1971 and later dates.]
4. On economic cosmologies, see Leijonhufvud (1985), pp. 187ff.
5. See Howitt (1986) and references listed there; also Tobin (1988), pp. 350ff.
6. Here I am referring obliquely to Leijonhufvud (1968).
7. For early comment on the origin of this differentiation, see Edgeworth (1925), vol. III, p. 169.
8. Marshall used as an aggregate output variable called 'gross real income' (g), in his (unpublished) 'Notes on the Theory of Economic Growth' (Whitaker, 1975, vol. 2, p. 309).
9. In Clower (1955), I called market makers and operators 'marketees' and called their customers 'marketors'. In keeping with the more familiar meaning of the words 'lessor' (owner of property leased) and 'lessee' (person using property of lessor), the earlier designations are here reversed.

10. It has been suggested that contemporary microtheory deals with 'pinhead' societies in which even 'large, square economies' (with a continuum of commodities and agents) can be 'placed' on the head of a pin. Of course, mathematical concepts can be 'placed' metaphorically as easily on a debating point as on a pin. But before this footnote becomes pointless, let me make its point: my 'definitions' of contiguity and connectedness are meant to be suggestive, not formal.
11. The association of 'brisk' trade with a 'thick' market is impressionistic. Formally, the concept of a 'thick' market requires continuity (or 'effective' continuity, somehow defined) of the marketor's sales function (a function that may be definable only as a Poisson process and, in any case, has no counterpart in received doctrine). See Robinson (1933); Triffin (1937) for extended discussion.
12. I use the term 'Marshallesque' rather than 'Marshallian' because the extension of Marshall's analysis from a single industry to output as a whole entails notions that have no literal counterpart in Marshall. Marshall evidently contemplated a similar extension in his early ruminations (1976–82) about J. S. Mill's Book IV (see Whitaker, 1975).
13. The word 'initially' has no temporal connotation; it just signifies the assignment of a first specific value to a 'free' parameter.
14. 'Price' may be thought of as a scalar representation of an index of a distribution of prices, perhaps a distribution over prices at a point in time, possibly a distribution over time of a single price. It might be a familiar function (mean, median, mode) or index number (CPI, WPI, GNP Deflator, etc.). On this, see Clower and Due (1972), pp. 50–1; also Bushaw and Clower (1957), p. 180.
15. Figure 4.1 and later diagrams represent magnified expansions (not necessarily on the same scale vertically as horizontally) of corresponding full-size diagrams that would be required if units were measured on an absolute scale; hence no significance attaches to slopes, curvatures, and so on.
16. There is a complication here that some writers appear to have overlooked (see Knight, pp. 151–2). Unless we suppose that unemployed workers are available for all levels of output less than Q_{MAX} (Keynes nowhere spells out the conditions on which this requirement implicitly depends), we cannot coherently treat the level of the supply price curve as given independently of the volume of output as a whole. For a single industry, perhaps; but for all industries taken together we can hardly suppose that increases in employment do not sooner or later put upward pressure on the money wage rate.
17. Elsewhere I have called $\bar{D}$ the '*mutatis mutandis*' demand curve. For details about the relationship between $\bar{D}$ and more familiar *ceteris paribus* demand [price] curves, (see Clower (1989), p. 136, esp. fn. 13).
18. The symbol $\mathring{A}$ denotes a 'shift parameter' which, depending on one's model, might refer to a *kx* 1 row vector representing such imponderables as 'the state of confidence', 'autonomous expenditure', 'uncertainty about future economic conditions' and so on. Thus $\mathring{A}$ is analogous to the 'portmanteau' shift parameter occasionally called 'animal spirits' (see Howitt and McAfee (1992), pp. 502–3). On this, see Keynes (1971–), JMKVII, pp. 148–9 (there are numerous implicit shift parameters in Keynes's analysis; see esp. Keynes (1971–), JMKVII, p. 24 and pp. 91–6).

19. See Clower and Leijonhufvud (1973), reprinted in Walker (1984), ch. 12.
20. See Mummery and Hobson (1889), p. 101; also Keynes (1971–), JMKIX, ch. 2 ('The End of Laissez-Faire'), pp. 272–87.
21. For a comparable treatment that emphasises the sometimes confused state of contemporary supply and demand analysis, see Haavelmo (1978), pp. 31 ff.).
22. Given Marshall's views on competition, it seems natural to suppose that actual market (transaction) prices lie generally in the neighbourhood of equilibrium points, and this 'naturalness' is precisely the source of potential confusion. For a definitive, non-technical discussion of this issue see again Haavelmo (1958).
23. The closest approach to such a construction is probably Davidson and Smolenski (1964), esp. ch. 10, p. 145 and fig. 10.3; but their construction treats employment as the sole independent variable and so leaves the connection between employment and output in the background. Phelps (1985, p. 525) draws a two-quadrant diagram that allows us to see directly the connection between employment and aggregate demand and supply, but Phelps's graph is in fact a fixed-point diagram similar to that developed later in this chapter, not a Keynesian cross in any customary sense. Hansen (1953, pp. 30–1) also attempts a graph, but either his discussion rests on some curious and unstated assumptions, or his diagrams are in error. For further discussion, and copious references to the literature on aggregate supply, see also the useful paper by Tarshis (1979), a former student of Keynes who sat in on Keynes' Cambridge lectures during the period when the *General Theory* was being written.
24. Patinkin (1982, pp. 142–50) identifies several ambiguities here. I shall not here attempt to deal with these and related issues in doctrine history. For further comment, see Clower (1988), p. 255.
25. I have substituted the symbol *E* for Keynes' *D* to make it clear that aggregate demand is an expenditure magnitude, not a quantity or price. It should be stated explicitly that my definitions of *Z* and *E*, unlike those of Keynes (Keynes (1971–), JMKVIII, p. 24, fn. 2) are *not net* of user cost – but my assumptions imply that all producers are fully integrated (the only variable or 'prime' costs are for factor services) – so producers purchase nothing from each other in any case. Keynes' procedure in this regard may help to account for the apparent confusion that emerges from his treating as synonyms the phrases 'what entrepreneurs expect' and 'what entrepreneurs can expect' (compare p. 30 with p. 25!).
26. This conjecture is based largely on Keynes' letter to Hawtrey of 18 November 1935 (Keynes (1971–), JMKXIII, pp. 602–3): 'The demand which determines the decision as to how much plant to employ must necessarily concern itself with expectations. And I am in this respect simply trying to put more precisely what is implicit in most contemporary economics.'
27. Compare Keynes' letter to Kahn, (Keynes (1971–), JMKXIII, pp. 422–3).
28. This conclusion supports E. J. Amadeo's thesis (Amadeo (1989), p. 150): 'the price versus quantity dichotomy should not be taken as the basis for a comparison between [the *Treatise on Money* and *The General Theory*'. It appears to conflict with Patinkin (1982), pp. 88–9. On this, see Clower (1991), pp. 254–5.

29. In 'The End of Laissez Faire' (Keynes (1971–), JMKIX), Keynes described the post-Smithian emergence of 'confidence in the automatic machinery of commerce' (mockingly mentioned in Keynes (1971–), JMKVII, p. 20, fn. 2, as a residue in the 1930s thinking of Lionel Robbins) as an excresence on economic thought created by mid-nineteenth-century ideologues, not economists – though J. B. Say no doubt helped by *ex cathedra* comments in his *Treatise*. So it is hard for a modern reader to take seriously the 'dust' about 'Say's Law' that Keynes throws in the eyes of prospective readers of chs 2 and 3 of *The General Theory*. Perhaps Keynes persuaded himself that economists in the 1930s *believed these things* even though he *could not formally show it*. However that may be, Keynes's polemical tactics undoubtedly served their intended purpose.
30. The proposition now called Say's Law in macroeconomic texts appears first to have been so stated by Keynes (1971–), JMKVII, p. 18), who apparently was unable to decide just what it meant (for his many remarks about the notion, see the index to Keynes (1971–), JMKXIV). An analytical version of the 'uncharitable' identity form of Says Law is proffered by Kalecki in a 1934 paper (published in 1990 in English translation as 'Three Systems') where he assumes 'all income must be spent immediately on consumer or investment goods'. (Kalecki (1990), p. 201).
31. See Andrews and Brunner (1975), ch. 2, esp. p. 29; also Okun (1981), pp. 178–81.
32. In Figure 4.4, I have set the parameter $\dot{A}$ to some initial level to fix the position of the aggregate expenditure curve.
33. See Samuelson (1939), p. 790 – p. 1115 in CSPII.
34. See Arrow and Hahn (1971), p. 28; Simmons (1963), pp. 337–43.
35. With apologies to Axel Leijonhufvud, whose *Keynesian Economics and the Economics of J. M. Keynes* (1968) distinguishes between 'Keynesian economics' and 'The economics of J. M. Keynes' in a manner quite different from my formulae [H] and [K].
36. See, again, Haavelmo (1958).
37. Klein (1947, p. 127) introduces a 'business sector' into his model by creating (with no preliminary discussion) a mechanical theory of output determination, according to which output increases if inventories decline and output declines if inventories increase. With this theory, one can talk about output 'equilibrium' in terms solely of equality between production and sales (forget prices, wages, revenues and costs!). The problem then is to account for inventory holdings and to square the associated view of output adjustment with the most elementary facts of everyday life. Parkin (1990) and Lipsey *et al.* (1990), arguably also Klein's MIT dissertation adviser, Paul Samuelson (Samuelson and Nordhaus, 1985), not to mention numerous others, have adopted, tacitly or explicitly, Klein's expedient 'solution' to what some students saw as an awkward simplification (thus, in Lipsey *et al.*, the possibility of price changes as an alternative to inventory adjustment is noted (p. 540, fn. 2), but quickly dismissed as 'excluded by assumption in this chapter'.
38. See, for example, Samuelson (1948, pp. 258–64), who prefaces one 'proof' with the section heading 'Income Determination: A Third Restatement'. This is reminiscent of Lewis Carroll's Bellman in 'The Hunting of the

Snark', who proclaims early in that poem: 'What I tell you three times is true' (a later line in the same poem reads 'A sentiment open to doubt').

39. Samuelson (1939), p. 790. L. R. Klein used the diagram a year earlier to decorate the title page of his best-selling *Keynesian Revolution*, and that gave the 'cross' additional prestige.
40. Bruce Littleboy (1990), p. 308, fn. 113, suggests 'Perhaps a less offensive and more modern term, "love-child Keynesians"' could be used instead.
41. I have proceeded thus because the original diagram (pp. 260) *y* along with much of the rest of the famous 1948 'green book', is truly an historic document of macroeconomics.
42. We might define Q and P as n-dimensional vectors ($P = (p_i, \ldots, p_n)$, $Q = (q_i, \ldots, q_n)$), and write E as the inner product $E \equiv \Sigma_{i=1}^{i=n} p_i q_i$. Then if the individual p_is were assumed always to vary in the same proportion, we could extract from this inner product a price-weighted scalar index of aggregate quantity, Q, together with a scalar index of the general price level, P, whence $E = PQ$. But it is simpler to label the output unit Q and its unit price P.
43. See Patinkin (1982), pp. 146–7.
44. This is first stated explicitly by Samuelson in the 2nd edn (1951) of *Economics* (p. 273), and is repeated through the 10th edn (1976, p. 221). But in the 11th edn (1980), the last with Samuelson as sole author, there is a change in wording. The phrase 'we shall initially neglect any changes in the price level' is extended to read 'in the price level – a neglect that will be only temporary' (p. 210).
45. But note that Peterson (p. 100), like many others (for example, Patinkin (1982), p. 152), identifies the 45° line with the 'aggregate supply curve'. In truth, the 45° line is nothing of the sort. As usually presented, the 45° line might be considered to indicate the 'aggregate proceeds' above which firms will be willing to supply any *demanded* quantity of output short of capacity ('full employment'); but even then, unless the line is explicitly assumed to be connected with business cost, it is not accurately described as a supply *function*.
46. For evidence supporting a more agnostic view, see Eckstein and Fromm (1968); and Okun (1980).
47. Axel Leijonhufvud probably does not recall this, but I first heard this phrase from him.
48. On this, see the revealing dissertation by Starr (1971); also Clower and Howitt (1992).
49. See Clower and Howitt (1992).
50. See Zuxxzs (2001), p. 2, a work I have not read, but which is listed conspicuously on p. 145 of Bushaw and Clower (1957).

References

Andrews, P. W. S. and E. Brunner (1975) *Studies in Pricing* (London: Macmillan).

Amadeo, E. J. (1989) *Keynes's Principle of Effective Demand* (Aldershot: Elgar).

Arrow, K. J. and F. Hahn (1971) *General Competitive Analysis* (San Francisco: Holden-Day).

Bishop, R. L. (1948) 'Alternative Expansive Fiscal Policies', in L. A. Metzler *et al.*, *Income, Employment and Public Policy: Essays in Honor of Alvin Hansen* (New York: Norton), part 3, ch. V; pp. 317–40.

Bushaw, D. W. and R. W. Clower (1957) *Introduction to Mathematical Economics* (Homewood,: Ill. Irwin).

Clemence, R. V. (ed.) (1950) *Readings in Economic Analysis*, vol. 1 (Reading, Mass.: Addison-Welsey).

Clower, R. W. (1955) 'Competition, Monopoly, and the Theory of Price', *Pakistan Economic Journal*, September, 1955, pp. 219–26 (reprinted in Clower 1995, pp. 181–8)

Clower, R. W. (1965) The 'Keynesian Counterrevolution', in Walker (1984), pp. 34–58.

Clower, R. W. (1975) 'Reflections on the Keynesian Perplex', in Walker (1984), pp. 187–208.

Clower, R. W. (1989) 'Keynes' *General Theory*: the Marshall Connection', in D. A. Walker (ed.), *Perspectives in the History of Economic Though*, vol. II, (Upleadon: Elgar), pp. 131–47.

Clower, R. W. (1990) 'Keynes' *General Theory*: A Contemporary Perspective', *Greek Economic Review*, vol. 12, supp., 73–84.

Clower, R. W. (1991) 'Ohlin and the *General Theory*', in *The Stockholm School of Economics Revisited* ed. Jonung (New York: Cambridge), pp. 245–62.

Clower, R. W. (1994) 'Economics as an Inductive Science', *Southern Economic Journal*, 60, pp. 805–14.

Clower, R. W. (1995) *Economic Doctrine and Method* (Aldershot: Edward Elgar).

Clower, R. W. and J. F. Due (1972) *Microeconomics* (Homewood Ill.: Irwin).

Clower R. W. and A. Leijonhufvud (1973) 'Says Principle', in Walker (1984), pp. 145–65.

Davidson, P. and E. Smolensky (1964) *Aggregate Supply and Demand Analysis* (New York: Harper & Row).

Davis, J. R. and F. J. Casey, (1977) 'Keynes' Misquotation of Mill'. *Economic Journal,* vol. 87, pp. 329–30; see also Davis (1979), further comment in *Economic Journal*, September, pp. 658–9.

Eckstein, O. and G. Fromm (1968) 'The Price Equation', *American Economic Review*, vol. 58, no. 5, part I (December), pp. 1159–83.

Edgeworth, F. Y. (1925) *Selected Papers in Political Economy* (London: Macmillan).

Hansen, A. H. (1938) *Full Recovery or Stagnation?* (New York: Norton).

Hansen, A. H. (1953) *A Guide to Keynes* (New York: McGraw-Hill).

Haavelmo, T. (1958) 'What Can Static Equilibrium Models Tell Us?', *Economic Inquiry*, April 1974, pp. 27–34.

Hicks, J. (1982) *Money, Interest and Wages (Collected Essays in Economic Theory*, vol. 2) (Oxford: Basil Blackwell).

Hicks, John (1989) *A Market Theory of Money* (Oxford: The Clarendon Press).

Howitt, P. (1986) 'The Keynesian Recovery', Innis Memorial Lecture, *Canadian Journal of Economics* vol. 19, November, pp. 626–41.

Howitt, P. and McAfee (1992) 'Animal Spirits', *American Economic Review*, vol. 82, no. 3, June, pp. 493–505.

Jaffe, W. (1956) English language translation of Walras's *Elements* (London: Allen & Unwin, for the Royal Economic Society).

Kalecki, M. (1990) 'Three Systems', in *Collected Works of Michal Kalecki*, vol. 1, J. Osiatynski (ed.) (Oxford: Clarendon Press).

Kahn, R. F. (1984) *The Making of Keynes' General Theory*, Raffaele Mattioli Foundation Lectures, June 1978 (Cambrdge University Press).

Keynes, J. M. (1934) 'Poverty in Plenty: Is the Economic System Self-Adjusting?', *Listener*, vol. 12, (November 21), pp. 850–1; reprinted in Keynes (1971–), *JMK XIII*, pp. 485–92.

Keynes, J. M. (1936) *JMK VII. (The General Theory)*.

Keynes, J. M. (1971–) *Collected Writings*; 30 vols, varying dates, 1971–90 (London: Macmillan, for the Royal Economic Society).

Klein, L. R. (1947) *The Keynesian Revolution* (New York: Macmillan).

Knight, F. (1938) 'Unemployment: And Mr. Keynes's Revolution in Economic Theory', *Canadian Journal of Economics and Political Science*, pp. 100–123.

Leijonhufvud, A. (1968) *On Keynesian Economics and the Economics of Keynes* (New York: Oxford).

Leijonhufvud A. (1985) 'Ideology and Analysis in Macroeconomics', in P. Koslowski (ed.), *Economics and Philosophy* (Tubingen: Siebeck), pp. 182–207.

Leijonhufvud, A. (1983) 'What was the Matter with IS–LM?', in Fitoussi (ed.), *Modern Macroeconomic Theory* (Oxford: Basil Blackwell, pp. 64–90.

Lipsey, R. G. (1972) 'The Foundations of the Theory of National Income: an Analysis of Some Fundamental Erros', in Preston and Corry (eds), *Essays in Honour of Lord Robbins* (London: Weidenfeld & Nicolson).

Littleboy, B. (1990) *On Interpreting Keynes* (London: Routledge).

Mankiw, N. G. (1992) *Macroeconomics* (New York: Worth).

Mankiw, N. G. (1990) 'The Reincarnation of Keynesian Economics', *European Economic Review*

Marshall, A. (1889) *The Pure Theory of Domestic Values* (London: University of London (Reprint of Scarce Tract, 1949).

Marshall, A (1920) *Principles of Economics* (London: Macmillan).

Mummery, A. F. and J. Hobson (1889) *The Physiology of Industry* (New York: Kelley) (1989 reprint).

Okun, A. (1981) *Prices and Quantities* (Princeton, NJ: Princeton University Press).

Parkin, M. (1990) *Economics* (Reading, Mass.: Addison-Wesley).

Patinkin, D. (1978) 'Keynes' Misquotation of Mill', *Economic Journal*, vol. 88, June, pp. 341–2.

Patinkin, D. (1982) *Anticipations of the General Theory?* (Chicago: University of Chicago Press).

Peterson, W. C. (1984) *Macroeconomics*, 5th edn (New York: Norton).

Phelps, E. S. (1985) *Political Economy: An Introductory Text* (New York: Norton).

Phillips, H. (1924) *'The Theory of Social Economy by Gustav Cassel'* (review of McCabe translation of 2nd edn (1921) of Cassels 1918 (Leipzig) *Theoretische Sozialokonomie, Economic Journal,* vol. 34, June, pp. 235–41.

Robinson, J. (1933) *The Economics of Imperfect Competition* (London: Macmillan).

Robinson, Joan (1962) 'The General Theory after 25 Years' (review of Harry Johnson's *Money, Trade and Economic Growth, Economic Journal*, vol 72, pp. 690–2.

Samuelson, P. A. (1939) 'A Synthesis of the Principle of Acceleration and the Multiplier', *Journal of Political Economy*, vol. 47, no. 6, December, pp. 786–97.
Samuelson, P. A. (1947) *Foundations of Economic Analysis* (Cambridge; Mass.: Harvard University Press).
Samuelson, P. A. (1948a) *Economics* (New York: McGraw-Hill).
Samuelson, P. A. (1948b) 'The Simple Mathematics of Income Determination', in L. A. Metzler *et al., Income Employment and Public Policy: Essays in Honor of Alvin Hansen* (New York: Norton), part 1, ch. 6; pp. 133–55.
Samuelson, P. A. (1951) *Economics*, 2nd edn (New York: McGraw-Hill).
Samuelson, P. A. (1955) *Economics*, 3rd edn (New York: McGraw-Hill).
Samuelson, P. A. (1973) *Economics,* 9th edn (New York: McGraw-Hill).
Schumpeter, J. A. (1936). 'Review of Keynes's General Theory', *Journal of the American Statistical Association,* December 1936, pp. 791–5.
Schumpeter, J. (1954) *History of Economic Analysis* (New York: Oxford University Press).
Simmons, G. F. (1963) *Introduction to Topology and Modern Analysis* (New York: McGraw-Hill), Appendix 1, pp. 337–43.
Starr, R. (1971) 'Notes on Microeconomic Monetary Theory', chapter 1 of Ph. D. dissertation, Stanford University.
Stigler, G. J. (1975) 'A Sketch of the History of Truth in Teaching', in Stigler, *The Citizen and the State* (Chicago: University of Chicago Press, pp. 189–93 (originally published in the *Journal of Political Economy*, March/April 1973).
Tarshis, L. (1979) 'The Aggregate Supply Function in Keyness General Theory', in M. Boskin (ed.), *Essays in Honor of Tibor Scitovsky* (San Francisco: Academic Press), pp. 361–92.
Tobin, J. (1988) 'The Future of Keynesian Economics', *Eastern Economic Journal*, vol. 12 October–December, pp. 347–58.
Tucker, R. S. (1938) 'The Reasons for Price Rigidity', *American Economic Review*, vol. 28, March, pp. 41–54.
Walker, D. A. (1984). *Money and Markets* (Cambridge University Press).
Walker, D. A. (1987a) 'Edgeworth versus Walras on the Theory of Tatonnement', *Eastern Economic Journal*, vol. 13, no. 2, April–June, pp. 155–65.
Walker, D. A. (1987b) 'Leon Walras', *The New Palgrave*, pp. 852–63.
Walras, L. (1874) *Elements of Pure Economics,* Jaffe translation, see Jaffe (1956).
Whitaker, J. (1975) *The Early Writings of Alfred Marshall–1867–1890*, 2 vols (New York: Free Press).
Zeuthen, F. (1954) *Economic Theory and Method* (London: Longman).
Zuxxzs, Y. Q. (2001) *Nostradamas was a Plker* (Pullman: Three Forks Fantasy Press), NYH (Not Yet Here).

6 Cash in Advance, Microfoundations in Retreat*

Peter Howitt

INTRODUCTION

Axel Leijonhufvud's *On Keynesian Economics and the Economics of Keynes* told of the strange transformations that macroeconomists can apply to a theory as they develop it. While the profession adopted the analytical tools of the *General Theory*, simultaneously it abandoned many of its central economic ideas. Thus Keynes' theory, which began as an attack on conventional ways of thinking, was turned into a special case of conventional equilibrium analysis. The proposition of 'classical' economics to which Keynes objected most fervently, namely that 'apparent unemployment (apart from the admitted exceptions) must be due at bottom to a refusal by the unemployed factors to accept a reward which corresponds to their marginal productivity' (p. 16), somehow became the distinguishing characteristic of Keynesian economics. And although Keynes had been an outspoken advocate of activist monetary policy, his ideas ended up being used to defend the proposition that monetary policy was unimportant. The tale is made even stranger by the fact that these outright negations of Keynes' central ideas were put forth not by his opponents but by his putative apostles.

The economics of Keynes was not the first body of economic theory to be perverted thus by our profession. Nor was it the last. Indeed, a similar fate has befallen some of the main ideas that Leijonhufvud employed in outlining an alternative to textbook income–expenditure Keynesian economics. He argued that Keynes' essential idea of less than infinitely rapid price adjustment could be derived endogenously using Alchian's[1] 'economics of information,' and the dynamic macro behaviour which Keynes was trying to deal with, especially the deviation-amplifying multiplier process, could be portrayed by taking into account the extra constraints

*This chapter was presented as a paper at the celebration in honour of Axel Leijonhufvud in Montevideo, 6 September 1993. It has benefited from numerous suggestions offered by my colleagues Joel Fried and David Laidler, and draws heavily on my ongoing work with Robert Clower, who also offered useful comments on an earlier draft.

which, according to Clower (1965, 1967), disequilibrium trading imposes on people in a money economy. In the years since the publication of *On Keynesian Economics and the Economics of Keynes*, the profession has made much use of the analytical tools of Alchian's economics of information and Clower's dichotomised budget constraint. Yet little is left, at least in the North American branch of the profession, of the economic ideas of disequilibrium dynamics which originally motivated the use of these tools.

What happened to the economics of information is well known.[2] As Leijonhufvud's book was being published, these ideas were being investigated intensively by Edmund Phelps, Dale Mortensen, Robert Lucas and others, whose works were about to appear, along with the first full-scale published version of Alchian's' ideas, in the famous Phelps volume (1970). Phelps's introduction to the volume made it clear that the objective which inspired this collective enterprise was that of finding a 'microfoundation for macroeconomics'. One can take it for granted that the macroeconomics in question was Keynesian macroeconomics, for at the time there were no serious competitors in macro theory. Thus the enterprise was just what Leijonhufvud had been advocating, especially in his summary statement (p. 397) to the effect that the economics of Keynes was economics without the Walrasian auctioneer. To many, the 'Phelps island' seemed a new paradigm to replace the Walrasian auction house.

Within ten years, macro theory had indeed been transformed by the economics of information, and the Phelps island had become the new conceptual model of macroeconomics. However, the macroeconomics that now filled the journals was not the economics of Keynes, but new classical economics, and the name most commonly associated with the island paradigm was not Phelps, but Lucas, the arch enemy of Keynesian economics. Instead of being a tool of disequilibrium analysis, the economics of information had become the basis of the 'Lucas aggregate supply curve', the central analytical construct of 'equilibrium business cycle theory'. The vision embodied in this theory was not that of a system with chronic co-ordination problems capable of amelioration by wise collective actions, but a system always in rational-expectations equilibrium, whose outcomes cannot be altered in any systematic way by the actions of the monetary authorities.

Less has been written on what happened to the other main component of Leijonhufvud's analysis, Clower's theory of budget constraints. In what follows, I argue that, as in the case of Keynesian economics and the economics of information, macroeconomists have adopted the analytical apparatus while at the same time negating the economic ideas underlying it. My purpose is partly to describe yet another example of a fascinating phenomenon in the development of economic knowledge; the transformation that our profession has applied to the dichotomised budget constraint has many points in common with its transformation of both the economics of Keynes and the economics of information. I also use this example to argue that a new conceptual framework will be needed if the kind of disequilibrium theory that Clower and Leijonhufvud were advocating is ever to fulfil the promise it once had.

CLOWER'S DICHOTOMISED BUDGET CONSTRAINT

What most writers have taken as Clower's contribution to the microfoundations of Keynesian economics, to what Leijonhufvud called the economics of 'income-constrained processes', is his (1965) theory of the 'dual-decision hypothesis.' According to this theory, a transactor formulating plans in one market will take into account not just the prices which enter the usual budget constraint but also the realised quantities from other markets. The result of this is that actual transactions quantities become arguments of the transactors demand functions, thus giving rise to something close to Keynes' consumption function. Barro and Grossman (1971) put this together with Patinkin's (1965, ch. 13) analogous treatment of the demand for labour by firms to formulate a general disequilibrium model of an entire economy.

Instead of focusing on Clower's 1965 analysis, however, I shall describe instead his (1967) analysis of the dichotomised budget constraint, which he proposed in order to recognise the role of money in the exchange process, a role which he argued was missing from general equilibrium theory.[3] Many writers who have commented on the relationship between the dual decision hypothesis and the dichotomised budget constraint have concluded that the latter is irrelevant for disequilibrium analysis.[4] However, this has not been the view of Clower himself, who has since maintained[5] that 'the two models are simply alternative formalisations of a single thought experiment,' and that the 1967 analysis supersedes the dual decision hypothesis, because it relaxes what was too tight an income constraint in the earlier model and provides recognition of a monetary constraint without which the dual decision hypothesis would make no sense.

The subject of Clower's analysis was the concept of 'effective demand', the key concept of the income-constrained process. He criticised general equilibrium theory for implying that 'all goods are indistinguishable from money as sources of effective demand,' and proposed an alternative formulation intended to capture 'the essential meaning of the traditional (but curiously nonmodern) contention that demand in a monetary economy is effective only if it involves a combination of desire with money purchasing power.' Although the importance of this monetary constraint for understanding effective demand has been denied by many, it was not lost on Benassy (1975), who showed that if trade were not restricted to being monetary in nature, the multiplier process which the dual decision hypothesis was supposed to account for could never arise. Nor was it lost on Leijonhufvud,[6] who pointed out that in order to understand how an economic system evolves over time we must describe how the distribution of assets evolves, and to do this requires, among other things, that we take into account the way trading in fact occurs. A minimum requirement for this is that we specify what constitutes the means of payment in an economy and how it enters into the constraints of demanders, for it is the fall in the availability of means of payment, both actual and predicted, that ultimately forces people to curtail their demands in a multiplier process.

Clower's 1967 analysis begins with a critical account of Patinkin's general equilibrium theory, taking as given Patinkin's conceptual framework of a sequence of Hicksian weeks. At the beginning of each week transactors receive manna endowments of the n goods. Each of them is also endowed with some part of the fixed supply of fiat money, which has been carried over from the previous week. At the next stage they all participate in a multilateral tâtonnement that enables them to formulate mutually compatible trading plans. Actual trades are executed after the conclusion of the tâtonnement. After trading, all goods are consumed, and all money is stored until the beginning of the next period, at which time the sequence of events is repeated.

The objective function of a transactor in Patinkin's model is the function $u(c, m/P)$, where c is an n-dimensional consumption vector, m the demand for money to be held at the end of the week, and P an index of money prices. The inclusion of real balances is rationalised by the transactor's desire to avoid inconvenience during a 'stochastic payments process' that follows the tâtonnement. The transactor maximises this function subject to the single budget constraint:

$$pc + m = py + m_{-1} \tag{1}$$

where p is the n-dimensional vector of money prices which the transactor takes as given by a market authority during the *tâtonnement*, y the vector of manna endowments and m_{-1} the holding of money at the beginning of the week.

Having sketched the Patinkin model, Clower's first task was 'to show that the conception of a money economy implicit in these constructions is empirically and analytically vacuous', which he undertook by conducting two conceptual experiments intended to reveal the fundamental flaw of Patinkin's analysis, specifically that every tradeable object plays the same role as money in defining the exchange opportunities available to a transactor. The second of Clower's conceptual experiments reveals the importance of disequilibrium and effective demand in his theory. He imagined that all prices were perfectly flexible except for the price of one good, which was in excess supply, and asked what would be the effect on demands for other goods, and on the set of other prices, if a set of transactors were to receive a sudden increase in their endowment of the good in excess supply. According to (1) the answer is the same as if the transactors had received an increase in their money holdings. In other words, supply could create its own demand, in that mere intentions to sell were as useful as money in enabling them to exercise an effective demand for other goods.

His next task was to suggest 'a reformulation of established microeconomic analysis suitable as a foundation for explicit analysis of the working of a money economy'. First he introduced the formal concept of an 'exchange relation' defined on the set of tradeable commodities. He argued that Patinkin had assumed implicitly an exchange relation appropriate to a barter economy, one that permitted every commodity to be exchanged directly for any other. His proposed reformulation was to assume an exchange relation at the other extreme, one appropriate to a pure money economy, according to which only the single commodity called money could be exchanged directly for a non-money commodity, and every commodity could be exchanged directly for money.

Going back to the conceptual framework of the Patinkin analysis he then made this abstract notion more concrete by dichotomizing the budget constraint (1). He argued that the exchange relation of the pure money economy imposed not one but two separate constraints. The first was an 'expenditure constraint,' to the effect that market value of positive acquisitions of goods were limited by 'the amount of money held by the transactor at the outset of the period':

$$p(c-y)^{+}-m_{-1} \leq 0 \tag{2}$$

where the notation x^{+} denotes the vector formed by substituting zero for all the negative elements of the vector x. The second was the 'income constraint,' which limited the gross acquisition of money during the period (the 'income demand for money') to the market value of planned sales of non-money commodities:

$$m^{y} = p(y-c)^{+} \tag{3}$$

He followed Patinkin in assuming that each transactor attached utility directly to the demand for money, with the twist that now there were two separate demands: the income demand m^y and the 'reservation demand' m^r, defined as the amount of initial cash holding which the transactor chooses not to spend (the negative of the left hand side of (2)).

Reformulating the opportunity set facing the transactor in terms of the pair of constraints (2) and (3), rather than the single constraint (1), avoided the anomalous results of Clower's conceptual experiments. In particular, the trading opportunities defined by this pair of constraints no longer assigned the same role to every commodity as they assigned to money, and it could now be claimed that 'supply of goods does not create demand for other goods'. Thus, for example, a transactor with no initial holding of money would have an effective demand identically equal to zero for every other commodity, regardless of his holdings of other commodities, and regardless of the vector of money prices he faced,[7] but a zero initial holding of any other commodity would not give rise to such a stark prediction.

UNFINISHED BUSINESS

Clower concluded his analysis by observing that his task of reformulating monetary theory was 'obviously unfinished'. Among the items of business left unfinished were some awkward questions which arise naturally when one examines his model, but which he did not even mention. One such question concerns the timing of transactions within the Hicksian week. The only money balances that could be used for the purchase of other commodities were those on hand at the 'outset of the period'. People had a demand for current income 'to replace cash currently scheduled for disbursement from initial holdings of money balances', but the analysis did not address the obvious question of why none of this current income could be used to make purchases before the end of the period. Certainly there was nothing in the formal exchange relation of a pure money economy that would rule out the use of intra-period receipts. Apparently some unspecified lag in the system prevented people from using their sales receipts until the next week. Just as apparently, if some such lag were not invoked, and the transactor could use receipts from current sales, then the dichotomised budget constraint would collapse to the single constraint of Patinkin. Clearly there was much left to be done, either in providing some convincing account of the black hole that absorbed sales receipts within the Hicksian week, or in providing an alternative account of what limited the transaction velocity of money.

Another awkward question begged by the analysis was how to account for the widespread use of credit in modern economies. The only reference to credit was an observation that a 'non-pure money economy' could 'contain numerous barter subsets (trade credit, blocked currencies, credit cards, demand deposits, etc.)'. But the analysis was directed almost entirely at the two polar extremes of a pure barter economy and a pure money economy. Many have interpreted Clower's analysis as describing 'capital market imperfections', and have seen the dichotomised budget constraint as a 'liquidity constraint',[8] which people face in addition to the single lifetime budget constraint that would pertain in a world of 'perfect capital markets'. Since the constraints facing someone wishing to make his/her demands effective in a modern economy have as much to do with the allocation of credit as the allocation of token money, this interpretation is at least consistent with the focus on effective demand. But, in truth, the 1967 analysis had almost nothing to say on the issue. Again, there was much left to be done, in providing some account of how credit is allocated and how this affects the monetary constraints facing transactors.

In addition to begging these questions, Clower's analysis clearly fell far short of providing a complete account of effective demand. An expenditure constraint such as (2) would eventually force a transactor to curtail his/her expenditures if his attempts to sell were repeatedly unsuccessful, and in that respect the analysis attained the objective of the dual decision hypothesis, namely that of having realised transactions affect the demand for goods independently of the effects of market prices, at least with a lag. Indeed, the effect could even be with a lead if sales difficulties were anticipated by the transactor. But to analyse such an effect would have required an account of the dynamic process governing the evolution of the transactor's cash holding over time, which Clower's single-period model was incapable of providing.

Indeed, the model did not even have anything to say about how the transactor's cash would evolve from the beginning of the period until the end, except under the inimical assumption that all planned trades would be realised. In Patinkin's terminology, it was a model of individual experiments, not market experiments. A complete account of effective demand would have required not just an analysis of how individual transactors lay their plans, but also an account of how actual transaction quantities would be determined and how the set of prices and transaction quantities would adjust when plans were not mutually consistent. To this end Clower provided no substitute for the Hicksian temporary equilibrium analysis in which effective demand plays no role.

MODERN CASH-IN-ADVANCE MODELS

Although the flurry of excitement over the dual decision hypothesis and disequilibrium analysis died out long ago, the formal analysis of the dichotomised budget constraint lives on, in the form of 'cash-in-advance' models. Following its adoption by Lucas (1980) and its vigorous defence by Kohn (1981), cash-in-advance has become one of the two or three most popular ways of introducing money into theoretical monetary analysis, along with Patinkin's money-in-the-utility-function approach and the overlapping-generations approach of Wallace (1980). While there are still many detractors of cash-in-advance[9] it has become a part of every graduate student's training. Sargent's (1987) graduate text, for example, devotes an entire chapter to cash-in-advance, while the Blanchard and Fischer (1989) text devotes most of its single chapter on money to models embodying the dichotomised budget constraint.

But while formal analysis of the dichotomised budget constraint lives on, the substantive ideas which originally motivated that analysis have been all-but abandoned. For the cash-in-advance constraint of contemporary monetary theory has been incorporated into the very general equilibrium approach which was the subject of Clower's attack, and which finds no role for the crucial concept of effective demand that he was trying to elucidate. Instead of pursuing the unfinished business of Clower's analysis, his followers have managed to build an entire approach to monetary theory around one of its least satisfactory aspects, namely the above-mentioned black hole that absorbs sales receipts within the Hicksian week. Without appearing to recognise it, even those who invoke Clower's criticism of Patinkin have produced a theory which, as will be shown below, constitutes little more than an economically uninteresting and empirically trivial special case of Patinkin's money-in-the-utility-function theory.

What I shall now proceed to sketch is an amalgam of the modern cash-in-advance models of Grandmont and Younés (1973), Wilson (1979), Svensson (1985), and Sargent (1987), based upon the conceptual framework which Kohn (1988) used in his survey of the literature, a framework which Kohn drew from my own early attempt at cash-in-advance theory (Howitt, 1974). An infinite sequence of Hicksian weeks is postulated, indexed by $t = 0, 1, \ldots$. During each week the same sequence of events occurs, as was described above in the account of Patinkin's theory, except that in addition to the n goods and fiat money the transactors can also exchange loans, in the form of bills promising to pay one unit of money the next week. In particular, the *tâtonnement* is assumed always to converge upon a temporary equilibrium vector of money prices and an equilibrium rate of interest on the loans before any trading occurs.

Individual transactors all live for ever, and have perfect foresight with respect to prices. The exogenous endowments, and hence the equilibrium prices, money holdings and amounts consumed, will generally vary from one week to the next. The intertemporal utility function of each transactor is assumed to depend only upon amounts consumed, with no separate utility attached to either a reservation or an income demand for money:

$$\sum_{0}^{\infty} \beta^t \, \overline{u}(c_t) \qquad (4)$$

where β is a discount factor between 0 and 1.

Where the cash-in-advance model departs from the analogous money-in-the-utility-function model is in its exact specification of what happens during the trading process that follows each week's *tâtonnement.* Suppose that there is a set of $n + 1$ shops, one for each good and one for loans, each presided over by a different shopkeeper. At the beginning of trading, everyone proceeds at once to the loan shop. Each transactor receives from the loan shopkeeper an amount of money equal to the transactor's net demand for bills in the last period minus the value of his net demand for bills in this period: $b_t - b_{t+1}/(1 + r_t)$ If this amount is negative he must pay it, in the form of money, to the loan shopkeeper.

Everyone then proceeds to the n goods shops, where trading takes place according to a set of rules that everyone must follow concerning the sequence of events. First, each transactor visits each shop to place the purchase or sell order for the good sold there, according to the net demand or supply as determined in the *tâtonnement.* No orders will be accepted at this stage unless the transactor makes a full deposit with the shopkeeper. That is, a net supplier must leave the amount of his sell order on consignment, and a net demander must pay in full for his purchase order. Next, the transactors make one more round of visits to the shops, at which time demanders can acquire the goods they have ordered and sellers can receive payment for the goods they have left on consignment. Since the *tâtonnement* has guaranteed that supply equals demand for each good, all purchase and sell orders will have been executed, and the shopkeepers will end the week with no money or unsold goods. Similarly, all net suppliers will have been paid in full, although too late for them to spend the proceeds in this week.

This account of the trading process implies that the transactor will be subject each period to two constraints. The first is that the holding of money at the beginning of the next week will be the holding at the beginning of this week, plus the amount received from the loan shopkeeper, minus the total amount paid to goods shopkeepers in payment for net demands: $p_t\,(c_t - y_t)^+$, plus the amount received from goods shopkeepers for net supplies: $p_t\,(y_t - c_t)^+$. That is:

$$m_{t+1} = m_t + b_t - b_{t+1}/(1 + r_t) - p_t(c_t - y_t) \qquad t = 0, 1, \ldots \qquad (5)$$

The next constraint is that the total value of goods the transactor can demand from the shopkeepers cannot exceed the amount of money with which he leaves the loan shop:

$$p_t(c_t - y_t)^+ \leq m_t + b_t - b_{t+1}/(1 + r_t) \qquad t = 0, 1, \ldots \qquad (6)$$

The cash-in-advance constraint (6) is a straightforward generalisation of the expenditure constraint (2) to cover the existence of a loan market. Constraint (5) similarly generalises Patinkin's single budget constraint (1). The two constraints (5) and (6) taken together thus generalise the dichotomised budget constraint formed by (2) and (3), since the latter pair is mathematically equivalent to (1) and (2) with $m = m^y + m^r$. Notice that the loan market does not eliminate the demand for money, because sellers are forced to hold their receipts in the form of cash until the beginning of the next period. This can be seen formally by noting that the pair of constraints (5) and (6) are mathematically equivalent to the pair (5) and:

$$p_t(y_t - c_t)^+ \leq m_{t+1} \qquad t = 0, 1, \ldots \qquad (6')$$

which puts a lower limit to the demand for money.

By themselves, however, (5) and (6) do not define a bounded set, since they do not rule out Ponzi games. Cash-in-advance models follow the now standard procedure of supposing that some unspecified mechanism is able to impose on the transactors the no-Ponzi-game restriction[10] that:

$$\lim_{t \to \infty} (1 + R_t)^{-t}\,(\mathrm{b}_t + m_t) = 0 \qquad (7)$$

where R_t is the implicit t-period interest rate.[11] As is easily verified, (5) and (7) are jointly equivalent to the single lifetime budget constraint:

$$\sum_{0}^{\infty}(1 + R_t)^{-t}\left[\left(\frac{r_t}{1 + r_t}\right)m_{t+1} + p_t(c_t - y_t)\right] = m_0 + b_0 \qquad (8)$$

Thus the transactor chooses a plan so as to maximise (4) subject to (6), (8) and initial conditions on m_0 and b_0.

Although this analysis preserves the formal structure of the dichotomised budget constraint, it contains no vestige of the notion of effective demand, being embodied in a neowalrasian system of general equilibrium with perfect foresight. Nor does the model make any real progress in dealing with Clower's unfinished business. Instead of elaborating the constraints on effective demand arising from the information and incentive problems of real world credit markets, the model has assumed 'perfect capital markets': people can borrow and lend all they want at the same structure of interest rates, at no transaction cost, subject only to the single no-Ponzi-game condition (7).

As for the black hole that absorbs sales receipts in the original model of the dichotomised budget constraint, although the shopkeeper story at least provides an explicit conceptual account of what was a mysterious unspecified lag in the original dichotomised budget constraint model, it makes it no less mysterious, and no more convincing. For why should the shopkeepers not allow sellers to return before the end of the period to see if their consignment has been sold, and, if so, to spend the proceeds immediately? And if this obvious possibility were allowed for, why would anyone plan to hold money at the end of the period if there were a positive rate of interest to be had on the loan market?

Instead of providing an alternative to the mysterious lag without which the model could not account for positive money holding, modern writers have made it the centrepiece of the cash-in-advance model. As Leach (1983) and Kohn (1988) have pointed out, the distinctive feature of cash-in-advance models is that they show anticipated inflation to impose an *ad valorem* tax on all transactions, at a rate that is increasing in the nominal rate of interest. Consider the position of someone selling good 1 at date t and buying good 2 at date t. The relative price faced by the transactor is not the ratio of current prices $p_{2,t}/p_{1,t}$, which it would be if current sales proceeds could be applied directly towards current purchases, but instead $(1 + r_t)\, p_{2,t}/p_{1,t}$, because the nominal sales proceeds will be available only after one period. By the same reasoning, the relative price faced by someone wishing to trade in the other direction is $p_{2,t}/p_{1,t}\,(1 + r_t)$, since that person must wait one period to receive the payment $p_{2,t}$ for each unit of good 2 sold at t. The relative price paid by the buyer is thus larger than that received by the seller, by a factor of $(1 + r_t)^2$, exactly as if the exchange of good 1 for good 2 were taxed at the rate $(1 + r_t)^2 - 1$. As long as inflation has a positive effect on the nominal rate of interest it will therefore raise this transaction-based tax rate. The reproduction by Grandmont and Younès (1973) of Friedman's result to the effect that the nominal rate of interest must be zero in order for the economy to achieve a first-best Pareto optimum is an immediate consequence of this result, a special case of the general proposition that all marginal tax rates should be zero to attain a first best optimal allocation. Similarly, Stockman's (1981) non-superneutrality result comes from this *ad valorem* tax being applied to the exchange of current for future consumption.

In short, cash-in-advance models have tried to make a virtue out of one of the least convincing aspects of the dichotomised budget constraint model, while denying the real essence of the model, namely the possibility that sales plans might not be realised. For if it were not for this possibility, there would be no apparent reason to worry about the fact that goods cannot be traded directly for other goods in a money economy. It would be the crudest form of money illusion not to see that you can indeed trade economics books for Big Macs by first selling the books for dollars and then bringing the dollars to McDonald's. To insist on representing separately the acts of purchase and sale in a transactor's constraints if these sales were always guaranteed of fulfilment, and if all that separated purchase and sale was the little time it took for money to go from the buyer's hand to the seller's, would be pedantic insistence on detailed descriptive reality, no matter how trivial the details. It would amount to requiring strict observance in the wearing of the veil of money. Such insistence is the very essence of cash-in-advance models.

To make this point quantitatively, consider first the size of the demand for money implied by the above model. Whenever the nominal rate of interest r is positive, the cash-in-advance constraint (6′) will hold with strict equality, implying that the aggregate demand for money in the economy will equal the volume of transactions during the Hicksian week. Even if we were to bend over backwards to interpret the model generously, by supposing that in real time it takes a day for a seller to acquire his/her receipts, it would still be the case that the demand for money would amount to no more than one day's income, less than 5 per cent of actual M1 holdings in Canada, where M1 is defined very narrowly. Similarly, the *ad valorem* tax rate at the heart of cash-in-advance models is just $(1 + r)^2 - 1$, where r is the simple nominal rate of interest per Hicksian week. If the Hicksian week were interpreted as being one real day, then this tax rate would be about one-fortieth of 1 per cent when the simple annual rate of interest was 5 per cent, and would not rise as high as 1 per cent until the simple annual rate of interest had risen to 515 per cent.[12]

In fact, what cash-in-advance models amount to is little more than a special case of Patinkin's money-in-the-utility-function approach. As we have seen, the conceptual framework is just a special case of the stochastic payment process that takes place after each tâtonnement,[13] one in which every transactor is required to make all payments before receiving any, and in which the 'inconvenience' of being unable to pay a bill entails infinite disutility. Formally, one can reinterpret the model in Patinkin's terms by imbedding the cash-in-advance constraint (6) into an indirect utility function. That is, the decision problem of a transactor in a modern money-in-the-utility-function model[14] is to maximise:

$$\sum_{0}^{\infty} \beta^{t} u(c_t, m_{t+1}, p_t) \tag{9}$$

where u is homogeneous of degree zero in $m_t + 1$ and p_t, subject to the lifetime constraint (8). The decision problem in a cash-in-advance model can be expressed in exactly the same terms, using the special form of $u(\ )$:[15]

$$\hat{u}(c, m, p, y) = \begin{cases} \bar{u}(c) & \text{if } m \geq p(y - c)^{+} \\ -\infty & \text{otherwise} \end{cases} \tag{10}$$

To be fair, the explicit representation of the constraints arising from the payments process in the cash-in-advance model reveals a feature not found in the money-in-the-utility-function literature, namely that the utility function should depend not just on consumption and money holdings but also on endowments.[16] If Patinkin had undertaken an explicit integration of the stochastic payments process with his analysis of the decision problem by which transactors arrived at their demands and supplies, he too would have found such a result. That is, a transactor in Patinkin's economy whose initial endowment was replaced by a bundle just equal to his desired consumption bundle would continue to demand the same bundle. But such a change would obviously affect the utility which the transactor would attach to the holding of money, for in the new situations he would no longer have any need to participate in the stochastic payment process.

The modification which cash-in-advance models reveal is needed in the money-in-the-utility-function approach also blunts Clower's original critique of the latter. For in the special case of cash-in-advance, the transactor will obviously not always respond to an increase in endowments exactly as if it had been an equal-valued increase in money holding.[17] Again, a thorough analysis of the more general stochastic payment process would have shown the same to be the case, and the difference would have been reflected in the way money and goods endowments enter the utility function. In short, since cash-in-advance models are just a contrived and empirically trivial special case of money-in-the-utility-function models, any claim that the latter is deficient applies with even greater force to the former.

EXPLANATION AND THE NEOWALRASIAN CODE

What we have described constitutes yet another example of the remarkable ability of conventional economics to absorb criticism. There are many points in common between the way it co-opted J. M. Keynes' *General Theory* and the way it co-opted Clower's dichotomised budget constraint, although, of course, Keynesian economics was of vastly greater significance than cash-in-advance. In both cases, what started out as devastating attack on the foundations of conventional theory was turned into a special case of that theory, even though accommodating the special case required some minor tweaking. In both cases, the novelty of the special case was made to rest on a part of the analysis that was ill-developed, and to which the original author had attached little importance – rigid wages in Keynes' case, and an arbitrary lag in acquiring sales receipts in Clower's. In both cases, work aimed at shedding light on a dynamic process of adjustment was absorbed into an equilibrium system in which all adjustments are instantaneous. In both cases, what the profession took from the original authors work was a mathematical skeleton without the substantive flesh. And in both cases the authors of this perversion were admirers of the original author, who saw themselves as extending and developing his original ideas.

The episode of cash-in-advance illustrates the difficulty of making progress in a science where adherence to an increasingly complex code of formal ideas has become the overriding criterion of success, rather than the fruitful modelling of observed phenomena. The code of modern economics has become for the most part that of neowalrasian analysis, with its rules for modelling all behaviour as the outcome of rational choice. Their frequent critiques of money-in-the-utility-function theory attest that the developers of the cash-in-advance literature were keen to account for some of the real world 'frictions', those underlying monetary exchange, that fit so poorly into neowalrasian analysis. But accounting for some phenomenon in a discipline dominated by an elaborate code consists not of telling stories designed to convince others that this is why the phenomenon exists, or why it appears the way it does, but of telling stories, no matter how *ad hoc*, that incorporate some aspect of the phenomenon, no matter how trivial, without violating the code.

Although there is a debate among purists as to whether or not the cash-in-advance approach really does conform,[18] its proponents at least tell a story in which the actions carried out by all participants are explicitly shown to be the outcome of maximising a tightly specified objective function of the conventional sort, subject to a consistent set of constraints. Because the model arising from the story generally possesses an equilibrium in which there is a positive demand for fiat money, it can be said by modern standards to account for the holding of money, even though most intelligent people who have not had a rigorous training in economics would not be convinced by an explanation that hinged on such trivial details, and which accounted for such an insignificant fraction of observed holdings.

In many ways, modern economic theory has gone the way of geometry, although without having accomplished what geometry did. That is, it has ceased to be a theory designed to make sense of the real world (measuring distances and angles, or accounting for the determination of prices and quantities transacted) and has become instead a purely logical discipline in which the objective is to follow a set of a priori rules with no connection to the external world. Economic theorists are meanwhile suffering from a similar kind of confusion[19] to that which afflicted mathematicians when they had not yet realised that what they were doing had no connection with any concrete interpretation, and that because of this, any one set of axioms was as justifiable as any other as long as it was logically coherent. Economists building 'rational models' to account for things not found in conventional theory think of themselves as seeking explanation in the usual sense, whereas in fact they are just addressing purely semantic question that do not even arise once one ventures out of the neowalrasian cloister. Only by the rarest fluke could someone working under such delusions come up with a convincing scientific explanation of anything.

It would be a mistake, however, to blame everything on Clower's followers. Much of what happened to the dichotomised budget constraint must be attributed to the above-mentioned shortcomings of Clower's own 1967 analysis. Indeed, that analysis itself seems in retrospect to have been much influenced by the neowalrasian code. The complaint that Patinkin's 'conception of a money economy ... is empirically and analytically vacuous' pertained not to the logic of choice but the actual working of organisations. A reformulation that addressed this complaint would clearly have to provide a convincing account of how the transactions institutions of actual economies operate, with emphasis on the *prior possession constraints*[20] imposed by those institutions. But Clower offered no concrete alternative to the Hicksian week, whose *tâtonnement* reduced the determination of transaction quantities down to the level of individual choice theory and obliterated the distinction between notional and effective demand. Instead, he merely proposed a quick fix for Patinkin's theory, one which tinkered with the budget constraint in the standard theory of choice without addressing the more fundamental issues of market organisation.

REPLACING THE HICKSIAN WEEK

What this suggests is that a more radical approach will be needed if Clower's original ideas are ever to produce a theory of effective demand. Since 1967 there has been at least a bit of progress in this direction, including some important contributions by Clower,[21] although as yet this progress has been little noticed in the profession. Still, it is possible to sketch a conceptual framework for monetary theory based on some of this work. In what follows I provide such a sketch, and indicate how the framework might deal with some of the above-described unfinished business.

I start by observing that in any advanced economy almost all trades involve business firms: wholesalers, retailers, brokers, jobbers, manufacturers, shopping malls, banks, commodity exchanges, auction houses, employment agencies, mail-order businesses, newspaper publishers, and so on. These business firms, unlike the ghosts of neowalrasian theory, earn profits by creating, maintaining and operating markets. It is business enterprises that enable households to acquire goods at a less than prohibitive transaction cost and, as Hicks (1932) emphasised, to engage in regular daily work without having to undergo constantly the inconvenience of finding buyers for their specialised labour services or for the fruits thereof. Thus a first step in replacing the conceptual apparatus of the Hicksian week is to return to John Hicks's earlier perspective and suppose that all transactions take place in markets organised by firms.

In effect, the firms that make markets perform the role of the shopkeepers in the above account of post-*tâtonnement* trading in cash-in-advance models. They also perform some of the services that the neowalrasian auctioneer is presumed to perform during the *tâtonnement*. That is, in most cases they administer prices, subject to the pressure of competition from other firms. Like the auctioneer, they do this not according to finely calculated estimates of what price will maximise the flow of profits, but according to simple rules of thumb. For the most part these rules can be rationalised only as a means of enhancing the prospect of survival in a world of competition and uncertainty. They cannot involve setting prices that harmonise instantaneously the plans of everyone in the economy, since no one has the miraculous powers of the Walrasian auctioneer.

What the discussion to this point suggests is that we model trade as in the analysis (Howitt, 1974) where I first introduced the shopkeeper device, which was almost exactly as in the above account of cash-in-advance, except that there was no pre-trade *tâtonnement*. Instead, prices were posted at the beginning of each period by the shopkeepers, who would not trade at any other prices. Transactors then went directly to the shops, without ever having assembled for a *tâtonnement*. The rules for visiting shopkeepers were as in the above account, except that there was no loan shop. Thus, in the first stage, buy and sell orders were placed, subject to a 100 per cent deposit; and in the second stage the transactors came by to acquire the goods and money from successfully executed orders. Because there was no tâtonnement, it was not generally possible for the shopkeepers to execute all orders, and the shopkeepers followed certain rationing rules for deciding which orders to leave unfilled. Thus, during the second round of visits, those whose orders had been filled would receive their goods or money, while others would reclaim their deposits. At the end of trading the shopkeepers would adjust the prices to be posted the following week, according to a rule similar to that followed by the neowalrasian auctioneer.

An important fact of real world markets is that typically they involve clientele relationships between firms and their regular customers. We can capture this by supposing that when transactors go to market to make purchases they visit only a small subset of all shopkeepers, changing that subset only infrequently. Underlying these relationships are set-up costs of transacting; irreducible costs incurred each time a transaction is made, regardless of its size. Because of set-up costs it pays a transactor to visit only those shops that have acquired a reputation for price, quality, availability, honesty, convenience, and so on. Set-up costs also provide shopkeepers with an incentive to develop this sort of reputation, or, in other words, to compete not for current profits but for repeat customers. Let us suppose accordingly that each transactor must pay a fixed cost with each visit to a shopkeeper.

The incentive to provide availability to their customers gives shopkeepers a demand for reliable sources of goods for sale. By the same token, the fact that transactors needs to sell goods in order to survive in a specialised economy gives them a demand for a stable output for their goods. Thus in contrast to the above story, in which each transactor placed sale goods on consignment with anonymous shopkeepers, transactors and shopkeepers here tend to enter into supplier relationships, in which the supplier agrees to provide a regular stream of goods for sale to the shopkeeper, in exchange for a regular stream of money income. Thus, while a transactor's purchase trips will take place at his/her own discretion, and can thus be varied on short notice, the supplier relationship will require his/her sale trips to be regular and predictable.

Incorporating set-up costs into the shopkeeper story entails a substantial formal modification. It implies that a transactor going out to acquire goods will limit his/her visits not only to a sub-set of all shops but also to a sub-set of all time periods. Furthermore, the fact that discretion has been granted to customers implies that it will be practically impossible for a shopkeeper to coordinate visits by different customers. Thus, at any time, the number of transactors in a given market will be random. The simplest way to model this is to suppose that time is continuous, and that shopkeepers stand open for business at all times, at their customers' disposal.

This way of thinking about markets also makes it clear that there is no opportunity for shopkeepers to use a tâtonnement to discover the price at which demand would just equal the amount which sellers have brought to market. For the arrival of customers appears to the typical shopkeeper as a Poisson process, with each arrival bringing a random quantity of demand by a single, possibly idiosyncratic, transactor. Varying the price might bring some information about the response of demand over a period of time long enough to sample a large fraction of the shop's customers, but over a long period *ceteris non sunt paribus*. Thus as Hicks (1982, p. 150) stressed, in practice it is not possible to estimate a demand curve with enough reliability to make use of it in pricing. Furthermore, as Okun (1981) noted, customer relations are damaged by excessive variation of prices.

In addition, to establish reliability of money income to regular suppliers, shopkeepers are encouraged to offer predictable purchase prices. To survive and cover overhead costs, shopkeepers will try to preserve a margin between wholesale and selling prices; so they will tend to regulate selling prices according to the principle of full cost pricing, except in times of extraordinarily high or low sales. This is why shops typically react to unexpected changes in demand, not by varying their prices, but by competing for customers on other fronts; advertising, offering better service, and so on.

Another modification needed to the original shopkeeper story is the holding of goods inventories on the part of shopkeepers. (Cash inventories will also play a big role in what follows.) A firm can provide availability only if it maintains buffer stocks that allow buyers to execute their purchase plans even when demand outruns supply. A firm that routinely refused to absorb day-to-day discrepancies between supply and demand would soon find itself unable to meet its fixed costs, as customers would learn to go elsewhere. Although buffer stocks will inevitably run out on occasion, at which time the firm will need to ration its customers, it will typically hold enough inventories to make rationing the exception rather than the rule.

Finally, we need to deal with the absence of credit in the original shopkeeper formulation. In a modern economy, most exchanges take place using not token money but either bank credit or trade credit. The most straightforward way is to assume that shopkeepers routinely grant a line of credit to each customer, with conventional repayment terms, subject to interest costs at a rate administered by the shopkeeper. A visit to a shopkeeper can then be used not just to buy goods but also to repay outstanding credit balances, or to be paid the amount extended to a shopkeeper. This is the way firms routinely deal with each other, and the way households deal with their employers, although, again, many real world complications are introduced by the existence of banks and other intermediaries that do not fit into any simplified conceptual scheme. We may suppose that shopkeepers allocate credit much the same way as they administer prices, that is, according to rules designed with the main purpose of enhancing the shopkeeper's survival as a profitable enterprise, in the face of competition from other shops offering similar products for sale. Visits to a shopkeeper by a customer repaying outstanding credit will be at the discretion of the customer, and hence subject to unpredictable fluctuations, whereas visits by a supplier claiming payment of outstanding credit will be more regular and predictable, in accordance with the supplier relationship.

FINANCE NOW

Although there remain many details to be filled into the above sketch of trading arrangements, enough has been said to indicate the nature of the constraints facing the typical transactor. In particular, they lead to a version of the dichotomised budget constraint that is free from some of the difficulties pointed out above with Clower's original formulation. An analogue to the expenditure constraint (2) can be derived as follows. Note that because of set-up costs, no transactor will trade continuously. Whenever trades occur, they will be trades of a discrete stock of money (now defined, as Clower[22] has long insisted, to include unspent trade credit as well as currency) for a discrete stock of goods (now assumed to be storable).

It follows from the above that demanders face a 'finance now' constraint of the form:

$$\sum_{p_i(t)D_i(t)>C_i^-(t)} (p_i(t)D_i(t) - C_i^-(t)) + \sum_1^n x_i(t) \leq R(t) + M^-(t) \quad \text{for all } t \geq 0. \qquad (11)$$

where $D_i(t)$ denotes the quantity demanded from the ith shopkeeper,[23] $C_i(t)$ the amount of unused trade credit with that shopkeeper, $R(t)$ the stock of money which the transactor is receiving at time t from shopkeepers to whom he is a supplier, and $x_i(t)$ the amount of credit being repaid at t to shopkeeper i. The superscript – denotes a left hand limit in continuous time; if t is a trading date for the transactor then either $M(t)$ or one of the $C_i(t)$s will have a jump discontinuity, but it is the amount leading up to the transaction that limits the transactor's purchasing power.

Constraint (11) is what Kohn (1981) calls a finance constraint, because while it requires that the transactor has adequate finance in order to exercise the demands D, the finance need not be in the form of cash. The first sum is taken only over those goods whose current demands cannot be financed with unused trade credit because, by assumption, credit, unlike cash, is not fungible across shops.[24] Thus (11) states that, at any given time, the value of what the transactor buys beyond what can be financed with credit, plus the amount of credit s/he chooses to repay, cannot exceed the amount of cash available for the purpose, the latter consisting of the predetermined stock $M^-(t)$ plus the amount $R(t)$ now being received from a shopkeeper.

Note also that the model does not involve any black hole absorbing sales receipts. The cash $R(t)$ received at time t is available to satisfy the finance constraint with no lag required. The constraint (11) thus requires not cash in advance but finance now. In doing so it redresses the two main shortcomings identified above in the dichotomised budget constraint, namely the absence of credit and the presence of an arbitrary positive lag in acquiring sales receipts.

An analogue to the other half of the dichotomised budget constraint, the income constraint (3), can also be derived from the above account of trading, although it is much more involved than (3). For simplicity, suppose that each transactor sells only one good. A trip to the shopkeeper who sells it for him can be made to deposit goods or to take possession of cash receipts. The latter activity is what results in positive values of the variable $R(t)$ in the finance constraint (11). What limits the receipts $R(t)$ is not just the supplier relationship but also the cumulative net receipts of the shopkeeper, which in turn will depend upon a number of considerations.

For simplicity, suppose that the costs of operating a shop are purely subjective and do not involve any outflow of cash. All outflow of cash from a shop is then the predictable amount dictated by existing agreements with suppliers. The inflow of cash will be dictated by the decisions made by the shop's customers as to the amount and timing of purchases and the method of financing them (since the shop's accounts receivable are assumed not to be usable as a means of payment in other markets), and the decisions that all transactors make concerning which shops to patronise. A shopkeeper's control over his/her shop's net cash receipts is thus limited mainly to the indirect control s/he exerts by varying prices, by changing the amount and terms of credit extended, by negotiating new supplier agreements, and by terminating old ones. At any given time the availability of cash on hand with the transactor's shopkeeper $S^{-}(t)$ can be taken as given by the shopkeeper, so that the demand for money income by transactors is limited by the constraint:

$$R(t) \leq S^{-}(t) \tag{12}$$

Although (12) takes a very simple form, underlying it is the above-mentioned complex set of considerations determining the evolution of $S^{-}(t)$.

Note that the above account, although it allows for the possibility that a transactor could acquire cash receipts instantaneously, will not generally imply a trivially small demand for cash holding on the part of transactors. In general, one would expect the costs of visiting different shopkeepers to involve a 'bunching cost'[25] that would prevent the transactor from being in more than one shop at a time, in which case only one item on the left-hand side of (11) could be positive at a time, and if one was positive, then $R(t)$ would be zero. However, we need not resort to a mechanical bunching cost reminiscent of the black hole of cash-in-advance theory. Even in the absence of bunching costs, shopkeepers would want to maintain substantial positive cash balances in order to guard against the possibility of being unable to fulfil their contractual obligations with suppliers: that is, to allow for slack in the income constraint (12). In short, they will try to use their limited control to regulate cash holdings much as in the standard theory of a firm's demand for precautionary balances.[26]

Furthermore, as Laidler (1974) has argued in a similar context, the transactors themselves will have a demand for precautionary balances of cash (and of unused trade credit) in order to deal with the uncertainties of the trading process.[27] That is, someone who always chose to make the finance constraint binding, as s/he definitely would in cash-in-advance models whenever the nominal rate of interest was positive, would not only pay high transaction costs and bunching costs, but would also run the risk of being unable to make future purchases if the shopkeeper ran into sales difficulties and had to abrogate the supplier relationship. The 'reservation demand' for money which Clower insisted should enter transactors' utility functions, and which has been all-but ignored by cash-in-advance models, is thus returned to the forefront by the above account of the trading process.

Finally, although this account is still sketchy, it sheds more than a glimmer of light on the logistics of trade, and so suggests in particular how ongoing trade could lead to something like Leijonhufvud's income-constrained process. Suppose there is an unusually large decline in demand for a large variety of goods on the part of a large number of transactors, for whatever reason. This will show up first as an increase in unsold inventories and a decrease in sales receipts at the shops. Although eventually shops may try to boost sales by reducing their selling prices, they will be reluctant to upset customer relations by such tinkering with prices, and also reluctant to cut into the profit margins necessary for their continued existence. Instead, they will try first to boost net receipts by finding new ways to reduce expenses, or by improving the quality of services provided to their customers, and advertising these innovations. Since inventories will now seem superfluous and cash scarce, shopkeepers may even cancel or fail to renew contracts with existing suppliers, paying the price of loss of reputation and the possible recruiting cost of replacing the suppliers in the future, in exchange for an enhanced ability to honour other contracts. This will, of course, result in a further shortfall in demand faced by shops being patronised by laid-off suppliers, thus amplifying the original fall in demand.

As Leijonhufvud (1973) stressed, the multiplier process will not be fully under way until buffer stocks of money have been largely depleted. Until then, the desire on the part of households to maintain accustomed standards of living, and on the part of firms not to lose their suppliers, will buffer the fall, and the precautionary stocks of money they hold will allow them to do this without violating the finance constraint (11). But buffer stocks are indeed likely to be depleted in the absence of any intervening shock. Since non-fungible trade credit enters the finance constraints, the attempt by everyone to use money to maintain previous expenditure levels will generally involve a running down of *aggregate* money holdings, something that would not happen in a cash-only world with a fixed stock of token money. Furthermore, the increased likelihood of future sales difficulties will induce many transactors to use credit rather than fungible cash, and to defer repayments, thus amplifying the running down of unused credit balances. As the process continues, the desire of shops to protect their net cash receipts, together with the increased likelihood of default by customers, will induce them to reduce the availability of credit even to customers who have not yet reached their limits. All these forces will amplify the endogenous contraction in the aggregate supply of money, broadly defined to include trade credit, unless there is a countervailing increase in the supply of token money. Because of this, the typical transactor will find that the finance constraint (11) is permitting a smaller and smaller amount of expenditure, and will be forced by this constraint to curtail his/her effective demands for goods regardless of the desire to maintain habitual consumption levels.

CONCLUSION

Although the above account of trading processes is seriously incomplete, something like it seems necessary if there is to be any hope of reviving the quest for microfoundations that Leijonhufvud helped to stimulate in the 1970s, the quest that was derailed by the new classical revolution. The microfoundation needed to understand the co-ordination problems with which Leijonhufvud was wrestling consists not of the formal translation of macro theory into the neowalrasian code of new classical economics, but the elaboration of more convincing conceptual accounts of how economies allow the highly complex interactions of modern economic life to continue in the face of disturbances. Until this need becomes more generally recognised, the best ideas are likely to go the way of the dichotomised budget constraint: that is, to be emptied of their substantive content and swallowed by conventional equilibrium analysis.

Notes

1. The unpublished paper cited by Leijonhufvud was entitled 'Unemployment and the Cost of Information.'
2. See, for example, Laidler (1986) or Howitt (1986).
3. As Kohn (1988) has pointed out, there are several historical antecedents to Clower's analysis, including the earlier analysis of Brunner (1951), and the similar contribution of Tsiang (1966) whose origins can be traced to Robertson (1926). None of these, however, had Clower's focus on effective demand, and none had a direct effect on the development on modern cash-in-advance models.
4. See, for example, Hahn (1975) or Drazen (1980).
5. Clower (1984), pp. 264–7.
6. See Leijonhufvud (1968, p. 80), and the discussion in Leijonhufvud (1973).

7. Assuming, that is, a strictly positive vector of money prices. Clower did not discuss the technical issue of zero prices but his analysis of the exchange relation presumes that all rates of exchange are strictly positive and finite, since it requires that all trades involve exactly two commodities.
8. This was the interpretation of Barro and Fischer (1976) in their prominent survey of monetary economics.
9. See, for example, Patinkin (1989, preface), Wallace (1980), or Kiyotaki and Wright (1989).
10. See Blanchard and Fischer (1989, p. 49).
11. That is, $R_0 = 0$, and for all $t \geq 1$ R_t is implicitly defined by the equation: $(1 + R_t)^t = (1 + r_0) \ldots (1 + r_{t-1})$.
12. Let x be the simple annual rate of interest. Then $1 + r = (1 + x)^d$, where d is the length, in years, of the Hicksian week. Thus the tax rate is: $(1 + r)^2 - 1 = (1 + x)^{2d} - 1 = T(x, d)$. Straightforward calculation yields: T (0.05, 1/365) = 0.000267, and T(5.146823, 1/365) = 0.01.
13. A similar story is told by Lucas (1980) of the divided household, in which the seller who sets out at the beginning of the post-tâtonnement trading process is out of touch with the buyer until the end of the week, and therefore cannot pass along any money received until it is too late for current purchases.
14. See, for example, Sargent (1987, ch. 5).
15. That is, as noted above, the pair of constraints (5) and (6) define exactly the same set as the pair formed by (5) and (6′). Since (5) and (7) together are equivalent to (8), therefore requiring (5) ~ (7) as in the cash-in-advance model is equivalent to requiring (6') and (8). But maximising (4) subject to (6') and (8) is clearly equivalent to maximising (9) subject only to (8), with u given by (10).
16. This has been pointed out by Ostroy and Starr (1990).
17. That is, according to (6) an increase in m_t would be equivalent to an equal-valued increase in endowment of a good for which the transactor has a net demand at time t, but not of one for which s/he has a net supply.
18. See, for example, the critique of Wallace (1980).
19. See Clower (1993).
20. See Howitt (1973).
21. See Walker (1984).
22. See, for example, Clower (1973).
23. Assume, for simplicity, that each shopkeeper deals in only one non-money good.
24. This would obviously not be true of the credit limits on widely accepted bank credit cards, or chequing deposits.
25. See Clower and Howitt (1978).
26. See, for example, Miller and Orr (1966). See also Howitt (1988) for a preliminary attempt to model the demand for money by a shopkeeper. The idea of eliminating the awkward lag in Clower's model by invoking set-up costs, and supposing that receipts are too frequent for people to find it optimal to spend them instantaneously, was first proposed by Fried (1973), and subsequently adopted by such writers as Jovanovic (1982) and Romer (1986).
27. See also Foley and Hellwig (1975).

References

Barro, R. J. and S. Fischer (1976) 'Recent Developments in Monetary Theory', *Journal of Monetary Economics*, vol. 2, April, pp. 133–67.

Barro, R. J. and H. I. Grossman (1971) 'A General Disequilibrium Model of Income and Employment', *American Economic Review*, vol. 61, March, pp. 82–93.

Benassy, J. P. (1975) 'Disequilibrium Exchange in Barter and Monetary Economies', *Economic Inquiry*, vol. 13, June, pp. 131–56.

Blanchard, O. and S. Fischer (1989) *Lectures on Macroeconomics* (Cambridge, Mass.: MIT Press).

Brunner, K. (1951) 'Inconsistency and Indeterminacy in Classical Economics', *Econometrica*, vol. 19, April, pp. 152–73.

Clower, R. W. (1965) 'The Keynesian Counter-revolution: A Theoretical Appraisal', in F. H. Hahn and F. P. R. Brechling (eds), *The Theory of Interest Rates* (London: Macmillan).

Clower, R. W. (1967) 'A Reconsideration of the Microfoundations of Monetary Theory', *Western Economic Journal*, vol. 6, December, pp. 1–9.

Clower, R. W. (1973) 'Theoretical Foundations of Monetary Policy', in G. Clayton, J. C. Gilbert and R. Sedgwick (eds), *Monetary Theory and Monetary Policy in the 1970s* (London: Oxford University Press), pp. 15–34.

Clower, R. W. (1984) 'Afterward', in D. A. Walker (ed.) *Money and Markets* (New York: Cambridge University Press).

Clower, R. W. (1993) 'Towards a Reconstruction of Economics', Address to the annual meeting of the Canadian Economic Association, Ottawa, 5 June, 1993.

Clower, R. W. and Peter Howitt (1978) 'The Transactions Theory of the Demand for Money: A Reconsideration', *Journal of Political Economy*, vol. 86, June, pp. 449–66.

Drazen, A. (1980) 'Recent Developments in Macroeconomic Disequilibrium Theory', *Econometrica*, vol. 48, March, pp. 283–306.

Foley, D. K. and M. Hellwig (1975) 'Asset Management with Trading Uncertainty', *Review of Economic Studies*, vol. 42, July, pp. 327–46.

Fried, J. S. (1973) 'Money, Exchange and Growth', *Western Economic Journal*, vol. 11, pp. 285–301. (now *Economic Inquiry*).

Grandmont, J.-M. and Y. Younès (1973) 'On the Efficiency of a Monetary Equilibrium', *Review of Economic Studies*, vol. 40, pp. 149–65.

Hahn, F. (1975) 'Money and General Equilibrium', *Indian Economic Journal*, vol. 23, October–December, Special Number in Monetary Economics, pp. 109–22.

Hicks, J. (1932) *The Theory of Wages*, (London: Macmillan).

Hicks, J. (1935) 'The Theory of Monopoly: A Survey', *Econometrica*, vol. 3, January, pp. 1–20. Reprinted in *Classics and Moderns*, vol. III of his *Collected Essays on Economic Theory* (Cambridge, Mass.: Harvard University Press, 1982), pp. 131–152.

Howitt, P. (1973) 'Walras and Monetary Theory', *Western Economic Journal*, vol. 11, December, pp. 487–99.

Howitt, P. (1974) 'Stability and the Quantity Theory', *Journal of Political Economy*, vol. 82, January/February, pp. 133–51.

Howitt, P. (1986) 'Conversations with Economists: A Review Essay', *Journal of Monetary Economics*, vol. 18, July, pp. 103–18.

Howitt, P. (1988) 'Money and the Timing of Transactions in Intermediated Exchange', in M. Kohn and S. C. Tsiang (eds), *Finance Constraints, Expectations, and Macroeconomics* (Oxford: Clarendon Press) pp. 54–70.

Jovanovic, B. (1982) 'Inflation and Welfare in the Steady State', *Journal of Political Economy*, vol. 90, June, pp. 561–77.

Kiyotaki, N. and R. Wright (1989) 'On Money as a Medium of Exchange', *Journal of Political Economy*, vol. 97, August, pp. 927–54.

Kohn, M. (1981) 'In Defense of the Finance Constraint', *Economic Inquiry*, vol. 19, April, pp. 177–95.

Kohn, M. (1988) 'The Finance Constraint Theory of Money: A Progress Report', Unpublished MS, Department of Economics, Dartmouth College, Hanover, New Hampshire.

Laidler, D. (1974) 'Information, Money and the Macroeconomics of Inflation', *Swedish Journal of Economics*, vol. 76, March, pp. 26–41.

Laidler, D. (1986) 'The New-Classical Contribution to Macroeconomics', *Banca Nazionale del Lavoro Quarterly Review*, vol. 156, March, pp. 27–55.

Leach, J. (1983) 'Inflation as a Commodity Tax', *Canadian Journal of Economics*, vol. 16, August, pp. 508–16.

Leijonhufvud, A. (1968) *On Keynesian Economics and the Economics of Keynes: A Study in Monetary Theory* (New York: Oxford University Press).

Leijonhufvud, A. (1973) 'Effective Demand Failures', *Swedish Journal of Economics*, vol. 75, March, pp. 27–48 (now *Scandinavian Journal of Economics).*

Lucas, R. E., Jr (1980) 'Equilibrium in a Pure Currency Economy', *Economic Inquiry*, vol. 18, April, pp. 203–20.

Miller, M. H. and D. Orr (1966) 'A Model of the Demand for Money by Firms', *Quarterly Journal of Economics*, vol. 80, August, pp. 413–35.

Okun, A. (1981) *Prices and Quantities* (Washington, DC: Brookings).

Ostroy, J. M. and R. M. Starr (1990) 'The Transactions Role of Money'; in B. M. Friedman and F. H. Hahn (eds), *Handbook of Monetary Economics*, Handbooks in Economics, no. 8, vol. 1 (Amsterdam: North-Holland), pp. 3–62.

Patinkin, D. (1965) *Money, Interest, and Prices*, 2nd edn (New York:Harper & Row).

Patinkin D. (1989) *Money, Interest, and Prices*, abridged 2nd edn (Cambridge, Mass.: MIT Press).

Phelps, E. S. *et al.* (1970) *Microeconomic Foundations of Employment and Inflation Theory* (New York: Norton).

Robertson, D. H. (1926) *Banking Policy and the Price Level: An Essay in the Theory of the Trade Cycle* (London: Staples Press).

Romer, D. (1986) 'A Simple General Equilibrium Version of the Baumol–Tobin Model', *Quarterly Journal of Economics*, vol. 101, November, pp. 663–85.

Sargent, T. J. (1987) *Dynamic Macroeconomic Theory* (Cambridge, Mass.: Harvard University Press).

Stockman, A. C. (1981) 'Anticipated Inflation and the Capital Stock in a Cash-in-Advance Economy', *Journal of Monetary Economics*, vol. 8, October, pp. 387–93.

Svensson, L. E. O. (1985) 'Money and Asset Prices in a Cash-in-Advance Economy', *Journal of Political Economy*, vol. 93, October, pp. 919–44.

Tsiang, S. C. (1966) 'Walras Law, Say's Law and Liquidity Preference in General Equilibrium Analysis', *International Economic Review*, vol. 7, September, pp. 329–45.

Walker, D. A. (ed.) (1984) *Money and Markets: Essays by Robert W. Clower* (New York: Cambridge University Press).

Wallace, N. (1980) 'The Overlapping Generations Model of Fiat Money', in J. Kareken and N. Wallace (eds), *Models of Monetary Economies* (Federal Reserve Bank of Minneapolis).

Wilson, C. (1979) 'An Infinite Horizon Model with Money', in J. Green and J. A. Scheinkman (eds), *General Equilibrium, Growth and Trade* (New York: Academic Press).

Part II

7 Viconomics: An Evolutionary Science of Mind and Society*

John McCall

> Vico was the first modern philosopher to discover in his own mind and in the European past, all human destiny. W. B. Yeats, *On the Boiler*

Axel Leijonhufvud is one of our profession's most prominent practitioners of *Viconomics*. Axel's writing is pregnant with metaphor. Seemingly unrelated ideas are linked effortlessly, illuminating the economic discourse. This Vichian talent is so natural that readers are sometimes unaware of the complexity of Axel's economic analysis. For example, in 1968 Axel[1] observed:

> The microbiological revolution that in recent years has begun to transform medicine, came about when the presumption that 'health is only natural' was abandoned and the basic question changes: 'How is it possible that such an improbable arrangement of molecules as a cell maintains itself?' 'How is it even conceivable that it could?' The changed formulating of the question reflects an exchange of 'presuppositions' at a very basic level. To find the system under study in an 'organized' or 'coordinated' state is seen as unexpected and, consequently, in need of explanation. The a priori probability of some breakdown in organization, of a prior 'coordination failure', is on the other hand seen as so large as not to deserve priority in the search for explanations. Nonetheless – and this, of course, is the point – the pursuit of answers to the question of how life is, improbably, maintained has proved more productive of ideas, hypotheses and results relevant to the important 'failure' or 'illness' questions than did the old 'direct' approach. As

*In his appraisal of Giambattista Vico ('one, of the greatest thinkers … in the field of the social sciences'), Schumpeter (1954) claims that 'His *New Science* is best described by the phrase "an evolutionary science of mind and society" … mind and society two aspects of the same evolutionary process'. This chapter has benefited from comments by A. DeVany, P. Burke, D. P. Verene and K. Velupillai.

> answers to the 'new' and 'indirect' question begin to come in, it becomes apparent how tenuous life is in the case of the cell – how many things must 'go right' for it to be maintained – and, consequently, at how many points it is possible for the maintenance mechanisms to break down.
>
> A modern economy is a highly improbable structure. Yet, by and large, economists have not, apparently, regarded it as such and have not assigned a high priority to the coordination question.

The importance of history, institutions and evolution characterise all of his research from his classic study of Keynes to his recent monograph (with Daniel Heymann) on inflation. In the latter, a society responds to relentless inflation by decomposing into smaller and smaller trading (exchangeable) groups. This 'unnatural' reduction in exchangeability[2] marks the onset of a collapsing hierarchical society. This analysis shows that trade is the language of economics and also reveals the importance of exchangeability in Viconomics.

The power and absurdity of artificial barriers to inter- and intra-disciplinary research is portrayed in Axel's elegant parable: 'Life Among the Econ'. Axel's critique of axiomatic economics is the exact counterpart of Vico's trenchant but constructive attack on René Descartes. Axel's ongoing research programme in Computable Economics is his constructive response to the 'current Cartesian status of economics'. There are many other parallels between Vico and Leijonhufvud, which I hope, will be evident as one reads this chapter.

THE LIFE AND TIMES OF GIAMBATTISTA VICO: A BRIEF SKETCH[3]

Vico was born in 1668. His father was a bookseller with a cheerful disposition, but Vico notes that his mother suffered from melancholia. Naples was his birthplace and for all practical purposes his entire life was spent in Naples, where he died in 1744. He was a poet and a lawyer. His life was that of a dedicated scholar at the University of Naples. He yearned for the chair of jurisprudence, but instead was appointed to a low-ranking chair in rhetoric, which he occupied from 1699 to 1741. The position was reflected in his salary. His large family and low income placed him well below the 'poverty line' of his time. A severe head injury in his youth left lifelong visible traces. Many of his contemporaries attributed his mercurial intellectual drive, cyclical emotions and constant revision of his philosophical system as new ideas were discovered, to his early accident. Vico acknowledges the gravity and influence of the injury on the first page of his *Autobiography*. His peculiar assessment of its results is most revealing:

> The surgeon, indeed, observing the broken cranium and considering the long period of unconsciousness, predicted that he would either die of it or grow up an idiot. However by God's grace neither part of his prediction came true, but as a result of this mischance he grew up with a melancholy and irritable temperament such as belong to men of ingenuity and depth, who, thanks to the one, are quick as lighting in perception, and thanks to the other, take no pleasure in verbal cleverness or falsehood. (Vico, 1944, p. 111)

Vico was Jesuit trained. His intelligence and diligence were apparent in grammar school from his own description of interactions with teachers and students. Based on extended absences from school and the use of private tutors, it may be best to say that he was 'self-trained'.

Tacitus and Plato were Vico's exemplars. They were harbingers of some of Vico's major contributions. Vico noted that

> Tacitus contemplates man as he is, Plato as he should be. And as Plato with his universal knowledge explores the parts of nobility which constitute the man of intellectual wisdom, so Tacitus descends into all the counsels of utility whereby, among the infinite irregular chances of malice and fortune, the man of practical wisdom brings things to good issue. Now Vico's admiration of these two great authors from this point of view was a foreshadowing of that plan on which he later worked out an ideal eternal history to be traversed by the universal history of all times, carrying out on it, by certain eternal properties of civil affairs, the development, acme and decay of all nations. From this it follows that the wise man should be formed both of esoteric wisdom such as Plato's and of common wisdom such as that of Tacitus. (Vico, 1944, pp. 138–9).

Francis Bacon is also proclaimed by Vico as 'a man of incomparable wisdom both common and esoteric, at one and the same time a universal man in theory and in practice', in short, Plato and Tacitus combined. Vico vows that all his own thinking and writing will be guided by the illumination flowing from this Trinity.

After reading *On the Law of War and Peace* by Hugo Grotius, Vico remarks that he

> found a fourth author to add to the three he had set before himself ... [Grotius] embraces in a system of universal law the whole of philosophy and philology, including both parts of the latter, the history on the one hand of facts and events, both fabulous and real, and on the other of the three languages, Hebrew, Greek and Latin; that is to say, the three learned languages of antiquity that have been handed down to us by the Christian religion. (Vico, 1944, pp. 154–5)

It is important to remember that Naples was controlled by Spain and Austria (1707–34) and the Inquisition was a force that must have constrained Vico's writing. Be this as it may, from my limited understanding of Vico, it seems clear that he was a deeply religious man whose silent critique of Catholicism was probably similar to that of many intellectuals who were eventually canonised.

A REVIEW OF VICO'S PHILOSOPHY

> To contemporary ears Plato's famous dictum in the *Laws* that a serious man keeps his seriousness for serious things, and 'does not waste it on trifles' such as human affairs may sound extreme; in fact, it was never challenged before Vico, and Vico had no influence or echo till the nineteenth century. (Arendt, 1978, p. 154)

The design and evolution of *social institutions* are receiving increased attention by the economics community. The early 1990s have witnessed a renewed interest in the *formation*, *function* and *demise* of economic institutions.[4] As anticipated, economic historians have been at the forefront of this revival. The research stimulated by this renewal has not been limited to historians and applied economists. It also includes economic theorists: mathematical economists and theoretical econometricians. The interest by economic theorists in history and institutions may seem surprising until one reflects on the relationship between the formation of institutions and the reduction of uncertainty.[6] Any serious study of the economics of uncertainty must consider the evolution of institutions as a major source of risk reduction.

Institutions include the customs, language and legal system of a society.[7] In addition, insurance, capital markets, information networks, the education system and industrial organisation are institutions. The manner in which they adapt to changing circumstances affects the well-being of each member of society. Uncertainty diminishes as institutions guide information flows and improve induction by the mind and society. There is a corresponding increase in predictability. Institutions not only reduce uncertainty and generate specialisation, regularity and civility. They also harbour the ethical qualities distinctive to humanity. Kumaraswamy Velupillai (private correspondence) says it best:

> Our sense of justice, goodness, etc., call forth institutions that are repositories of them. Laws, moral codes, ethical conduct, etc. These must have a relative permanence so that behaviour can be in accordance with those senses that were the backdrop for the emergence of these institutions. Hence the reduction of uncertainties and the enhancing of predictabilities. In this way *symmetry* plays the fundamental role ... The manner in which institutions adapt to changing circumstances must, somehow, be compatible with our sense of justice, goodness, beauty, truth and kindness. When the adaptive process violates any of these entities then one will detect asymmetries ... Till now one was always theorizing about the symmetric–asymmetric state. Thus the importance of exchangeability.

Properly functioning institutions are indiscernable. Individuals convert complex and risky behaviour into regular routines that are virtually effortless and riskless. This permits each of society's members to specialise in those activities yielding the greatest personal rewards. If society is organised *smoothly*, the welfare of its members coincides with society's welfare.[8]

The genetic evolution that was necessary for the emergence of *Homo sapiens* is now fairly well understood and provides a satisfactory explanation of *physical* evolution. Clearly, nature would not endow man with self-consciousness until the mind had devised methods for increasing the predictability of his behaviour above some threshold. That is, humans would survive only if inherent, 'raw' risk was reduced by the development of 'insurance' institutions. They enabled individuals to rely on routines or habits for resolving problems, while they concentrated on specialised activities. A highly conscious organism would be unable to accomplish this resolution if it was thrust into a dangerous, uncertain and unpredictable environment. Dread would induce inaction and extinction. Thus society, mind and the brain evolved concurrently.

The prime mover in Vico's evolutionary theory of society was the quest to control uncertainty.[9] Uncertainty manifested itself in 'sheer fright' about the inexplicable behaviour of natural phenomena. According to Vico, responses to this uncertainty included marriage, monogamy, family, tribes and, finally, nations. The evolution of language was a necessary prerequisite to this *nation-building*. Vico's theory of linguistics was highly original. Communication was first by sign language and grunting, then by a guttural singing refined into Homeric poetry. Rhetoric, eloquence and jurisprudence emerged from this primitive poetry and *were absolutely essential* to the diminution of uncertainty and the formation of a civilised society. Just as primitive man was certain of the sunrise, so civilised man must be certain of a stable legal system: a system that defined the sphere in which he could operate without fear of violence by either individuals or government. Religion explained nature's erratic behaviour. Language was the cement that held groups together. Eloquence was the method used to persuade society to undertake new and risky enterprise such as forming a treaty with contiguous strangers or changing an essential technology, such as adopting farming. Eloquence merged into jurisprudence that 'protected' the individual and group, during these enterprises.

Vico is associated frequently with the Platonists and idealists. Yet he was a vehement critic of Descartes' idealism. His emphasis was on *the real, the tangible*, and *the frailties and flux of human nature*. Most Vico scholars seem to agree that Bacon, Plato and Tacitus were the strongest forces in Vico's philosophy. But the most decisive influence on Vico was Descartes.[10] He provided the stimulus that ignited Vico's creative genius. The provocative function of conflicting ideas is undervalued in their genesis of novelty. The sparks generated by the collision of the Cartesian and Vichian systems may be the most significant contribution to the intellectual evolutionary process.

Vico repeatedly attacked the abstract concept that was never enriched by concrete experience. Descartes' profound mistake was to abstract from concrete reality before perceiving the relationships among facts. Vico anticipated the limitations of a purely mathematical social science. By withdrawing from the richness of experience one was left with a distorted and desiccated view of social phenomena.

The decline of altruism as one moved away from the family was one manifestation of Vico's non-Utopian view of human nature. Another was his clear recognition of the primitive aggressiveness that reigned before the formation of institutions and the onset of civilisation. Only the simultaneous *growth of mind and civility* could keep destructive instincts in check. These instincts persisted in civilised societies and therefore the return to chaos was always a threat. He was painfully aware of the Heraclitian flux and, indeed, it is essential to his theory of a *recurrent civilised society*.

The Well-known Contributions of Vico

Vico is best known for his trenchant critique of Cartesian philosophy. He refused to acknowledge the bifurcation of man into mind and matter, the supremacy of mathematics as a science, and the deterministic system that behaved with clockwork precision. The Cartesian position was very popular among eighteenth-century philosophers, who regarded Descartes' philosophy as the humanistic counterpart of the scientific discoveries of Sir Isaac Newton and Johann Kepler. But Vico would have none of this! Concrete reality was too unpredictable to be captured by any deductive system. This was especially true of the social sciences.

Vico's writings were well known to Auguste Comte and Karl Marx, but not to Adam Smith or David Hume. One of the most famous expositor of Vico's philosophy was Benedetto Croce who, of course, selected those parts that agreed with his philosophy. As M. Mooney sagely observes:

> Croce could only be embarrassed by Vico's link with the rhetorical tradition. For nothing is quite so foreign to romanticism as any thought of utility, and nothing is further from the spirit of oratory than the romantic idea of self-expression.
>
> Although poetry represents, as Croce saw, a form of knowledge that is imaginative and not intellectual, it does *not* mark, as he also held, the boundary between arts and sciences. Not only does Vico maintain, with Bacon his mentor, the traditional understanding of 'arts', but he also assigns a 'poetic' origin to the entire cycle of learning – sciences as well as arts. Poetry – or 'poetic wisdom', as Vico more commonly called it – designated that earliest phase of human culture which was spontaneous, synthetic, and constructive, in contrast to a later phase, which was reflective, discursive, and critical. Vico had no interest, as Croce so vainly hoped, in 'forms of the spirit' independent of the course of concrete history. He was totally absorbed in the question of how humanity arose, and once established, how it advanced from its crude beginning to an age critical intelligence. Within his theory of culture, poetry is the crucial reality. (Mooney, 1985, pp. 27–9)

The power, rhetorical splendor, and in some cases, strident assertiveness of Vico's writing would not attract most economists, and certainly not the Scots. He attempted to mould a language encompassing the wisdom of Homer, Tacitus and Bacon that would promote his own deep insights. Homer came to dominate this mythical 'Irish stew' which was savoured by James Joyce and W. B. Yeats, but proved repugnant to many 'rational' economists.

The main point is that the other social sciences were influenced directly by Vico. Indeed, many attractive ideas usually attributed to Comte, Marx, Emile Durkheim, or Marcel Mauss are easily traced to Vico. These include: the network of contractual relationships[11] that comprise the individual, influence individual decision-making, and reduce uncertainty; the basic institutions – family, community, state and nation that create homogeneity (exchangeability) when combined with a common legal system that protects and enforces contracts, a common moral code that has become *instinctual* by reason of parental educational and religious training. The process orientation of many social scientists manifested by an evolutionary approach to social problems and an affinity for classification that is shared by many scientists, especially biologists, are also Vichian. Indeed, as we will see later, all the physical sciences and social sciences are characterised by adaptive classification processes.

Vico recognised clearly that the thirst for order and stability in everyday affairs was genetic and first quenched by the Homeric myths, then by religion and, finally, by science. These 'mysterious' forces were used to explain nature's disorderly conduct, while at the same time providing the impetus for the evolution of civilised societies.

Economists sometimes ignore the 'magical' aspects of their creative models. However, Friedman (1953), observes:

> The construction of hypotheses is a creative act of inspiration, intuition, invention; its essence is the vision of something new in familiar material. The process must be discussed in psychological, not logical, categories; studied in autobiographies and biographies, not treatises on scientific method; and promoted by maxim and example, not syllogism or theorem.

The third conceit

Vico distinguishes between two errors that afflict intellectuals. The first is called the 'conceit of nations': if an idea or invention is used in every country, it must have originated in one's own country. Since Vico regarded the creation of a civilised society as *the* most spectacular invention, his decisive example of this conceit was the belief that this singular creative event occurred in one's own country and was copied by the rest of the civilised world. This conceit was anathema to Vico's evolutionary theory. The civilised society that evolved in one country reflected the peculiarities of its historical unfolding. Each civilised country invented its *own* social structure. Diffusion of a unique society invented in one country was the clearest and most egregious form of the 'conceit of nations'. A modern version of this phenomenon is the transfer of technological innovations among countries. This diffusion is relatively easy when the invention is culture-free. However, costs rise dramatically as the cultural properties of the invention become more pronounced. Computer software is a good example of the latter.

Vico referred to the second error as the 'conceit of scholars: if I am able to solve a particular puzzle, my ancient ancestors could also solve it. The manifestation of this error most offensive to Vico was the common belief in a 'golden age' or the 'wisdom of the ancients'. This was mystical anhistorical nonsense. The problems that confronted the ancients emanated from an environment entirely different from those of the twentieth century. If the problems were unique, they would never be reconsidered. On the other hand, many recurring problems that were threatening presumably were solved so frequently that the solutions became routine features of civilised behaviour. These routines would go unnoticed by modern people, who are confronting new problems. The second conceit belonged to the classic quarrel between ancient and modern wisdom that was raging during Vice's life. The contributions of each, imagination of the ancients and abstraction of the moderns, were acknowledge by Vico. However, in his reaction to the Cartesian dismissal of the poetic wisdom of the ancients, Vico developed a theory of sequential learning anchored in the imagination and memory. The modern (Cartesian) abstract approach to learning was egregiously defective because of its disdain for the poetic, rhetorical and subjective composition of wisdom. In this version of the quarrel, Vico was on the side of the ancients. See Verene (1981) for a lucid discussion of this feud.

We may add a third conceit not mentioned by Vico: the 'Vico conceit'. The error is to label Vico a prolific, but disorganised thinker – the producer of many *disconnected* ideas. Some of these ideas were highly original, some were incomprehensible, and some were just plain wrong. This is an error in that it does not agree with either Vico's philosophy or his personality. To be ingenious was his mission. Ingenuity measured his intelligence, and his pride would never allow him to construct a disconnected tree. This Vico tree spans all Vico's ideas. There is a unity here that forbids sampling. How, then, can Vico be the patron for Marxists, Pragmatists, Idealists, Romanticists and so on? One must read all of Vico before selecting an especially attractive branch; and the branch may be less appealing and even repulsive after the entire tree has been envisioned.

Having said this, I must confess that ignorance and indolence have conspired to render me an active member of the Third Conceit Club.

Evolution

Social evolution progressed from marriage to family to community to nation. At the same time, language and jurisprudence were evolving. Language began with pointed grunts and evolved through a poem-like singing to a simpler, more conventional form. The key to Vico's philosophy is the development of this common language. *Rhetoric merged into eloquence and these culminated in jurisprudence.*

Contracts

Thus, language gave rise to the legal system which protected individuals from one another, guaranteed property rights, and enforced contracts. Contracts (treaties) with other nations were enforced by the threat of war. Vico considered war to be contractual sanction that *reduced* its likelihood. Preparedness for war was thus a deterrent to its occurrence: 'The origins of war and peace, whereby all commonwealths were brought into being by force of arms and then composed by laws. From the nature of these human institutions derives their external property: that wars are waged so that peoples may live secure in peace.'

The symmetry that Newton and Sir Francis Bacon discovered in nature was also present in civilised societies.[12] While uncertainty was not eliminated, it was reduced substantially. There were costs associated with these homogeneous societies. The outsider was disdained, innovations were not encouraged, conformity was applauded,[13] ant the reflective, solitary individual was suspect.

Co-operation and exchangeability

Becker (1976) has observed that co-operative behaviour is most likely within families, while competitive behaviour characterises non-family relationships. As usual, he devises an ingenious economic argument that is compatible with the Hamiltonian–genetic explanation of this phenomenon.[14]

Both these explanation can be augmented by an exchangeability approach that underlies B. de Finetti's attempts to study social phenomena. The homogeneity that resides within families has been promulgated by nature, common culture and familiarity with one anther's idiosyncracies: that is, a smoothing of heterogeneity. This homogeneity is similar to exchangeability and is the major source of co-operation.

But let us stop and listen to Vico:

> We thereby establish the fact that man in the bestial state desires only his own welfare; having taken wife and begotten children, he desires his own

welfare along with that of his family; having entered upon civil life, he desires his own welfare along with that of his city; when its rule is extended over several peoples, he desires his own welfare along with that of the nation; when nations are united by wars treaties of peace, alliances, commerce, he desires his own welfare along with that of the entire human race. In all these circumstances man desires principally his own utility.

As one moves from the family to the 'tribe', many of these homogeneous forces remain, and hence a 'hardball, tit-for-tat, competitive' environment need not dominate tribal relations. Common moral codes, common language and common legal systems engender homogeneity and co-operative relationships. As the group enlarges, heterogeneous factors tend to outweigh the commonalities and co-operation is less likely. But one need not remain in this anonymous, alien land. One learns quickly how to discriminate and classify individuals, given their key background variables. This gives rise to exchangeability and instils the confidence required for co-operative arrangements. In short, there is a hierarchy of exchangeable environments with the family at the centre and unexplored regions at the outer boundary. As the cost of achieving contractual relations increases uniformly, the outer boundary shrinks and the hierarchy begins to collapse on to the family. Vico would interpret this collapse as the decline of society. At the limit, not even the family is an exchangeable entity – the initial feral state has recurred.

High inflation exhibits this collapsing phenomenon (see Heymann and Leijonhufvud (1994). Vico's language mixed with de Finetti's is apt: the 'unnatural' reduction in exchangeability marks the initiation of a collapsing society.

Mancur Olson (1982): *The Rise and Decline of Nations* New Haven, Yale University Press argues that the decline of nations is accompanied ('caused by') the formation of self-interest groups which tend to divide society. This division of society into 'exclusive rent-seeking clubs' increases transaction costs, reduces trade, interferes with the efficient operation of the economy, induces conflict, and, in short, undoes the homogeneity process brought about by common culture, common laws and common language. It remains to be seen how powerful this Balkanisation process is relative to the forces of homogenisation that have been remarkably potent in the United States of America.

Processes

Perhaps Vico's most important contribution lies in his abhorrence of static systems. Of course, this Heraclitian realism was not novel in itself. The novelty manifested itself in two ways. First, societies are processes and are subject to cyclical decline and renewal: 'Men first feel necessity, then look for utility, next attend to comfort, still later amuse themselves with pleasure, thence grow dissolute in luxury, and finally go mad and waste their substance'. More succinctly, 'the nature of peoples is first crude, then severe, then benign, then delicate, finally dissolute'. Thus no civilisation can persist without adapting to these 'natural phenomena'. So, too, no mathematical model of nature is invariant to the flux of concrete phenomena. Unless a science adapted to these fluctuations, it would soon be obsolete. Just as a static society was doomed, so too was any mathematical model that remained remote from the concrete.

Many economists have interpreted the myriad asymmetries in economic processes as proclamations of market failure with some sort of mending required by non-market interventions. Closer scrutiny reveals that many of these asymmetries have been symmetrised by natural forces.[15] These forces have produced a variety of institutions that are sometimes quite subtle and difficult to observe. Nevertheless, they may be sufficiently strong and durable to reduce asymmetry. This, of course, does not mean that all is well and no interventions are justified. Asymmetries may be mended after a diligent search for symmetrising mechanisms comes up empty. However, the absence of 'natural' institutions *suggests* that their implementation is difficult and *perhaps* more costly than *doing nothing*.

Insurance

Another uncertainty reducing institution that Vico advocates is *insurance*:

> In the every birth of [domestic] economy, they fulfilled it in its best idea, which is that the fathers by labor and industry should leave a patrimony to their sons, so that they may have an easy and comfortable and secure subsistence, even if foreign commerce should fail, or even all the fruits of civil life, or even the cities themselves, so that in such last emergencies the families at least may be preserved, from which there is hope that the nations may rise again. (Vico, 1948, p. 181 [525])

This description shows Vico's concern for *renewal* when societies collapse.

A stable society is much less vulnerable to disintegration than is a fluctuating society.[16] There is a balance between innovation and tradition that resides in all surviving institutions. This natural range of flexibility *permits* the institution or organism to adapt to changes that would extinguish a brittle entity, while at the same time *controls* endogenous change by preserving the 'wisdom of the past'. Thus, internal heterogeneity never becomes self-destructive. Observe that a wildly fluctuating heterogeneity increases the costs of partitioning groups into homogeneous sub-sets. For example, it becomes more and more difficult for firms to rank their work-force, for contracts to be formed and sustained, and for the associated institutions to survive.

Vico's primitive society is characterised by three principles:

> And in no nation, however savage and crude, are any human actions performed with more elaborate ceremonies and more sacred solemnity than the rites of religion, marriage, and burial. For, by the axiom that 'uniform ideas, born among peoples unknown to each other, must have a common ground of truth', it must have been dictated to all nations that from these three institutions humanity began among them all, and therefore they must be most devoutly guarded by them all, so that the world should not again become a bestial wilderness. For this reason we have taken these three external and universal customs as three first principles of this Science. (Vico, 1948, p. 97 [333])

Many anthropologists discount Vico because he denies diffusion – each civilised society began in isolation – and he ignores the polyandry and other aspects of primitive life that are now well known. Vico might claim that these primitive 'societies' were still in the feral state. It is more likely, however, that he would have adapted his theory to these concrete findings. On the other hand, Martin and May (1982) conclude: 'Much work on birds, mammals and other creatures indicates that monogamy often emerges when large investments in offspring necessitate cooperation between the parents. It is possible that the increasing dependence of the human infant, associated with progressively increasing brain size and cultural complexity, has favored culturally determined monogamous tendencies in various human societies, even though these have not led to the complete suppression of biological indicators of a polygynous society.' There is also much truth in Vico's *burial principle*. Scarre (1984) observes that the genetic composition of bones found in ancient graves show that families were buried in the same grave site.

Ingenuity

Vico's notion of ingenuity is 'the faculty of bringing together things that are disparate and widely separated'. Mooney (1985) notes that 'ingenuity' is used consistently throughout his work and 'is a lodestone of his thought, drawing to itself so many of his original ideas':

> One does not need to call on ingenuity; one either has it or does not, sees connections or misses them utterly. Vico was a child of acute ingenuity, he claimed, and so, too, are children generally, if only we will recognize it and train them accordingly. (Mooney, 1985, p. 151)

Vico's passionate concern for education anticipates much of John Dewey's philosophy, as does his attitude toward uncertainty. His emphasis on the concrete and the richness of experience that must replace the abstract mathematics of the Cartesian curriculum reminds us of Ernst Mach:

> the object of science is the connection of phenomena; but the theories are like dry leaves which fall away when they have ceased to be the lungs of the tree of science. (Mach, 1906)

Consider the following quotations:

> In the main, if a phenomenon is important for our welfare, it interests and excites us ... Dangerous things fill us with involuntary fear; poisonous things with distaste; indispensable things with appetite. Mind and world in short have been evolved together, and in consequence are something of a mutual fit.
>
> However man may have acquired his faculty for divining the ways of nature, it has certainly not been by a self-controlled and critical logic ... It resembles instinct too in its small liability to error, for though it goes wrong oftener than right, yet the relative frequency with which it is right is on the whole the most wonderful thing in our constitution.

They certainly could have been said by Vico, but the first was by William James in 1890, and C. S. Peirce wrote the second in 1903. Both quotes appear in a basic paper by Anderson (1983) on neural networks.

Poetry and mathematics

The sensuous aspects of language enhance its ability to convey information and to perform induction; that is, to extract symmetry, pattern and meaning from the vast variety of waves emanating from concrete phenomena. Vico was the first to see poetry as the primal source of human intelligence. The evolution of language fortified induction, not deduction. Induction is the basis of both Bacon's physical science and Vico's 'new' science. In his attack on rigid Cartesian formalism, Vico overlooked the practical beauty inherent in mathematics. Just as poetry is the language of Vico's 'New Science', so too is mathematics the language of Bacon's physical science. The transmission and interpretation of information – *induction* – began with grunting and counting. These basic activities evolved into poetry and mathematics. Both processes are refinements of their original functions – learning and trading.

Creativity is common to both poetry and mathematics. However, the pathos and ethos characteristic of poetry is quite different from the mechanical and perfunctory aspects of mathematics. The human spirit cannot be captured by a mathematical process.

Induction

Berlin (1976) considers Vico's reconstruction of the past to be one of his most important contributions. He refers to Vico as the 'true father both of the modern concept of the culture ... cultural pluralism, according to which each authentic culture has its own unique vision, its own scale of values'. Furthermore, these values and visions change – culture is a process. With hard work and imagination, individuals living in period $t_0 + t$ can infer the culture of their forefathers at t_0. I. Berlin sees this backward induction as being similar to the studies of modern anthropologists, who search for meaning in the unfamiliar and, frequently, grotesque myths, rituals, visions and 'natural responses of extant primitives'. This spatial induction is quite difficult. Decoding observed primitive behaviour requires insight, tenacity and luck. The attempt by Vico to *reconstruct* the behaviour of his early ancestors and to *explain* the myths and other features of their reconstructed behaviour was truly breathtaking in its novelty and precariousness. It shattered the static concept of human nature. Yet Vico thought the poetry of these early barbarians possessed a simplicity and linguistic purity superior to that of Vico's contemporaries. Vico contemplates a *process* when the seventeenth-century notion of time was completely static?[17] In the midst of the Inquisition, how was Vico able to evade persecution? P. Rossi claims that:

> The tortuous, difficult path that had led humanity out of barbarism to civil order was, for Vico, decisive proof of the presence in history of a Providence that 'disposes to a universal end' the particular ends of human actions and that uses those narrow ends, beyond the conscious purposes of individuals and of peoples 'to preserve the human race upon this earth'. To acknowledge that brutish state of barbarism and to assert a generalized dehumanizing of men seemed to Finetti,[16] on the contrary, an impious denial of the work of Providence:

> What could be harder to accept, more difficult to harmonize with the idea that we have of Providence, than to imagine that the Lord, after having made the supreme gesture of preserving mankind from total destruction through the Ark, should then abandon it so completely that men reached the point where they became like beasts, all religion lost, the use of reason lost, all language lost, all companionship lost, in few words, humanity lost, falling into a bestial state, rather, worse than that of the beasts themselves? (Rossi, 1984, p. 256)

Vico's preference for induction is explicit in his discussion of medicine:

> In medicine it would be safer to adhere to concrete particulars, and to refrain from assigning to the sorites a role more important than that which, in this field, it deserves to play; it would be safest to lean firmly on induction. (Vico, 1990, pp. 32–3)

and

> we should pay a great deal of attention to symptoms and diagnostics. Let us cultivate the practices of preventive medicine as applied by the Ancients, chiefly gymnastics and dietetics, along with the curative procedures devised by us. (Vico, 1990, p. 33)

Vico reiterates his belief in the uniqueness of the individual and the difficulty of learning in a fluctuating heterogeneous environment. No two sick people are ever the same and the diseases are always novel.

Those using the syllogistic method in this milieu almost never come close to the real source. They 'are led astray by their fondness for that strictly deductive form of reason which the Greeks called *sorites* define. The person who uses the syllogism brings no new element, since the conclusion is already implied in the initial proportion or assumption'. The most we can hope for in our assiduous application of inference to a longitudinal data base is a 'probable approximation to truth'.

On probability

Vico's notion of probability or 'luck' is seminal. It differs substantially from earlier concepts.[19]

> A private evil may be a public good; and just as in the republic founded by men the safety of a nation is the supreme law, so in the universe

> established by God luck would be the queen of all. To put it another way, luck is God's will by which He dominates over the private goods of all men and their peculiar natures, while He looks to the safety of the universe. And just as private safety yields to public safety, so the good peculiar to each person must take second place to the conservation of the universe, and in this way the adversities of nature are its goods. (Vico, 1988, p. 108)

The environment where prudence is essential resembles the landscape of economic decision-making.

> When it comes to the matter of prudential behavior in life, it is well for us to keep in mind that human events are dominated by Chance and Choice, which are extremely subject to change and which are strongly influenced by simulation and dissimulation (both pre-eminently deceptive things). As a consequence, those whose only concern is abstract truth experience great difficulty in achieving their means, and greater difficulty in attaining their ends. Frustrated in their own plans, deceived by the plans of others, they often throw up the game. Since, then, the course of action in life must consider the importance of the single events and their circumstances, it may happen that many of these circumstances are extraneous and trivial, some of them bad, some even contrary to one's goal. It is therefore impossible to assess human affairs by the inflexible standard of abstract right; we must rather gauge them by the pliant Lesbic rule, which does not conform bodies to itself, but adjusts itself to their contours. (Vico, 1990, p. 34)

On myth, language and art

According to Cassirer (1955):

> modern philosophy has in this point turned more and more to man's subjectivity. Myth became a problem of philosophy insofar as it expresses an original direction of the human spirit, an independent configuration of man's *consciousness*. Anyone aiming at a comprehensive system of human culture has, of necessity, turned back to myth. In this sense, Giambattista Vico, founder of the modern philosophy of language, also founded a completely new philosophy of mythology. For Vico the true unity of human culture is represented in the triad of language, art, and myth. (Cassirer, 1955, p. 3)

COMMON SENSE[20]

Common sense glues the institutional bricks of Vico's 'New Science'. This notion reveals the economic character of Vico's philosophy.

Following the spectacular discoveries of Newtonian science and the Cartesian attack on the humanities, Vico announces that man can only know what he has made, society and mind. Furthermore, the physical sciences, in 'learning' and 'controlling' what God has made cannot acquire true knowledge. The entire enterprise of deducing the behavior of the *material universe* from mathematical models is *stochastic* at its core. Vico then proceeds to construct an axiomatic model of social processes. The axioms of common sense are:

> XI. Human choice, by its nature most uncertain, is made certain and determined by the common sense of men with respect to human needs or utilities, which are the two sources of the natural law of the gentes.
>
> XII. Common sense is judgment without reflection, shared by an entire class, an entire people, an entire nation, or the entire human race ... This axiom, with the following definition, will provide a new art of criticism concerning the founders of nations, who must have preceded by more than a thousand years the writers with whom criticism has so far been occupied.
>
> XIII. Uniform ideas originating among entire peoples unknown to each other must have a common ground of truth ... This axiom is a great principle which establishes the common sense of the human race as the criterion taught to the nations by divine providence to define what is certain the natural law of the gentes. And the nations reach this certainty by recognizing the underlying agreements which, despite variations of detail, obtain among them all in respect of this law. Thence issues the mental dictionary for assigning origins to all the diverse articulated languages. It is by means of this dictionary that the ideal eternal history is conceived, which gives us the histories in time of all nations. The axioms proper to this dictionary and to this history will shortly be proposed ... This same axiom does away with all the ideas hitherto held concerning the natural law of the gentes, which has been thought to have come out of one first nation and to have been received from it by the others. (Vico, 1948, pp. 63–4)

Common sense is the basis of decision-making. The instruments used in the evolution of mind are the topos, rhetoric and metaphor drawn from common sense by ingenium. The evolution of society occurs simultaneously as the mind, fortified by ingenium, designs institutions and arranges them so that society is most likely to survive:

> In reasoning of the origins of institutions, divine and human, in the gentile world, we reach those first beginnings beyond which it is vain curiosity to demand others earlier; and this is the defining character of [first] principles. We explain the particular ways in which they come into being; that is to say, their nature, the explanation of which is the distinguishing mark of science. And finally [these origins] are confirmed by the eternal properties [the institution] preserve, which could not be what they are if the institutions had not come into being just as they did, in those particular times, places, and fashions, which is to say with those particular natures.
>
> In search of these natures of human institutions our Science proceeds by a severe analysis of human thoughts about the human necessities or utilities of social life, which are the two perennial springs of the natural law of the gentes. In its second principal aspect, our Science is therefore a history of human ideas, on which it seems the metaphysics of the human mind must proceed. This queen of the sciences, by the axiom that 'the sciences must begin where their subject matters began', took its start when the first men began to think humanly, and not when the philosophers began to reflect on human ideas ....
>
> And the criterion our criticism employs is that taught by divine providence and common to all nations, namely the common sense of the human race, determined by the necessary harmony of human institutions, in which all the beauty of the civil world consists.
>
> the first indubitable principle posited above is that this world of nations has certainly been made by men, and its guise must therefore be founded within the modifications of our own human mind. And history cannot be more certain than when he who creates the things also narrates them. Now, as geometry, when it constructs the world of quantity out of its elements, or contemplates that world, is creating it for itself, just so does our Science [create for itself the world of nations], but with a reality greater by just so much as the institutions having to do with human affairs are more real than points, lines, surfaces, and figures are. (Vico, 1948, pp. 103–5 [346–9])

As one might guess, Leonardo da Vinci developed a very modern version of Thomas Aquinas's *Common Sense* (see Leonardo da Vinci, pp. 109–09).

Two centuries after Vico, *psychophysics* was inaugurated by Haeckel, Helmholtz, Weber, Fechner *et al.*[21] Psychophysics is a mixture of Vico's and Leonardo da Vinci's ideas of common sense. Ironically, it sought and obtained respectability by its nexus with the physical sciences. Both neural science and artificial neural networks have their roots in psychophysics. Of greater significance for our purposes, the similarity between the Weber–Fechner law and the Cobb–Douglas production (utility) function reveals the profound effect of *common sense* on economics.[22] Of course, neural nets are used routinely by some economists, but they are not yet an integral part of economic science. Note that, if Vico were correct, Axiom XI implies that probabilistic economics is unnecessary. Does this ludicrous implication mean that Vico's 'New Science' and his concept of common sense had no merit? Not at all. Vico anticipated Adam Smith in observing the unintended and felicitous consequences of human actions which were attended by 'ferocity, avarice, and ambition, the three vices which run throughout the human race'. Reiterated in Smithian style:

> It is true that men have themselves made the world of nations ... but the world without doubt has issued from a mind often diverse, at times quite contrary and always superior to the particular ends that men had proposed to themselves; which narrow ends, made means to serve wider ends, it has always employed to preserve the human race upon this earth. (Ibid., p. 425)

The unintended, but momentous, consequence of Vico's 'New Science' has been to convert physical science into a *process* comparable to Vico's historical process. Both processes are *stochastic*. Common sense unifies the humanities and the sciences. Plausible reasoning is as important in mathematics and the sciences as it is in poetry and jurisprudence. Communication is conducted by language and counting, the two activities that distinguish man from the other animals. Finally, both these activities have their origin in economics, the unifying discipline.

DISCOVERY OF THE TRUE

Homer and Poetic Wisdom

Vico tells us that *the discovery of the true Homer* was his key insight and the product of 'twenty years of continuous hard thought'. This discovery was simply that Homer was not a single individual, but a series of Greek authors writing over several generations. This insight was the 'master key' of his 'New Science'. According to Mooney (1985), Vico found the master key in a 'blaze of ingenuity', exactly the manner in which many new arts and sciences are discovered. Vico recalled:

> the line from Aristotle's *Rhetoric* that simple minds like to hear stated as universal truth the opinions they hold about particular cases, and the lines from the *Poetics* that poetic statements, unlike those of history, have the nature of universals; he remembered as well that poetic maxims, the *sententiae* ingeniously grasped in a flash, were thought by the ancients to be, as their named implied, rather felt by the senses than seized by the mind. These disparate, seemingly unrelated things he recalled, and all these he combined, in typically transmuted form, with his philological research and – in one of the most singular instances of *ingenium* on record – drew his famous conclusion: Poetry is the native tongue of the race, the clarion call of humanity's birth. Poetry is not a product of mind, but the logic of the mind's development. Not philosophy or human wisdom, but *poetic* wisdom was the wisdom of the ancients. Here in a single phrase adapted from the Stoics, he had the perfect thesis of wisdom and eloquence, of *res* and *verba*, for which he had striven his whole life. (Mooney, 1985, pp. 207–8)

Notice the graphical (network) structure of Mooney's beautiful description of Vico's creativity. The connectivity of seemingly diverse nodes (ideas) occurs in a combinatoric blaze. George Polya has remarked that ideas are like mushrooms. It may take a long time to find one, but if you find one you usually see many others. The suddenness and spontaneity of *ingenium* is exactly the threshold phenomenon discovered by Erdos and Renyi (1961).

Two of Vico's main insights were: poetry is the native tongue of the race and the basic institutions of social evolution, transmitted by a startling metaphor – a bolt of lightning that inarticulately revealed the three genetic institutions of social development: religion, marriage and burial. Know God, know your children, and know your ancestors. These institutions are the genes of an evolving society. Language, the logic of mind, is the communication institution evolving concurrently with mind and society. 'Poetic Wisdom' – the title of Part II in Vico (1948) was the *creative* wisdom containing the template for building human institutions. The discovery of this master key of the 'New Science' occurred simultaneously with the discovery of the true Homer.

Vico describes the nature of 'poetic wisdom':

> In such fashion the first men of the gentile nations, children of nascent mankind, created things according to their own ideas. But this creation was infinitely different from that of God. For God, in his purest intelligence, knows things, and, by knowing them, creates them; but they, in their robust ignorance, did it by virtue of a wholly corporeal imagination. And because it was quite corporeal, they did it with marvelous sublimity; a sublimity such and so great that it excessively perturbed the very persons who by imagining did the creating, for which they were called 'poets', which is Greek for 'creators'.

The generative power of 'poetic wisdom' encompassed not only the three basic institutions, but also the arts and sciences. Vico's outline of Book II *Poetic Wisdom* in Vico (1948) conveys this creative power in a tree metaphor:

> for all these reasons we must trace the beginnings of poetic wisdom to a crude metaphysics. From this, as from a trunk, there branch out from one limb logic, morals, economics, and politics, all poetic; and from another, physics, the mother of cosmography and astronomy, the latter of which gives their certainty to its two daughters, chronology and geography – all likewise poetic. We shall show clearly and distinctly how the founders of gentile humanity by means of their natural theology (or metaphysics) imagined the gods; how by means of their logic they invented languages; by morals, created heroes; by economics, founded families, and by politics, cities; by their physics, established the beginnings of things as all divine; by the particular physics of man, in a certain sense created themselves; by their cosmography, fashioned for themselves a universe entirely of gods; by astronomy, carried the planets and constellations from earth to heaven; by chronology, gave a beginning to [measured] times; and how by geography the Greeks, for example, described the [whole] world within their own Greece. (Vico, 1948, p. 112)

and then defines his 'New Science':

> Thus our Science comes to be at once a history of the ideas, the customs, and the deeds of mankind. From these three we shall derive the principles of the history of human nature, which we shall show to be the principles of universal history, which principles it seems hitherto to have lacked. (Vico, 1948, p. 112)

Metaphorical learning is modified as new experiences unfold. Eventually, imagination gives way to reason and poetry yields to a more efficient method of communication – language disciplined by the logic of grammar. Language connects individuals and ideas. It is *the* informational institution. Communication among society's members is prerequisite to co-ordination of activities. Hence, society can respond to new ideas and environmental challenges in a concerted manner. Adaptability, survival and growth are thereby promoted. Language continues to be a source of creativity in so far as its nexus with poetic wisdom is preserved. Each member of a society contributes to creativity every time a conversation occurs.

One of Vico's greatest insights was his anticipation of the neural network. His metaphorical description of the mind, its connectivity and the occurrence of sudden ingenious acts are all compatible with modern neuroscience. Ingenuity also has a modern interpretation in terms of combinatorics. Roughly speaking, the mind arranges the pieces of a puzzle until a sensible, and frequently novel, pattern emerges. The emergence is spontaneous – like a flash of lightning – and often induces actions that inherit the novelty of the thought process. The value of inventive behaviour varies enormously. Most of it proves useless or harmful. However, the small fraction that survives differentiates a dynamic, adaptive society from one that is static and, hence, moribund in the face of a fluctuating environment.

The knowledge acquired by the information flowing through these institutions over many generations (learning) was essential for the evolution of mind and society. The evolution of economy was indistinguishable from the evolution of society. In Vico's scheme, religion invested parents with the authority to teach, the institution of marriage was maintained by the 'stern restraints of frightful religions', which also forced parents to acknowledge their offspring and held them responsible for education. In turn, the children yielded to the 'stern discipline' of the household, converting their learning into productive uses. Finally, the burial instituted property rights: 'Thus by the graves of their buried dead the giants showed their dominion over the lands.'

Successful families are stable and attract wanderers. The growth of communities begins as the benefits from co-operation and trade become evident. Contracts evolved and, in the absence of money, trade was conducted by barter and contract. This was essentially the state of the economy contemplated by Hicks (1969) before the rise of the market economy.

The invention of money was *ingenium* par excellence. In one stroke, all commodities become comparable. The 'invisible hand' spontaneously assigned relative values to these *comparable entities*. The institutions *money* and *private property* and *trade* gave rise to the open economy, just as marriage, burial and child-rearing gave rise to society. Creation of new commodities such as the creation of new ideas is based on combinatorics, a mixing or shuffling process driven by changing tastes and technologies.

A child's learning is characterised similarly by seizures of ingenuity. In his discussion of the learning process, Vico uses 'children' and 'primitive man' interchangeably. Like children, early man's learning relied on imagination. Learning is both gradual and spontaneous. The child moves at a positive rate until a connection is spotted and the process jumps to a higher level. Some insights cause negative jumps. A set of ideas which was previously seen as being tightly related suddenly falls apart. Learning drops until the puzzle is resolved. A learned procedure is stored in memory until the recurrence of the environmental conditions which produced it.

RHETORIC AND ECONOMICS

It seems odd that a Professor of Rhetoric should be interested in evolutionary processes. Mooney (1985) observes that Vico envisages rhetoric as a practical endeavour:

> Like the general at war or the doctor at a patient's bedside, the orator is concerned with results. His task is 'to persuade, to induce in the hearer a mind agreeable to the oration'. His metier is the debating of specific courses of action, matters that can be framed by the question, *quid sit agendum*. Abstract, theoretical question, (*universa argumenta*), such as the nature of law or virtue, are properly left to philosophers; rhetors are concerned with concrete issues, particularly ones that are neatly circumscribed (*questions peculiares finitae*), such as whether peace should be offered to a treacherous enemy who violates treaties. Issues of this kind are proper to all types of oratory – forensic, political, and epideictic – but for Vico, as for Cicero, the forensic is paradigmatic, for in a court of law, the fate of an accused hangs fully on the ability of opposing attorneys to sharpen the issue in a way favorable to their interest. (Mooney, 1985, pp. 71–2)

Rhetoric in this Aristotelian sense is a process in which the orator (lawyer) must adapt continuously to the changing mental arrangements of the audience (jury) by selecting arguments persuasive to the momentary configuration of opinion. His ability to perceive configurational fluctuations from the peculiar language (signals) emitted by the audience (jury) and to select the matching argument that is most persuasive determine his success. But this is exactly how survival occurs in evolutionary processes. The successful orators (lawyers) survive because their behaviour is determined by winning genes (abilities).

Vico saw that the process just described exemplifies the evolution of mind and society. The creative use of language is poetic wisdom. Survival in this practical arena requires poetic economics – the creative order continuously maintained among the arguments (family) and the changing configurations (environment).

The skilled orator senses the currently held opinions and designs rhetorical arguments that are most plausible. Following Aristotle and de Finetti, 'opinion' is another way of saying probability. Plausible arguments are compatible with opinion, whereas persuasive arguments direct it towards the desired outcome. The metaphor is a powerful ally of persuasion. Aristotle's definition of metaphor resembles Vico's notion of *ingenium*: It cannot be learnt from others; and it is also a sign of genius, since a good metaphor implies an intuitive perception of the similarity in dissimilars. The probability held by an audience can be enhanced by pointing out its resemblance with the probability of another event, where the latter probability is much higher than the former.

To summarise, rhetoric is the logic of eloquence. Plausibility measures the strength of opinion. Opinion is simply subjective probability. Persuasion alters opinion by revealing new information contained in metaphorical connections. The truly eloquent speaker conveys wisdom.

Many of the original and influential *economic* applications of rhetoric elaborated in McCloskey (1983) are compatible with Vico's vision. In this sense, D. N. McCloskey's essay is a modern corollary of Vico (1990).[23]

According to the eminent Vico scholar, Donald Verene:

> Vico's conception of rhetoric is the key to his construction of the inner form of the human world. His concern with rhetoric and its relation to truth reaches from the earliest work of his original philosophy to his last, his 1737 address to the Academy of Oziosi on the connection between philosophy and eloquence. Latin eloquence, or rhetoric, was the subject in which Vico was completely learned. To approach the *New Science* without attention to Vico's conception of rhetoric is to miss one of the bases on which it is founded. (Verene, 1981, p. 163)

Verene links Rhetoric to Vico's notion of *ingenium*, the unifying concept of this paper. He does this by interpreting Vico's statement that 'topics has the function of making minds inventive'.[24]

> ... the power of invention, or *ingegno*, was crucial in the times of the first men when out of necessity everything had to be invented. In his composite notion of memory consisting of *memoria, fantasia*' and *ingegno*, Vico says that *ingegno* occurs within memory when in relation to things 'it gives them a new turn or puts them into proper arrangement and relationship'. *Ingegno* does not function alone, but is always connected with the other two powers of memory, ... Specifically, *ingegno* is the power to extend what is made to appear from sensation beyond the unit of its appearance and to have it enter into connection with all else that is made by the mind from sensation. (Verene, 1981, p. 167)

Ingenuity is a combinatoric process – it rearranges the sensual flux so that the mind can extract meaning and sometimes discover new relations. The discovery of novelty is analogous to Friedrich Hayek's discovery process. Verene has a more general interpretation. Ingenuity is the discovery process that transforms nature into culture.

Vico also shows that much of what most economists refer to as 'imperial economics' is, in fact, a renewal process. 'Household economics', which is considered as an invasion of sociology, is compatible with the original meaning of economics – an orderly household. The household or family is one of Vico's basic institutions. Obviously, economics in this original sense has since become more orderly as society has evolved. However, once this connection is made, history is recognised as being indispensable to comprehending the origin and evolution of economics. *There are no ideas independent of the historical context comprising their origin and evolution.*

We claim that language and counting are the fundamental institutions and that both are driven by economic forces. Vico's genetic trinity – marriage, child-rearing and burials – gives rise to families and communities. The order of these interacting institutions is essential to their survival and, as we saw, is exactly what *economics* means. Co-ordination and co-operation depend on the information flowing through the community's networks. That is, conversation resolves tensions and conflict, creates harmony and alerts the community to environmental changes. The response to change sometimes requires reorganisations based on crude combinatorics. The maintenance of order in a fluctuating environment, the essence of the evolutionary process, would be impossible without language and counting.

In Viconomics, it is precisely the creativity of language that explains the evolution of society and the evolution of mind.[25] Vico's concept of language must encompass counting.

With one foot in the humanities and the other in the physical sciences, economics cannot only improve its foreign relations, but also 'prove' by its vitality that the humanities and physical sciences are complementary. Finally, economists should learn to respect their subject matter – *all* creatures follow economic principles. Respect will become reverence when academic economists realise that many of their ingenious insights were discovered and implemented by practical economists.

METHODS OF OUR TIME

The economic relevance of Vico's combinatoric vision of science reaches its zenith in *On the Study Methods of Our Time* (Vico, 1990). The power of his constructive approach to wisdom is hazardous to both friend and foe. What one previously considered as novel and seminal in one's own constructive methods must be recognised as being *at least* 300 years old. However, there is exhilaration in this experience of humility. On the other hand, if one's method of achieving wisdom relies on an axiomatic, deductive approach, then Vico must be subverted or dismissed in order to avoid humiliation. This attack on the abstract is too withering to be *true*. There certainly is a deductive niche in the creative process. But never again can the serious reader of Vico identify creativity with abstract deductions.

Vico's aesthetics is drenched by the practical. A constitution, an invention, a language, a legal defence is beautiful in so far as it reflects the natural flux, the heterogeneous, and *the vulnerable*[26] in an economical manner.

The idiosyncratic response of Benedetto Croce verifies Vico's assertion that each individual is unique:

> ... The fact is that his philosophy of history, his *Scienza nuova d'intorno alla communa natura delle nazioni*, does not concern the concrete empirical history which unfolds itself in time: it is not history, it is a science of the ideal, a Philosophy of the Spirit. (Croce, 1909, p. 231)

and

> In conclusion, we remark that fundamental terms are not always used by Vico in the same sense: it is not always clear how far 'sensation',

'memory', 'imagination', 'wit' are synonymous or different. Sometimes 'sensation' seems outside the spirit, at others one of its chief moments; poets are sometimes the organ of 'imagination', sometimes the 'sensation' of humanity; and imagination is described as 'dilated memory'. These are the aberrations of a thought so virgin and original that it was not easy to regulate. (Croce, 1909, pp. 233–4)

As we suggest[27] 'dilated memory' is a measure of creativity compatible with a constructive approach to knowledge.

It is not only consistent with the most recent advances in neuroscience *and* number theory, but also satisfies the practical (economical) criteria of mechanical design.

In his perceptive monograph on design, French (1988) uses similar criteria for measuring creativity:

Perhaps the most important evidence about the nature of invention has come from mathematicians and physicists, and it supports two principal ideas. The first is that the essential process is one of choice, and that the choice is made on principles which are best described as aesthetic. The mathematician Poincare wrote of aesthetic sensibility as the 'delicate sieve' with which the most useful combinations of ideas are separated from all the others which might be studied. (There is an interesting parallel here with the crude sieve, natural selection, with which nature separates the most useful combinations of genes from all the others which sexual reproduction generates.)

The second is that the process of invention usually falls into stages, of which the first three may be described as preparation, incubation, and illumination.

The inventive man is marked out by the range of elements he considers combining, and the skill and judgement with which he selects or converges upon his preferred combination. (French, 1988, pp. 278–9)

Vico would agree with French's assessment of invention of the printing press,[28] but adds a *caveat*, exemplified by the printing press:

When books were written by hand, only works composed by authors of tested and well-established reputation were reproduced, since they were the only ones worth the labor spent on the task of copying them; and since the price of such works was, at times, very high, students were obliged to copy them in their own handwriting. Now, there is no better exercise than this; we meditate on the text and write without haste or interruption, calmly and with continuous order. By copying, we gain, not a perfunctory knowledge, but an intimate familiarity with the original, and we are , so to speak, transformed into the author's very self. That is why second-rate authors palled in the copying; but superior ones were held in great esteem and celebrated, to the great profit of the general public. (Vico, 1990, pp. 72–3)

In his insightful monograph on the importance of being 'fit', Calder (1984) notes that nature's inventions must satisfy Stahl's conservation principles: (1) volume must be preserved and (2) time must be synchronised. The Ramanujan graph is a modular form and satisfies (1).[29] The timing constraint is a sequential decision rule and is satisfied when the process is 'fair'. Vico is cognizant of both of these constraints on inventiveness: 'We need to train young minds for the practice of mechanics by means of a close study of visual geometrical figures' and

But the sage who, through all the obliquities and uncertainties of human actions and events, keeps his eye steadily focused on eternal truth, manages to follow a roundabout way whenever he cannot travel in a straight line, and makes decisions, in the field of action, which, in the course of time, prove to be as profitable as the nature of things permits. (Vico, 1990, p. 35)

Language is a pillar of creativity. Sensuous languages increase the mind's inductive prowess. These languages prepare the mind for its contact with reality via sight, hearing, taste and touch. The thinking, writing and conversation that precede and coincide with sensual contact are less likely to overlook crucial, but subtle, connections when they are guided by a sensuous language. The reconstruction process (induction) will see the invariants associated with recurring patterns when it is conducted with a sensuous instrument – language. This process is unlikely to be corrupted by noise when its instruments possess a symmetry akin to that inherent in nature's recurring patterns. Finally, individuals who are able to extract nature's signals from the apparent turbulence of reality achieve satisfactory rewards. Hence language, aesthetics and utility conspire to decode nature's mysteries.[30] Vico says it best:

We Italians, instead, are endowed with a language which constantly evokes images. We stand for above other nations by our achievements in the fields of painting, sculpture, architecture, and music. Our language, thanks to its perpetual dynamism, forces the attention of the listeners by

> means of metaphorical expressions, and prompts it to move back and forth between ideas which are far apart. (Vico, 1990, pp. 40–1)

In his jingoistic passage,[31] Vico confirms his theory spontaneously! It is too instinctive to be a contrivance. After reading this ode to the Italian language one is tempted to insert 'Q.E.D.'[32]

VICO AND SMITH

On the surface, one would be hard pressed to find two intellectual giants as different as Adam Smith and Giambattista Vico. Smith was blessed with a balanced personality. His seminal contributions as the founder of economic science were recognised during his lifetime. He was a leading member of the Scottish Enlightenment – David Hume was his best friend – and his writing and thinking were unconstrained by any strict religious ideology. He was a prosperous bachelor and enjoyed a tranquil lifestyle conducive to intellectual activity. George Stigler's comment is, as usual, most apt: 'his life was remarkable only for its placidness'.

As we have seen, Vico had a mercurial personality, lived in poverty amidst a large, boisterous family, and was considered by many of his contemporaries to be a bit mad. His extraordinary originality went unappreciated during his lifetime. Indeed, his ideas were so novel, they gave credence to negative opinions and diminished his influence. Finally, the closest readers of his work may have been religious censors searching for heresy.

We should also note the similarities between these two intellectual giants. Both were Professors of Rhetoric, and had wide-ranging interests.[33] The social consequences of probability were emphasised by both. For Vico, society was harmonised by information flows within and among institutions, while economic forces achieved the Smithian Harmony. Many claim that Smith was not interested in co-operation or social interactions: they have not read *The Theory of Moral Sentiments*. Others hold that Vico was an idealist with little interest in the practical. This absurdity survived even the scathing attack by Beckett *et al.* (1929):

> G. Vico was a practical roundheaded Neapolitan. It pleases Croce to consider him as a mystic, essentially speculative ... It is a surprising interpretation, seeing that more than three-fifths of his *New Science* is concerned with empirical investigation ... His treatment of the origin and functions of poetry, language and myth ... is as far removed from the mystical as it is possible to imagine. (p. 3)

Finally, both agree that all plans have unintended consequences. That is, while the pursuit of selfish ends does create harmony is a principle shared by Vico and Smith, neither claims to understand how it works.

In spite of their sharp 'environmental' differences, Smith and Vico produced intellectual offspring which together comprise the principles of the modern open society.

Of course, everyone acknowledges that the evolutionary open economy is founded on Smithian principles with a little help from Charles Darwin. Smith's key insight was to show how the natural pursuit of individual welfare by each member of society gave rise, spontaneously, to a system of prices. These prices conveyed exactly the information necessary for success in the pursuit of welfare. As the environment changes, so do prices, and economic agents adapt accordingly.

Vico's evolutionary society is also founded on information flows and learning. These flows occur in the myriad of institutions and groups comprising the society. Some of these groups are society's durable pillars – family, school, job and church. Each group contains an information network. Pertinent Smithian prices flow through these nets, as well as an enormous amount of informal information. Some groups are ephemeral and anonymous, such as commuters from home to job, yet members are constantly exchanging information as they conform to the rules of the network.

The intimate practical bond joining Smith's 'invisible hand' and Vico's networks has received little attention. It may be the most important subject in the social and biological sciences and will be considered in a subsequent paper.

VICO'S IMPACT ON THE SOCIAL SCIENCES

Marx discovered Vico and transformed him into an historical materialist. Leon Trotsky quotes Vico on the first page of his *History of the Russian Revolution*. E. Wilson's *To the Finland Station* begins by referring to Michelet's discovery of Vico. Michelet translated an abridged version of Vico into French, where it achieved great popularity. Weber translated Vico into German in 1822, but without Michelet's success.

Given this background, it is not surprising that most American economists are unfamiliar with Vico's work. The sparkling combination of Vico, de Finetti and modern law and economics in Posner (1981) is unaware of its debt to Vico. In the Superb Viconian essays by Hayek (1967), there is no mention of Vico. On the other hand, in 1878,

F. B. Sanborn, one of the founders of the American Social Science Association referred to Vico, along with Smith and Auguste Comte, as comprising the intellectual foundation of the social sciences.

In his *Opus Magnus*, Mayr (1982) does mention Vico's devastating attack on Descartes and observes that 'For Vico, the various periods of human history were not different aspects of essentially the same story; rather they were successive stages of a continuing process, a process of necessary evolution'. Toulmin (1982) also notes that: 'In biology as in the human sciences structural perspectives – in the spirit of Descartes and Newton – can at most complement, never displace, the temporal perspectives of Vico or Darwin.

REFLECTIONS

There are, of course, flaws in Vico's philosophy.[34] His creative and spontaneous writing is truly extraordinary, but it does lend itself to misinterpretations that a more disciplined, less exciting prose would avoid. Nevertheless, many of his positions are crystal clear. The attack on Descartes' abstract analysis and his pedagogical thesis that the concrete should illuminate analysis retain their vitality. However, the durability of a society depends on a subtle mixture of uniformity and diversity. It would seem that a common language and a common legal system are essential. The individuality that is the hallmark of Smithian-de Finetti's philosophy is conspicuous in Vico, but not really embraced. The equivalence between certainty and uncertainty is not realised by these great men.

It is unlikely that Vico would agree that abstraction in one discipline *when sufficiently deep* encounters the concrete base unifying the arts, social sciences and physical sciences.[35]

His position does seem compatible with the view promulgated by Peirce that the symmetries present in nature are reflected in the composition of the evolving minds of all animals.

Recent neurological research has confirmed Peirce's hypothesis, while augmenting it in a fundamental way. The symmetry persists up to a threshold, at which point the neural behaviour becomes erratic.

This sort of symmetry-breaking is analogous to the maintenance of a contractual arrangement. As long as the behaviour of both parties remains within a critical region, the contract is followed. If a boundary is broken, the contract disintegrates. Replace contract by treaty, interpret boundary literally, and replace broken by invasion. The result is the transition from the symmetry of peace to the chaos of war.

CONCLUSION

The main point of this chapter is to show that Vico anticipated many modern developments in economics and the other social sciences. Exchangeability clarifies these remarkable insights. When one combines Vico's theory of renewal with his network analysis, a whole new circle of ideas is revealed. They begin with Peirce's combinatorial analysis extend to Markov's stochastic networks, and then to the seminal work of Irving Fisher and Eugen Slutsky on business cycles, of Hebb (1949) and Hopfield and Tank (1986) on neural networks, of MacArthur on ecology, of Konrad Lorentz on animal behavior, and to the fundamental insights of Gottfried Leibniz, Alan Turing, and artificial intelligence. The key to these relationships is some form of symmetry like exchangeability with birth and death Markov processes and no-arbitrage conditions being major actors. These relationships will be the subject of a subsequent essay. For now, note that the structure of networks is characterised by their spectral properties. The sensual roots of the inductive process is nicely conveyed by spectral analysis. In a classic article, Kac (1966) captured this delicate and intricate intertwining of the abstract and the concrete. The senses are able to learn novel information from the process of emitting signals, merge this information with the accumulated pool (memory) and make wiser decisions. In particular, one can detect structural change and mend the methodology accordingly. The analytical power of this inductive process resides in group theory and its symmetric patterns, recurrences and modularity. The synthetic strength is in the sensual contacts with concrete phenomena.[36] The combined wisdom of Vico and de Finetti has led us to this rich inductive process. Thus the practical (trade), the beautiful (poetry), and the abstract (mathematics) are united in the *Economy of Nature*.[37]

Notes

1. See Leijonhufvud (1981), ch. 11.
2. Roughly speaking, exchangeability and indistinguishability are synonymous. In an inflation-free economy, money creates prices such that all goods are exchangeable.
3. The reader is urged to consult Vico (1944) and the recent study of the autobiography by Verene (1991).
4. Excellent portraits of this rebirth are contained in Alchian (1984), Matthews (1986) and Williamson (1985). An economic version of the evolution of institutions is presented in Hicks (1969).
5. The terms 'risk', 'uncertainty', 'probabilistic', and 'stochastic' are used interchangeably throughout this paper.
6. Recall that the Latin root for 'institution' means 'to teach'. More specifically: L. *instituere* means to teach or arrange, and L. *institutum* means design.
7. This does not mean that society is tranquil and homogeneous. Buchanan-like clubs partition and generate tensions and conflict. Nevertheless, when common sense prevails among all members, these tensions are productive. As common sense lapses, destructive behaviour begins.
8. The close resemblance with Dewey (1929) marks Vico as a forerunner of Pragmatism.
9. As one might guess, Vico was a devout Cartesian before building his 'New Science'.
10. It was Peirce who formalised these relations in his seminal contribution to Graph Theory. See McCall (1993).
11. Martin Gardner (1980) notes that: 'wherever there is symmetry there is a group'. The permutation is the group we are focusing on. That is, exchangeability is closely related to the permutation of groups. An excellent discussion of the manifold applications of group theory and graph theory to the social sciences is presented in Hage and Harary (1983). See the recent monograph by Coleman (1990), its review by Frank (1992), and the earlier monograph by White (1970). The classic study of symmetry is Weyl (1952).
12. The human kind is naturally impelled to take delight in uniformity.
13. The brilliant contributions of W. Hamilton are exposited with wit and wisdom in Richard Dawkins (1976): *The Selfish Gene* Oxford, Oxford University Press. A subtle critique of the sociobiological industry, is contained in the last chapter of the monumental study by Eisenberg (1981).
14. Just as advances in medical science have accompanied the study of the sick, so, too, many insights have been gleaned from economists' study of asymmetric behaviour.
15. The trade-off between stress and durability has received much attention in the ecological literature. For an extensive discussion, see Lippman and McCall (1984).
16. An excellent discussion of these points is contained in the scholarly monograph by Rossi (1984).
17. Bonifazio Finetti, an astute Dominican, considered Vico's writings to be impious and dangerous. The reader should not confuse Bruno de Finetti with his earlier namesake.

18. L. M. Palmer has a nice discussion in Vico (1988).
19. See Galanter and Miller (1960) for its influence on modern psychology.
20. For an excellent discussion of the relationship between Weber–Fechner and Cobb–Douglas, see Coleman (1990). Recall that the Weber–Fechner law is: the change in response induced by a change in sensual stimulus is proportional to the level of the stimulus.
21. See Vico (1990), and 'Methods of Our Time' in this chapter.
22. Recall that the theory of topics is an integral part of rhetoric.
23. One of Vico's many innovations was to recognise language as *the* creative process.
24. it is an error to apply to the prudent conduct of life the abstract criterion of reasoning that obtains in the domain of science. A correct judgment deems that men – who are, for the most part, but fools – are ruled, not by forethought, but by whim or chance. The doctrinaires judge human actions as they *ought* to be, not as they actually are (i.e., performed more or less at random). Satisfied with abstract truth alone, and not being gifted with common sense, unused to following probability, those doctrinaires do not bother to find out whether their opinion is held by the generality and whether the things that are truths to them are also such to other people. (Vico, 1990, p. 35)
25. See McCall and Velupillai (1991b), where an analogue of an individual's creativity is a Ramanujan graph. As more ideas (nodes) enter memory and are connected by the imagination with the current inventory, creativity is enhanced. This process continues as long as the individual resides in a stable environment and can specialise in that combination of 'jobs' most agreeable with his tastes and talents.
26. See French (1988) pp. 249ff.
27. This conservation principle was emphasised by Mach. We call it the *economy of nature*: see McCall and Velupillai (1991).
28. This is a glimpse of the process whereby nature's recurring patterns are embedded into the form and structure of survivors. It is well known that some seeds will not germinate unless they have passed through the digestive tract of predators; other seeds require the intense heat of a 'forest fire' to trigger their reproductive mechanism. Hence, induction or learning is nature's sculpting process. The selected statues 'fit' in their *habitat*. In short, they are *economical*.
29. The interested reader should read the entire section in Vico (1990) to appreciate the label *jingoism*. In his cogent introductory essay, E. Gianturco disputes the 'jingoistic' interpretation. The essay also contains many other perceptive remarks about Vico's contributions; see Vico (1990).
30. Mach is another brilliant and persuasive advocate of the *sensuous foundations of knowledge*; see Mach (1906).
31. The intellectual breadth of Smith is portrayed nicely in Luna (1993).
32. This portrait of Vico ignores the contributions of the 'analytic mind' to society's pool of wisdom. Any balanced appraisal of epistemology must weight both synthetic and analytic contributions. Vico's lack of balance is necessary to capture the attention of a society dominated by Descartes. Similarly, our position with respect to economics emphasises synthesis because of the reign of the deductive.

33. An attempt at unifying these diverse disciplines was begun in McCall (1986).
34. For a beautiful extension of Kac's insights, see Berard (1986).
35. See McCall and Velupillai (1991).

References

Alchian, A. A. (1950) 'Uncertainty, Evolution and Economic Theory', *Journal of Political Science*, vol. 58, pp. 211–21.

Alchian, A. A. (1984) 'Specificity, Specialization and Coalitions', *Journal of Institutional and Theoretical Economics*, vol. 140, pp. 34–49.

Aldous, D. J. (1985) *Exchangeability and Related Topics* (New York: Springer-Verlag).

Anderson, J. A. (1983) 'Cognitive and Psychological Computation with Neural Networks', *IEEE TRANS. on Syst., Man, and Cybern*, SMC-13, pp. 799–815.

Arendt. H. (1978) *The Life of the Mind* (New York: Harcourt Brace Jovanovich)

Arrow, K. J. (1974) *The Limits of Organization* (New York: Norton).

Becker, G. S. (1976) *The Economic Approach to Human Behavior* (Chicago: University of Chicago Press).

Beckett, S. *et al.* (1929) *James Joyce/Finnegans Wake: A Symposium* (New Directions).

Berard, P. H. (1986) *Spectral Geometry: Direct and Inverse Problems* (Springer-Verlag).

Berlin, I. (1976) *Vico and Herder* (Hogarth Press).

Berlin, I. (1991) *The Crooked Timber of Humanity* (New York: Alfred A. Knopf).

Buchanan, J. M. and V. J. Vanberg (1991) 'The Market as a Creative Process', *Economics and Philosophy*, vol. 7, pp. 167–86.

Burke, P. (1985) *VC* (Oxford University Press).

Burks, A. W. (1977) *Change, Cause, Reason* (Chicago: University of Chicago Press).

Calder, W. A. III (1984) *Size, Function, and Life History* (Cambridge: Harvard University Press)

Caponigri, R. (1953) *Time and Idea: The Theory of History in Giambattista Vico* (Notre Dame: University of Notre Dame Press.)

Cassirer, E. (1955) *The Philosophy of Symbolic Forms: Mythical Thought* (New Haven, Conn.: Yale University Press).

Cavalli-Sforza, L. L. and M. W. Feldman (1981) *Cultural Transmission and Evolution* (Princeton, NJ: Princeton University Press).

Chapman, M. (1988) *Constructive Evolution: Origins and Development of Piaget's Thought* (Cambridge University Press).

Coase, R. H. (1937) 'The Nature of the Firm', *Economica*, vol. 16, pp. 386–405.

Coase, R. H. (1960) 'The Problem of Social Cost', *Journal of Law and Economics*, vol. 3, pp. 1–44.

Cohen, M. A. and S. Grossberg (1983) 'Absolute Stability of Global Pattern Formation and Parallel Memory Storage by Competitive Neural Networks', *IEEE Trans. on Syst., Man, and Cybern*, SMC-13, pp. 815–25.

Coleman, J. S. (1990) *Foundations of Social Theory* (Cambridge, Mass.: Harvard University Press).
Collingwood, R. G. (1946) *The Idea of History* (Oxford University Press).
Cornes, R. and T. Sandler, (1986) *The Theory of Externalities, Public Goods, and Club Goods* (Cambridge University Press).
Croce, B., (1909) *Aesthetic* (London: Macmillan).
Dawkins, R. (1976) *The Selfish Gene* (Oxford: Oxford University Press).
de Finetti, B. (1964) 'Foresight: Its Logical Laws, Its Subjective Sources', in H. E. Kyburg and H. E. Smokler (eds), *Studies in Subjective Probability* (New York: John Wiley).
de Finetti, B. (1974) *Theory of Probability I and II* (New York: John Wiley).
Dewey, J. (1929) *The Quest For Certainty* (New York: G. P. Putnam).
Eco, U. (1988) *The Aesthetics of Thomas Aquinas* (trans. H. Bredin) (Cambridge, Mass.: Harvard University Press).
Eisenberg, J. F. (1981) *The Mammalian Radiations* (Chicago: University of Chicago Press).
Erdos, P. and A. Renyi (1960) *On the Evolution of Random Graphs* (Hungarian Academy of Sciences).
Frank, R. H. (1992) 'Melding Sociology and Economics: James Coleman's *Foundations of Social Theory*', *Journal of Economic Literature*, vol. 30, pp. 147–70.
French, M. J. (1988) *Invention and Evolution: Design in Nature and Engineering* (Cambridge University Press).
French, S. (1987) 'The Analysis of Multiple Choice Tests in Educational Assessment', in R. Viert (ed.), *Probability and Bayesian Statistics* (Plenum Press).
Friedman, M. (1953a) *Essays in Positive Economics* (Chicago: University of Chicago Press).
Friedman, M. (1953b) 'Choice, Chance, and the Personal Distribution of Income', *Journal of Political Economics*, vol. 61, p. 277–90.
Galanter, E. and G. A. Miller (1960) 'Some Comments on Stochastic Models and Psychological Theories', ch. 19 in K. J. Arrow, S. Karlin and P. Suppes (eds), *Mathematical Methods in the Social Sciences* (Palo Alto, Calif.: Stanford University Press).
Haddock, B. A./ (1979) 'Vico's Discovery of the True Homer', *Journal of History Ideas*. vol. 40, pp. 583–60.
Hage, P. and F. Harary (1983) *Structural Models in Anthropology* (Cambridge University Press).
Hayek, F. A. (1967) *Studies in Philosophy Politics and Economics* (Chicago: University of Chicago Press).
Hayek, F. A. (1978) *New Studies* (Chicago: University of Chicago Press), pp. 264–5.
Hebb, D. O. (1949) *Organisation of Behavior*, J. Wiley and Sons, 1949.
Heymann, D., and A. Leijonhufvud (1994) *High Inflation* (Oxford University Press).
Hicks, J. R. (1969) *A Theory of Economic History* (Oxford,).
Hirshleifer, J. (1985) 'The Expanding Domain of Economics', *American Economic Review*, December, pp. 53–68.
Hobbes, T. (1651) *Leviathan* (Harmondsworth: Panguin).

Holland, J. H., K. J. Holyoak, R. E. Nisbett and P. R. Thagard (1986) *Induction* (Cambridge, Mass.: MIT Press).

Hoppfield, J. J. and D. W. Tank (1986) 'Computing with Neural Circuits: A Model', *Science*, vol. 233, pp. 625–33.

Joyce, J. (1961) *Ulysses* (London: Random House).

Kac, M. (1966) 'Can One Hear the Shape of a Drum', *American Mathematical Monthly*, vol. 73.

Kenner, H. (1984) *A Colder Eye*, (Harmondsworth: Penguin).

Kimura, M. (1983) *The Neutral Theory of Molecular Evolution* (Cambridge University Press).

Leijonhufvud, A. (1967) *On Keynesian Economics and the Economics of Keynes* (Oxford University Press).

Leijonhufvud, A. (1981) *Information and Coordination: Essays in Macroeconomic Theory* (Oxford University Press).

Leonardo da Vinci, (1952) *Notebooks*, selected and ed. by I. A. Richter (Oxford University Press).

Levinthal, D. (1984) *A Survey of Agency Models of Organizations*, Tech. Rep. 443, Department of Economics, Stanford University, California.

Lippman, S. A. and J. J. McCall (1984) *Ecological Decisionmaking and Optimal Stopping Rules* (UCLA).

Luna, F. (1993) 'From the History of Astronomy to the Wealth of Nations: Wonderful Wheels and Invisible Hands in Adam Smith's Major Works', unpublished UCLA Working Paper No. 6915.

Mach, E. (1906) *Space and Geometry* (Open Court).

Martin, R. D. and R. M. May (1982) 'Outward Signs of Breeding', in J. Maynard Smith (ed.), *Evolution Now* (W. H. Freeman).

Matthews, R. C. O. (1986) 'The Economics of Institutions and the Sources of Growth', *Economic Journal*, vol. 96, pp. 903–18.

Mayr, E. (1982) *The Growth of Biological Thought* (Cambridge, Mass.: Harvard University Press).

McCall, J. J. (1986) *The Bridge Across the Disciplines* (UCLA).

McCall, J. J. *The Smithian Self and Its Bayesian Brain*, UCLA Working Paper, 1989.

McCall, J. J. (1991a) 'Exchangeability and Its Economic Applications', *Journal of Econ. Dyn. and Cont.*, vol. 15, pp. 549–68.

McCall, J. J. (1991b) 'Exchangeability and the Structure of the Economy: A Preliminary Process Analysis', in H. Brink (ed.), *Advances in Business Cycle Theory* (London: Macmillan).

McCall, J. J. (1991c) 'The Random Foundations of Economic Processes I', in K. Velupillai (ed.), *Proceedings of the International Economic Association* (London: Macmillan).

McCall, J. J. (1993) *Graphical Economics: A Survey*, UCLA Working Paper.

McCall, J. J. and K. Velupillai (1991a) *The Economy of Nature: I. A Survey*, UCLA Working Paper.

McCall, J. J. and K. Velupillai (1991b) 'A Neuroscientific Approach to Economics', Abstract *Math. Soc. Sciences*, vol. 22.

McCloskey, D. N. (1983) 'The Rhetoric of Economics', *Journal of Economic Literature*, vol. XXI, pp. 481–517.

Minnis, P. E. (1985) *Social Adaptation to Food Stress* (Chicago: University of Chicago Press).

Mooney, M. (1985) *Vico in the Tradition of Rhetoric* (Princeton, NJ: Princeton University Press).

Olson, M. L. (1982) *The Rise and Decline of Nations* (New Haven: Yale University Press)

Pompa, L. (1990) *VC: A Study of the 'New Science'*, 2nd edn (Cambridge University Press).

Posner, R. A. (1981) *The Economics of Justice* (Cambridge, Mass.: Harvard University Press).

Priest, G. L. (1987) 'The Current Insurance Crisis and Modern Tort Law', *The Yale Law Journal*, vol. 96, pp. 1521–90.

Rossi, P. (1984) *The Dark Abyss of Time*, trans. L. G. Cochrane (Chicago: University of Chicago Press).

Salmon, W. C. (1984) *Scientific Explanation and the Causal Structure of the World* (Princeton, NJ: Princeton University Press).

Scarre, C. J. (1984) 'Kin-Groups in Megalithic Burials', *Nature*, vol. 311, pp. 512–13.

Schotter, A. (1981) *The Economic Theory of Social Institutions* (Cambridge University Press).

Schumpeter, J. A. (1954) *History of Economic Analysis* (Oxford University Press).

Shepherd, G. M. (1988) *Neurobiology* (Oxford University Press).

Shils, E. (1983) *Tradition* (Chicago: University of Chicago Press).

Smith, A. (1976) *The Theory of Moral Sentiments* (1853) (Liberty Classic).

Smith, A. (1978) *Lectures on Jurisprudence* (Oxford University Press).

Smith, A. (1983) *Lectures on Rhetoric and Belles Lettres* (Oxford University Press).

Snow, C. P. (1959) *The Two Cultures and the Scientific Revolution* (Cambridge University Press).

Suppes, P. (1984) *Probabilistic Metaphysics* (Oxford: Basil Blackwell).

Toulmin, S. (1982) *The Return to Cosmology* (University of California Press).

Toulmin, S. (1972) *Human Understanding* (Princeton, NJ: Princeton University Press).

Toulmin, S. and J. Goodfield (1965) *The Discovery of Time* (London: Hutchinson).

Turnbull, H. W. (1951) *The Great Mathematicians*, 4th edn (London: Methuen).

Verene, D. P. (1981) *Vico's Science of Imagination* (Ithaca, NY: Cornell University Press).

Verene, D. P. (ed.) (1987) *Vico and Joyce* (State University of New York Press).

Verene, D. P. (1991) *The New Art of Autobiography* (Oxford University Press).

Vico, G., (1944) *Autobiography*, translated by M. H. Fisch and T. G. Bergin, Cornell University Press, 1944.

Vico, G. (1948) *The New Science*, trans. M. H. Fisch and T. G. Bergin (Ithaca, NY: Cornell University Press).

Vico, G. (1988) *On the Most Ancient Wisdom of the Italians*, trans. L. M. Palmer (Ithaca, NY: Cornell University Press).

Vico, G. (1990) *On the Study Methods of Our Time*, trans. E. Gianturco (Ithaca, NY: Cornell University Press).

Vico, G. (1993) *On Humanistic Education* (Six Inaugural Lectures, 1699–1707), trans. G. R. Pinton and A. W. Shippee, (Ithaca, NY: Cornell University Press).

Weyl, H. (1952) *Symmetry* (Princeton, NJ: Princeton University Press).

White, H. C. (1970) *Chains of Opportunity* (Cambridge Mass.: Harvard University Press).

Williamson, O. E. (1985) *The Economic Institutions of Capitalism* (The Free Press).

Wilson, E. O. (1975) *Sociobiology* (Cambridge, Mass.: Harvard University Press).

Wittgenstein, L. (1969) *On Certainty* (New York: Harper and Row)

Young, J. Z. (1988) *Philosophy and the Brain* (Oxford University Press).

8 From the *History of Astronomy* to the *Wealth of Nations*: Wonderful Wheels and Invisible Hands in Adam Smith's Major Works*

Francesco Luna

> From Zeus let us begin; him never do we men pass by
> In silence. Full of Zeus are all the streets,
> And all the market-places of men... (Aratus)

INTRODUCTION

Adam Smith's works are particularly dear to Axel Leijonhufvud. Well before the spreading interest in endogenous growth, Professor Leijonhufvud revived 'an old parable, Adam Smith's theory of manufacturing production, which has been shunted aside and neglected because it has not fitted into the formal structure of either neo-classical or neo-Ricardian theory'.[1]

However, one may ask, what is the use of a past that has no present? Somewhere in the past – Professor Leijonhufvud will answer – faced with a fork in the road, researchers directed their efforts along the route that promised to be viable and successful, whereas the other path remained untried. The few pioneers who took some steps into the untamed wilderness were abandoned to their destiny. It is when the main road ends up in a swamp that scouts are sent back to look for Doctor Livingstone. In this vein, Axel Leijonhufvud reconsidered the Smithian division of labour, and he should be satisfied to see that once more he has pointed at a road that has recently led to some 'progress' in economic theory. Nevertheless, that very parable 'can illuminate many and diverse problems' that are different from the sustainable endogenous growth in the styles of Lucas or Paul Romer. In particular, very convincing answers can be given to a series of questions concerning the institutional nature of the capitalist firm.

This essay deals with a different Smithian parable or – better – metaphor: namely, the invisible hand. However, the reason for undertaking such a project and the approach followed are in syntony with Professor Leijonhufvud's research. Pushed in part by dissatisfaction with the dominant rationalist approach to economics and in part by pure aesthetic pleasure, I have tried to backtrack along the decision tree[2] that describes the evolution of economic thought, to find the latest economist who adopted induction as the preferred technique for his/her scientific investigations. Having settled for Adam Smith as the relevant empiricist, we have concentrated on references to the invisible hand metaphor in two of Smith's works written before *Wealth of Nations.* Such references establish a continuity in Smith's adherence to induction.

This chapter proposes a re-reading of Smith's major works in the light of his early, and certainly less applauded, essay tracing the history of astronomy from antiquity to the middle of the eighteenth century. The methodology employed to explain the evolution of these theories goes beyond the specific work and becomes Smith's own guideline in his overall philosophical investigation. Our analysis leads to an all-encompassing interpretation of the *invisible hand* as it appears in *History of Astronomy, The Theory of Moral Sentiments,* and *Wealth of Nations.* Smith uses the invisible hand as a rhetorical device, as an expression denoting an unknown value or functional form which solves the problem at stake. The number of interpretations of such a metaphor proposed in the literature reflects the implicit attempt at finding *the solution to the problem.* Hence it should not be surprising that sometimes seemingly irreconcilable contradictions are considered to be the effects of chance and as such are dismissed, or otherwise settled at a metaphysical level. Smith, we maintain, believed in and applied induction in his philosophical investigation of ethics in *The Theory of Moral Sentiments*, as well as of economics in *Wealth of Nations.* His goal was that of reducing *wonderful* phenomena to a well-connected sequence of familiar objects following the infinite regression up to the first of the sensible (visible) causes.

In the first two sections we analyse Smith's methodology as presented in *History of Astronomy.* Drawing a parallel between the sentiment of *wonder* – the basis of any philosophical investigation – and the concept of the *sublime* as theorised by Edmund Burke in 1757, we suggest that the acquisition of knowledge in Smith's major works is founded upon aesthetic categories. The third section gives a critical account of several interpretations of the invisible hand, starting with those given by Viner (1928) and Robbins (1952), to conclude with the ones proposed by Ahmad (1990) and Davis (1990). Dissenting from these standpoints, in the fourth section we introduce our own position which draws from and at the same time complements some of Leijonhufvud's ideas on the division of labour. The fifth section concludes the chapter.

SMITHS'S WONDER AND BURKE'S SUBLIME

The Principles which Lead and Direct Philosophical Enquiries; Illustrated by the History of Astronomy (*HA*) was published in 1795, after Adam Smith's death. However, a passage from this essay provides a precious *terminus ante quem* which helps us identify the position of such a work in the chronology of Smith's literary career:

> His [Newtons] followers have, from his principles, ventured even to predict the returns of several of them [comets], particularly of one which is to make its appearance in 1758. (*HA*, IV, pp. 74)

Moreover, Smith inserts a footnote concerning Halley's Comet, which clearly supports our assumption:

> It must be observed, that the whole of this Essay was written previous to the date here mentioned; and that the return of the comet happened agreeably to the prediction.

To be able to ascertain the date of the *HA* and to affirm, more specifically, that it precedes Smith's major works[3] is crucial to our present purposes, since we maintain that this essay is the *manifesto* of Smith's methodology. The very philosophical approach designed by Smith before 1758 will guide his investigations of ethics in *The Theory of Moral Sentiments* (*TMS*) and of economics in *An Inquiry into the Nature and Causes of the Wealth of Nations* (*WN*). Because of such a unique thread linking his works, the notorious 'Adam Smith problem' should cease to exist.[4] In fact, Smith himself never perceived any inconsistency within his whole project. In the *advertisement* to the seventh edition of *TMS*, Smith writes that the *WN* partially accomplishes the original plan he had anticipated in the introduction to the first edition of the *TMS*, namely that of giving an account of *general principles* of law and government.

In addition, the continuity of methodology and intents that we emphasise throughout the corpus of Smith's writings acquires further significance when related to similar projects of his time. Among the possible connections between Adam Smith and his contemporaries, we will examine one of the least exploited – probably because it goes beyond the realm of economics to which Smith is usually confined – but, on the other hand, quite rich in implications.

In 1757 Edmund Burke published *A Philosophical Enquiry into the Origin of our Ideas of the Sublime and Beautiful*: the links that can be made between this book and Smith's *History of Astronomy* are very interesting and suggestive of a pervasive research programme common to the period which we may indicate, on Smith's part, as the Scottish Enlightenment. To present Smith as a member of the Enlightenment and at the same time to associate him with one of the most outstanding theoreticians of the violent passions arising from the sublime might at first seem contradictory. Nevertheless, we argue, on the one hand the Scottish Enlightenment is unquestionably less dogmatic than the orthodox tradition of the French *philosophes*; on the other, Burke himself only anticipates some Romantic ideals, but still clings to a doctrine of moderation – of reform and not of revolution – both in philosophy and in politics. Therefore, both Smith and Burke can be seen as transitional thinkers who reflect the substantial cultural transformations occurring in Great Britain in the second half of the eighteenth century.

Some comparative studies of these two figures have been already accomplished – for example by Nicholson (1910), Dunn (1941), Fay (1960), as well as Jacob Viner (1977) in his introduction to John Rae's *Life of Adam Smith*. However, they all stress the relationship that existed between Smith and Burke when the latter was an active member of the Whig Party. Instead, the parallel we propose is between Smith and Burke the philosopher and literary scholar rather than the politician.

It may be interesting to know whether Smith had read Burke's *Philosophical Enquiry* before writing the *History of Astronomy*, but we do not have any conclusive argument in this sense. The *Catalogue of the Library of Adam Smith* compiled by James Bonar (1932) records a copy of the *Philosophical Enquiry*, but this was already its ninth edition published in 1782. Furthermore, Burke wrote a review[5] of *The Theory of Moral Sentiments* after having received a copy of the book which David Hume had sent him. In his letter[6] to Smith to thank him for the present, Burke formulates the desire of 'being made personally known to' him. However, beyond any punctual distinction between precursor and epigone – which is not the objective of this discussion – we will be content with arguing that the similarities in the concepts and ideas used by the two authors are the result of their common intellectual and historical background.

The opening section of the *History of Astronomy* is devoted to the definition and exemplification of three concepts: wonder, surprise and admiration. As Smith explains:

> what is new and singular, excites that sentiment which, in strict propriety, is called Wonder; what is unexpected, Surprise; and what is great or beautiful, Admiration. (*HA*, Introduction, p. 1)

Wonder is the most relevant to our analysis. One cause of this sentiment is the novelty or unfamiliarity of an object or situation:

> The imagination and memory exert themselves to no purpose, and in vain look around all their classes of ideas in order to find one under which it may be arranged. They fluctuate to no purpose from thought to thought, and we remain still uncertain and undetermined where to place it, or what to think of it. It is this fluctuation and vain recollection, together with the emotion or movement of the spirits that they excite, which constitute the sentiment properly called *Wonder* (*HA*, II, p. 3)

This description can be related to Burke's analysis of the sublime and, in particular, to the effect that the sublime in its highest degree has on the spectator:

> astonishment is that state of the soul, in which all its motions are suspended, with some degree of horror ... Hence arises the great power of the sublime, that far from being produced by them, it anticipates our reasonings, and hurries us on by an irresistible force. (*PE*, part 2, sec. I)

Smith recognises a second important cause of wonder, such as an unfamiliar succession of objects:

> Such is the nature of this second species of Wonder, which arises from an unusual succession of things. The stop which is thereby given to the career of imagination, the difficulty which it finds in passing along such disjointed objects, and the feeling of something like a gap or interval betwixt them, constitute the whole essence of this emotion. Upon the clear discovery of a connecting chain of intermediate events, it vanishes altogether. What obstructed the movement of the imagination is then removed. (*HA*, II, p. 9)

Through the whole *History of Astronomy, imagination* is the faculty responsible for the connection of ideas, for the reduction of sensorial data to familiar categories. It is significant that Smith uses the word *imagination* rather than *reason*; the former does not only evoke the rational component of the mind that is involved in the process of 'familiarisation', but also intuition: in Smith's connotation of the term, imagination includes some sort of creativity. Not accidentally, the role that Smith attributes to such a faculty seems to anticipate the concept of *secondary imagination* as used by Samuel Taylor Coleridge in his *Biographia Literaria,*[7] when the Romantic movement was already in full swing. The secondary imagination, which according to the English poet co-exists with the conscious will, 'dissolves, diffuses, dissipates, in order to recreate', but where this is not possible, still 'struggles to idealise and to unify'. According to Smith, philosophy should be counted precisely among those arts that address themselves to the imagination.

> Philosophy is the science of the connecting principles of nature. Nature, after the largest experience that common observation can acquire, seems to abound with events which appear solitary and incoherent with all that go before them, which therefore disturb the easy movement of the imagination ... Philosophy, by representing the invisible chains which bind together all these disjoint objects, endeavours to introduce order into this chaos of jarring and discordant appearances, to allay this tumult of the imagination, and to restore it ... to that tone of tranquillity and composure, which is both most agreeable in itself, and most suitable to its nature. (*HA*, II, p. 12)

It may be argued that, for a romantic artist, imagination is the catalyst for creation, whereas for Smith it has the task to discover the links of a chain of events which – though perhaps not visible – already exist as connecting principles or laws of nature. In this respect, Smith is clearly a figure of the Enlightenment. On the other hand, some evidence equally suggests that Smith possesses a sort of historical relativism which prevents him from accepting any kind of dogmatism, no matter whether founded on superstition, religion or reason. For example, he writes the following lines about the history of the philosophy of nature:

> Let us endeavour to trace it, from its first origin, up to that summit of perfection to which it is at present *supposed* to have arrived, and to which, indeed, it has *equally been supposed to have arrived in almost all former times.* (*HA*, II, p. 12. My emphasis)

Even though at various stages of human history the same question is asked, it will receive different answers according to the specific institutional framework and to the scientific knowledge attained.

HIDDEN CHAINS IN SMITH'S PHILOSOPHY

Nevertheless, Smith's relativism does not hinder him from posing the existence of a universally valid epistemological tool. By abstracting from the specificity of the various sociocultural contexts, he attempts to explain the dynamics underlying any philosophical investigation. In fact, the latter is triggered by the sentiment of wonder which causes deep discomfort by preventing the smooth passage of the imagination between two seemingly disjointed observations. Smith therefore feels it necessary to found such an investigation upon a sentiment which is more powerful and persistent than mere curiosity. In this, again, we find an echo of Burke's analysis, according to which, curiosity is 'the most superficial of all the affections':[8] and, even though its appetite is very sharp, it is also very easily satisfied. Burke maintains that, to exert itself the mind has to be affected by powers and passions other than curiosity:

> But whatever these powers are, or upon what principle soever they affect the mind, it is absolutely necessary that they should not be exerted in those things which a daily and vulgar use have brought into a stale unaffecting familiarity. Some degree of novelty must be one of the materials in every instrument which works upon the mind. (*PE*, part I, sec. I)

Therefore both authors agree on the role that novelty plays in arousing passions. Similarly, they both attribute to familiarity a dampening effect: for Smith, the latter is an obstacle to philosophical investigation since it prevents the feeling of wonder; for Burke, by inhibiting the production of the sublime, it deprives the mind of that peculiar pleasure that he defines as *delight*.[9] More significantly, Smith realises that, if the insurgence of wonder or the production of the sublime are natural effects (since they are linked to the innate faculty of imagination) philosophy, on the other hand, is an artificial construct, which could blossom only after the 'establishment of law, order, and security' (*HA*, III, p. 1). Nevertheless, the uneasiness caused by wonder can be solved by merely *supposing*[10] the existence of a train of familiar events which abridges the disconnected observations, thus allowing the smooth passage of the imagination. Anthropomorphic deities are created as a result of this very need to reduce to everyday experience some most irregular and terrifying occurrences. The acceptance of a superhuman power accounts for the grandiosity of the event; simultaneously, the earthly passions and feelings – which reveal the gods' human component and to which the savage is mostly accustomed – can explain the irregularity and unpredictability of the wonderful phenomenon.

The same ideas about the nature and goals of philosophy emerge surprisingly with almost specular words, in the *Wealth of Nations* when Smith is stressing the importance of the education of youth:

> The great phenomena of nature ... are objects which, as they necessarily excite wonder, so they naturally call forth the curiosity, of mankind to enquire into their causes. Superstition first attempted to satisfy this curiosity, by referring all those wonderful appearances to the immediate agency of the gods. Philosophy afterwards endeavoured to account for them, from more familiar causes, or from such[11] as mankind were better acquainted with than the agency of the gods. (*WN*, V, ch. I, III, art. II)

The philosopher is he who, because of some natural talent or through prolonged training, has developed a sensibility that makes him feel some incongruence, 'some interval betwixt objects' where common people would not find any uneasiness. Smith offers the simile of the 'nicer ear of the musician', which will find a want in rhythm and consonance in those sounds which to the greater part of men seem perfectly agreeable to measure and harmony.[12] However, the philosopher also has the active role of finding the 'invisible chain' to link seemingly disconnected occurrences. This concept is also presented and somehow generalised in *WN*: 'philosophers or men of speculation ... are often capable of combining together the powers of the most distant and dissimilar objects'[13] to bring about the 'improvements in machinery' which were never invented by those 'who had occasion to use the machines'.

This quotation suggests that the source and aims of philosophy are not limited to the investigation of astronomy or of what Smith calls *natural philosophy*. Smith maintains that, since natural phenomena must have been the first ones to strike the imagination of mankind, this type of philosophy was the first one to emerge. However, *moral philosophy* is founded on similar principles, and addresses similar needs. In Smith's opinion, the fact of leading a social life necessarily made individuals conscious of the 'characters, designs, and actions of one another', so that a number of 'rules and maxims for the conduct of human life' was established. Later, the introduction of writing promoted the multiplication of such maxims, proverbs, apologues and fables. Eventually, after a similar exercise had been *beautifully*[14] accomplished by the essays dealing with natural philosophy:

> the maxims of common life were arranged in some methodical order, and connected together by a few common principles, in the same manner as they had attempted to arrange and connect the phenomena of nature. The science which pretends to investigate and explain those connecting principles, is what is properly called moral philosophy. (*WN*, V, ch. I, part. III, art. II)

It is important to stress that the links between the various moral sayings are to be found in a 'few common principles'. These connecting causes must be few in number for the sake of simplicity, but the adjective 'common' hides some ambiguity. It can, first of all, be interpreted as implying some similar property among the elements of a sub-group of maxims, so that the 'methodical order' suggested here would be simply the result of a classification;[15] however, Smith may as well have used 'common' with the same connotation as in 'common life', that is, with the meaning of 'customary, familiar'. Obviously, these two interpretations are not exclusive; rather, they complement each other since a researcher, in order to be successful in his philosophical investigation, must be acquainted with the categories under which s/he groups the objects s/he is examining. Only in this case will the imagination once more 'glide smoothly' over the sequence of connecting objects that compose – or at least, are supposed to compose – the mechanism of nature .

The metaphor of the machine and its mechanism appears several times in Smith's works. In *The History of Astronomy*, the various systems created by the ingenuity of natural philosophers are 'imaginary machine[s] invented to connect together in the fancy those different movements and effects which are already in reality performed' (*HA*, IV, p. 19). Furthermore, an early system is as rude and complex as the first machine designed to produce a particular effect; but eventually the philosophical system is simplified in the same way in which a newer machine obtains the original effect 'with fewer wheels'. The *History of Astronomy* gives a detailed account of the evolution of the cosmological paradigms along this line. The Newtonian system is the latest and the most successful one at connecting the 'phenomena of the Heavens' with only one common principle. Among all the qualities of matter (after its inertness), gravity is indeed the one which is most familiar to us (*HA*, IV, p. 76). However, in the *Theory of Moral Sentiments* there seems to be a shift in the use of such imagery: the universe itself becomes the great machine endowed with 'secret wheels and springs' (*TMS*, I.i.4.2). We might interpret this change hurriedly in emphasis as an involution, as a return to the key-metaphor of the Enlightenment project, namely that of a mechanistic world. Nevertheless, Smith reinstates the meaning proposed in the *HA* by depicting human society as 'a great, an immense machine' only because 'we contemplate it in a certain abstract and philosophical light' (*TMS*, VII.iii.1.2). Once again, it is the theoretical model of reality and not reality itself which functions according to mechanistic relations.

This is not the only point of departure from the canonical principles of the Enlightenment. The most famous of Smith's metaphors – that of the invisible hand – reinforces precisely the notion of an organic universe to the detriment of a mechanistic one. As a synecdoche for a living body, the hand points at the need for connections between disjointed, and thus wonderful, events in a world already rich in Romantic nuances.

THE CONJURING TRICK OF THE INVISIBLE HAND

The image of the invisible hand has captured the attention and aroused the interest of generations of economists, and is usually taken as a synonym for the *laissez-faire* doctrine. For Viner (1928), Smith's main originality was that of introducing into economics the concept of a unified natural order, operating according to natural laws, and if left to its own course producing results beneficial to mankind.[16] Philosophers and theologians had already made a similar attempt for the world in general. Some of them – among whom was Smith's teacher, Francis Hutcheson – had extended the approach to the economic realm, at least sketchily. According to Viner, then, Smith took an important further step in analysing the economic process at large with the explicit purpose of 'discovering the nature of the order which underlay its surface chaos'. He was trying – to employ the same terminology adopted in the *History of Astronomy* – to find the connecting principles that would allow the imagination to move smoothly along the sequence of those most wonderful phenomena.

In *The Theory of Moral Sentiments*, the harmonious order of nature was the effect of a divine guidance, whereas in *The Wealth of the Nations* this benevolent deity is almost completely absent. In the latter work, harmony is the result of more familiar causes, so that Smith 'is free to find defects in the order of nature without casting reflections of the workmanship of its Author'.[17] Hence, the *Harmonielehre* of the *TMS* is universal, but in the *WN* 'it is as a rule a sort of average or statistical harmony',[18] which allows for cases in which a generally beneficent effect may be detrimental to some individual. In this respect, Viner sees the invisible hand in the *Theory of Moral Sentiments* as one of Smith's many synonyms for 'the great Director of Nature', 'the all-wise Author of Nature', expressions that stress the role played by the beneficent deity. However, in the *Wealth of Nations*, the invisible hand is just an isolated reference and a much weaker piece of evidence in favour of a complete reconciliation of the 'doctrines expounded in the two works'.[19]

Viner correctly interprets the *WN* as Smith's lay attempt at explaining economic phenomena. Hence the irreconcilability of the two books stems from the fact that Viner considers the *TMS* to be almost a religious work. An interesting experiment, however, would be to try a comparison of both books from a lay perspective. After the publication of the *TMS*, David Hume wrote a letter to Smith to congratulate him on the editorial success that the book was already having. In a humorous phrase, after announcing the great favour encountered by the book among several bishops and clergymen, Hume says that the work will not be received positively by the 'true Philosophers ... when these Retainers to Superstition praise it so highly'.[20] It is not so obvious that the *TMS* was designed to be an apology for Christian morality. The book was based originally on a series of lectures that Smith gave in Glasgow, and the class he addressed included a large portion of pupils destined for the Church. However, James Wodrow, who had attended those lectures, wrote:

> [Smith's] Theory of Moral Sentiment founded on sympathy, a very ingenious attempt to account for the principal phenomena in the moral world from this one general principle, *like that of gravity in the natural world*, did not please Hutcheson's scholars so well as that to which they had been accustomed. The rest of his lectures were admired by them and by all especially those on Money and Commerce, which contained the substance of his book on the *Wealth of Nations*.[21]

However simplistic, Wodrow's view adds some evidence in favour of the hypothesis that Smith had embarked in one grand enterprise. He wanted to reduce to a series of familiar causal or logical relations those phenomena of the social realm that had aroused his curiosity. One of Smith's *Lectures on Rhetoric* confirms the pivotal role of the Newtonian methodology in Smith's explanation of reality. Comparing it with the Aristotelian method, he considers Newton's approach to be 'the most Philosophical, and in *every science* ... vastly more ingenious'.[22] It is not only on the cognitive level, but also on the aesthetic one that Smith extols Newton over Aristotle:

> It gives us *pleasure* to see the phenomena which we reckoned the most unaccountable all deduced from some principle (commonly a well-known one) and all united in one chain. (*LR*, ii. p. 134)

Therefore Smith's own investigation from natural philosophy to social science can be seen as a search for some equivalent to the gravitation principle that could explain how, in general, a moral judgement may emerge from self-loving individuals, or that could justify a society based on division of labour and on exchange, and yet be composed of self-interest-seeking agents.

As noted by Viner, Smith's work was part of the project of his time, and the parallel can still be made with Burke's study of the beautiful and the sublime. In particular, Burke states explicitly that he is looking for the equivalent of attraction as an *efficient cause*. He is seeking certain affections of the mind, that cause certain changes in the body; or certain powers and properties in bodies, that work a change in the mind.[23] This *efficient* cause is strictly related to the immediately sensible qualities of things; it is not the *first* cause which, on the other hand, would force us to entertain some metaphysical discourse beyond our understanding:

> When Newton first discovered the property of attraction, and settled its laws, he found it served very well to explain several of the most remarkable phenomena in nature; but yet with reference to the general system of things, he could consider attraction but as an effect, whose cause at that time he did not attempt to trace. (*PE*, part IV, sec. I)

Smith, the philosopher, was mainly interested in these *efficient* causes and followed the same inductive methodology that he had praised so highly in the *History of Astronomy*, in the *TMS*, as well as in the *WN*. This is, in our opinion, the common thread linking the different works of the Scottish author. Viner (1928), however, stress that induction is the methodology used mainly in *WN*, and that it is employed there to find a natural harmony making inference from specific data rather than assuming its existence from metaphysical beliefs, as in the *TMS*. Therefore, the 'invisible hand' of the *WN* is nothing more than a relic of the superimposed divine harmony of the previous work. It should simply be considered as a *lapsus linguae*, although unfortunate, and should be dismissed as such without much consequence for the correct reading of Smith's major work.

Robbins (1952) follows a similar line of reasoning to that proposed by Viner. However, he takes it one step further, to state that the harmony of the system described in the *WN* is very limited and depends on certain preconditions. In particular, the government must not favour any subsection of society, and the market has to function freely without the constraints imposed by monopolies or other market imperfections. Only in this case will the individual self-interest be brought into line with the common good. Smith is in line with the other classical economists when he considers harmony as 'never ... arising in a vacuum but always very definitely within a framework of law'.[24] The legal framework that Robbins is advocating must be a precondition: it is logically – if not chronologically – anterior to the recommendation of economic freedom. What is the role of the invisible hand in this picture?

> [It] is not the hand of some god or some natural agency independent of human effort; it is the hand of the law-giver, the hand which withdraws from the sphere of the pursuit of self-interest those possibilities which do not harmonize with the public good.[25]

Robbins' *market-plus-framework* approach, as Samuels (1966) calls it, is founded upon a static notion; the idea of a framework involves a truism to discriminate what is part of it from what remains out of it. Furthermore, every aspect of a social structure within which interactions take place must be considered to be part of this framework. Self-interest-seeking individuals act in a habitat where the best rules of the game have been written by this benevolent law-giver (a dictator?), and efficient institutions are in place. There is no evolution of the structure and no attempt is made at explaining why these laws and institutions should emerge from the original anarchy.

Samuels (1966) criticises Robbins' interpretation, for three main reasons. First of all, no attention is paid to non-legal aspects of social interaction; second, the role of the government in establishing or maintaining the framework is not completely specified; third, the classical view of the law as an instrument or mode of change is not correctly characterised. As far as the first criticism is concerned, we must agree that such non-legal forces as morals, religion and customs, are essential components of the social fabric, and Smith was certainly aware of them. He had already stressed the importance of custom in Part V of the *TMS*, and in the *WN* itself he devotes an important part of Book V to the institutions designated to the education of the young. Robbins well recognises that government participation is necessary to the success of the *laissez-faire* system, and that the state is an important agent in the economic decision-making process; yet he does not analyse its framework-provision functions explicitly. In particular, Samuels points at two such functions: that of establishing the spheres of competence of the public and private sectors, and that of structuring the private decision-making participation. However, he maintains that:

> legislation (via court and parliament) was a mode of determining the scope of legislation and the role of government, and thereby, also, the overall division of power[26] ... Law thus participated in the determination of its own scope, and in the scope and role of private enterprise *writ large*.[27]

It is clear that Samuels is here complementing or making Robbins' interpretation more precise rather than presenting a serious criticism; in fact, the stress is once more on the legal aspects of the framework, even though it is at a meta-level. The third point of dissent that Samuels expresses is centred on the immutable nature of the market-plus-framework hypothesis as proposed by Robbins. As noted above, his analysis is synchronic and therefore does not consider the possibility of a clash of interest between groups of agents feeding back to the institution design process. Rather than being guided by an invisible and infallible hand, the law-giver closely resembles the Walrasian auctioneer who, in this case, records the excess discontent and gropes towards a social equilibrium. Simultaneously, non-legal institutions are born, evolve and disappear in a similar trial-and-error process. Hence, the idea of an *invisible hand*, as a pre-existent order, was

> epistemologically either mere assertion or hypothesis. The history of subsequent economic thought may be viewed as efforts aiming at the specification of the conditions of mutuality or harmony of interests.[28]

If Robbins was interested in describing a *status quo*, Samuels shifts the attention to the need for a continuous *recreation* of harmony. However, he does not try to isolate the *efficient cause* that first led to the creation of the original institution; he does not look for the gravitational principle that could bring together a social organisation.

Viner's, Robbins' and Samuels' treatments of the *invisible hand* are based on an analysis of this expression as it appears in the *TMS* and in the *WN*. Smith, however, had already used such a metaphor in a previous occasion, namely, as the invisible hand of Jupiter in the *History of Astronomy*. As noted above, this essay was published posthumously in 1795, but had probably been written before 1758, prior to the *TMS*. Macfie (1971) is apparently the first[29] who calls attention to this early passage and compares it with the later uses that Smith makes of it. The original passage reads:

> For it may be observed, that in all Polytheistic religions, among savages, as well as in the early ages of Heathen[30] antiquity, it is the irregular events of nature only that are ascribed to the agency and power of their gods. Fire burns, and water refreshes; heavy bodies descend, and lighter substances fly upwards, by the necessity of their own nature; *nor was the invisible hand of Jupiter ever apprehended to be employed in those matters.* But thunder and lightning, storms and sunshine, those more irregular events, were ascribed to his favour, or his anger. (*HA*, III, p. 2, my emphasis)

According to Macfie, 'Smith … enjoyed pithy, forceful phrases'[31] so when he was writing the *Theory of Moral Sentiments* he remembered the metaphor and employed it, this time, to 'express *his own* view as to the relation between divine guidance, the system of nature, and human behavior'.[32] In fact, the role that the invisible hand plays in the two early works is reversed. Heathen gods act to disturb, thwart and stop the ordinary course of things, whereas in the *TMS*, as well as in the *WN*, the Deity acts to preserve and develop the purposes of Nature when they are disturbed by men.[33]

> [The rich] consume little more than the poor, and in spite of their natural selfishness and rapacity … they divide with the poor the produce of all their improvements. They are led by an invisible hand to make nearly the same distribution of the necessaries of life, which would have been made, had the earth been divided into equal portions among all its inhabitants, and thus without intending it, without knowing it, advance the interest of the society. (*TMS*, IV.I.10)

> [Every individual] generally, indeed, neither intends to promote the public interest, nor knows how much he is promoting it. By preferring the support of domestic to that of foreign industry, he intends only his own security; and by directing that industry in such a manner as its produce may be of the greatest value, he intends only his own gain, and he is in this, as in many other cases, led by an invisible hand to promote an end which was no part of his intention. (*WN*, bk IV, ch. II)

Macfie's reading is religious and supports the metaphysical interpretation of the metaphor. However, his comment on a different function attributed to, or role played by, the invisible hand has been the first in a series of articles reviving interest in this particular aspect of Smith's literary production.

Ahmad (1990) recognises four different invisible hands, possibly reducible to one invisible power. Davis (1990) proposes that the reconciliation between individual and social interest is performed by a sleight of hand, suggesting, perhaps, that the invisible hand is nothing but a conjuring trick. Macfie (1971) maintains that the nuance contained in the expression as it appears in the *TMS* is somewhat different from that of the *WN*: the former refers to distribution and the latter to maximisation of wealth. Persky (1989) seems content to define the invisible hand simply as Adam Smith's simile for the market system[34] even though the title of his piece is 'Adam Smith's Invisible Hands'.

THE INVISIBLE HAND: *UNA, NESSUNA, CENTOMILA*

The question has not changed: what is it? The title of a novel by Luigi Pirandello, *Uno, Nessuno, Centomila (One, None, One Hundred Thousand)* seems appropriate to characterise the riddle of the invisible hand. One: Smith employs *this* metaphor on three different occasions. None: after all, the hand is *invisible*. One hundred thousand: as many *different* interpretations have been given for it as there have been serious scholars investigating it. Does the puzzle have a solution? Once more, of course, God may be the easy way out, but we believe there is another option which satisfies the constraints.

In the first section of this chapter, we tried to show the mechanism that, according to Smith, leads to the accumulation of knowledge. The *imagination* is disturbed, thwarted, and stopped by the occurrence of a *wonderful* event. This state of mind is so unbearable that it may cause lunacy and delusion: this is Burke's notion of sublime with terror. The only way to end the 'vain recollection' of the memory and imagination is to abridge or to fill the gap in order to allow the smooth passage along those that once were such 'different objects'. The bridge is given by an explanation or, as we stressed above, by the mere supposition of an explanation. When a conjuring trick is performed for the first time, we *know* it is just a trick and that it can easily be explained: perhaps we are satisfied by recalling the saying 'the hand is faster than the eye'. In sections II and III of the *History of Astronomy* the adjective *invisible* emerges nine times in as many pages. In particular, when *wonder* is analysed from a 'modern' philosophical perspective, the expressions used are: 'chain of intermediate, though *invisible* events', 'invisible effluvia', 'chain of invisible objects', *invisible* chains. On the other hand, in the section devoted to describing the origin of philosophy, the word *invisible* appears in 'intelligent, though *invisible* causes', 'invisible and designing power', 'intelligent, though invisible beings', 'invisible hand of Jupiter'. The last time it is adopted, it reconciles the different uses with the original lay connotation of *hidden, unknown*. In fact, the members of a somewhat developed society will become 'more desirous to know what is the chain that links' together some irregular phenomena, and the level of strength and security they have reached

> renders them less disposed to employ, for this connecting chain, those *invisible beings* whom the fear and ignorance of their rude forefathers had engendered. (*HA*, III, p. 3, my emphasis)

The 'hidden chains of events', 'the invisible chain', and 'the invisible hand of Jupiter' all stand for the want of an explanation.

We believe that Macfie is right when he says that Smith later remembered the expression and employed it on purpose. He used *the invisible hand* as a modern scientist would use x to denote the unknown – and not necessarily knowable – solution of a functional equation: it is just a rhetorical device. This is the other possible way to solve the riddle. One: the rhetorical phrase is always the same. None: it is just a signifier,[35] an empty box to be filled with meaning. One hundred thousand: the plethora of solutions to different problems at hand.

Why is there an implicit idea of harmony and order when the expression is adopted in the *TMS* and the *WN*, but it is definitely lacking in the *HA*? It is because the notion of an *acceptable* explanation, the set of admissible solutions, has changed over time. In the *History of Ancient Physics*, Smith traces this evolution. He writes that 'in the first ages of the world ... the idea of a universal mind ... who originally formed the whole, and who governs the whole by *general laws*'[36] was extraneous to mankind, but was later accepted

> by those philosophers, when, upon a more attentive survey, they discovered, or *imagined they had discovered*, more distinctly, the chain which bound all [nature's] different parts to one another. (*HAP*, p. 9, my emphasis)

The *invisible hand* is something that makes wonder disappear, thus allowing the disturbed imagination to rest once more. However, if the wonder arises in the savage as a result of some strange, peculiar event and dissolves with the *ad hoc* explanation – no matter how contradictory with respect to the explanation of another similar event – of an invisible and intelligent being, later the philosopher will not be satisfied. S/he will not manage to get rid of the wonder until s/he finds a logical chain of events that makes the irregular observation fit into a harmonic system.

In the *History of Astronomy*, the savage, struck by a wonderful phenomenon, invokes the invisible hand; however, the rich in the *Theory of Moral Sentiments*, and the 'investor' in the *Wealth of Nations* are unconscious agents led by the invisible hand to promote the general welfare. Also from this perspective there seems to be an inconsistency, but it is easily solved. The alter ego of the savage terrified by lightning is not the self-loving agent of the *TMS* or of the *WN*; rather it is Smith himself. In the same way in which the scared primitive tries to give himself a 'reason' for the irregular event, Smith the philosopher is full of wonder when he considers that a complex society can work reasonably well in spite of the fact that each of its members seeks explicitly only his/her own interest. Hence, Smith has set for himself a problem whose solution – let us call it *x*, or, why not, *invisible hand*[37] – will reconcile individual egotistical actions with a higher level of harmony.

This chapter would not be complete if we did not try to give *our* solution to the problem, just to make it one hundred thousand *and one*. As we stated above, Smith's set of admissible solutions is composed of *efficient* causes which, in Burke's definition, are the closest to the *first* cause without entertaining any metaphysical discourse. Burke points at the Newtonian gravitational principle as an example of such an efficient cause, gravity being an observable phenomenon which becomes the foundation of a coherent philosophical system. Smith, for his part, praises the Newtonian system as the only one that drew him insensibly to make use of language expressing [its] connecting principles, as if they were the real chains which Nature makes use of to bind together her several operations.[38] As far as the problem at hand is concerned, Robbins (1952) stressed the importance of the market, and of institutions in general, in limiting the individual to exercise his/her freedom in the pursuit of such objectives that could be reconciled with the common good. Samuels (1966) complemented this view by arguing that harmony is the dynamic result of a dialectical process involving institution design and historical contingencies. However, the original problem remains; the efficient cause has not been isolated explicitly.

Let us rephrase the question: what is it that makes individuals band together as the first example of an institution? What else if not *self-interest*? An important step in this direction is taken by McCall (1991) in a paper[39] on Giambattista Vico, an Italian philosopher and one of Smith's contemporaries. McCall writes:

> Any serious study of the economics of uncertainty must consider the evolution of institutions as a major source of uncertainty reduction ... The reduction of uncertainty occurs because institutions guide information flows and improve induction by the mind and society, with a corresponding enhancement of predictability. Institutions not only reduce uncertainty and generate specialization, regularity and civility. They also harbor the ethical qualities distinctive to humanity.[40]

Institutions are a source of 'uncertainty reduction', not only because they offer security in the form of the law and the police, but also because they provide a very regular, hence predictable, environment. This allows each individual to concentrate on those activities that s/he can perform best because of natural talents or acquired skills. Not surprisingly, we are back to the division of labour as the efficient cause – mediated by the creation of institutions – for the wealth of nations. The argument is made poignantly by Leijonhufvud (1989):

> I believe that it is better to conceive of economic development as the creation of increasingly more complex structures of the division of labor. The standard of living of the average person is thus seen to depend on our ability to maintain the conditions – the extent of the market – that make these complex patterns of co-operation feasible.

Smith himself stressed that the division of labour is a necessary consequence of the natural human propensity to truck and barter; however, let us reiterate once more our point in order to complement Leijonhufvud's analysis. How could institutions – complex patterns of co-operation – arise if not because each member gave up part of his/her freedom in order to benefit from the uncertainty reduction the institution could offer? What is this if not self-interest, again?

CONCLUSION

The *Theory of Moral Sentiments* and the *Wealth of Nations* are parts of a single project that Smith – as he states in the advertisement to the seventh edition to the *TMS* – had initially planned to complete in his lifetime. The leading thread is given by the methodology he employs in each case: induction. This crucial aspect of Smith's philosophy is not stressed sufficiently in the contemporary literature which, dealing with co-ordination and policy issues, rather invokes Adam Smith as the prominent theorist of the *laissez-faire* doctrine. General equilibrium models seem to have solved the theoretical riddle of the invisible hand by attributing to prices the power to co-ordinate a decentralised economy characterised – as a first approximation – by complete and perfect markets. However, these models are based on deduction, which is typical of the neoclassical paradigm. Since the economic theory emerging from this tradition is often criticised for being far apart from everyday experience, we welcome a renewed and well-informed interest in Adam Smith's methodology, one which starts precisely with observed phenomena in order to build a connected and coherent theoretical system.

Notes

* An earlier version of this paper appeared as UCLA Working Paper No. 691S. The author would like to thank Professors Kumaraswamy Velupillai and Michele Luna for their unconditional support, Professors L. Punzo, J. McCall, A. Leijonhufvud, R. Clower for their valuable comments, and Dr N. Pireddu for her suggestions and insights on the philosophical and literary aspects of this paper. Any remaining errors are, of course, the author's.

1. Leijonhufvud (1986) p. 203.
2. See Leijonhufvud (1987).
3. Some evidence suggests that this essay was written even before Smith was appointed to the Chair of Logic in Glasgow in 1751. In his introduction to *HA*, Wightman, commenting on a statement Smith makes on chemistry (*HA*, II, p. 12), says: 'Smith could hardly have written this in Glasgow, where William Cullen in 1748 began his epoch-making "popularization" of chemistry'. Furthermore, in a letter Smith sent to Hume in 1773, he defines *HA* as a juvenile work. In support of this hypothesis, Macfie (1971) maintains that Smith would hardly call any essay written in Glasgow 'juvenile'.
4. A thoughtful analysis of the debate on the consistency of Adam Smith's two major works, that took place toward the end of last century, may be found in Oncken (1897).
5. *Annual Register*, 1759, pp. 484 ff.
6. Quoted in Fay (1960), p. 10.
7. Coleridge (1983) On the Imagination, (ch. 13).
8. *PE*, part I, sec. I.
9. As I make use of the word *Delight* to express the sensation which accompanies the removal of pain or danger; so when I speak of positive pleasure, I shall for the most part call it simply *Pleasure* (*PE*, part I, sec. IV).
10. *HA*, II, p. 8.
11. Another example of moderation and of historical perspective shown by Smith.
12. *HA*, II, p. 11.
13. *WN*, bk I, ch. I.
14. The term is not fortuitous since, at least implicitly, Smith is suggesting that the *beauty* of a systematical arrangement of different observations (*WN*, bk

V, ch. I, III, art. II) of natural phenomena was one of the causes for a similar exercise in morals.

15. In Smith's opinion, classification is an important tool which accompanies advance in knowledge and experience. The example he refers to in *The History of Astronomy* (*HA*, II, p. 2) is that of botany. Smith, as is his habit, gives a very detailed account which may be somewhat familiar to the reader acquainted with the author's description of the pin factory in *The Wealth of Nations*. Since the philosopher is the one who has the ability to connect the characteristics of 'distant' objects or ideas, it is not completely absurd to hypothesise that the theory of the division of labour – that is, the idea of a much greater efficiency in production which derives from the recognition of a variety of particular tasks within a single process – was suggested to him by his 'distant' knowledge of classification in botany.
16. Viner (1928), p. 118.
17. Viner (1926), p. 127.
18. Viner (1926), p. 128.
19. Viner (1926), p. 127.
20. Letter 31, 12 April 1759. Partially quoted in the Introduction to D. D. Raphael and A. L. Macfie (eds) (1982), *The Theory of Moral Sentiments*, Liberty Press.
21. As quoted in the Introduction to Raphael and Macfie (eds) (1982), as Note 20 above. My emphasis.
22. *Lectures on Rhetoric* (*LR*), ii, p. 133. My emphasis.
23. Burke (1990), part IV, sec. I.
24. Robbins (1952), p. 191.
25. Robbins (1952), p. 56.
26. Samuels (1966), p. 103.
27. Ibid., p. 107 [sic].
28. Samuels (1966), p. 2.
29. Viner (1928), however, quotes a passage from the *History of Ancient Physics*, one of the essays on *The Principles which Lead and Direct Philosophical Enquiries. The History of Astronomy* is part of the same work and precedes only by few pages that on the *History of Ancient Physics*. It is strange that such a peculiar phrase did not capture the attention of an otherwise very meticulous scholar as Viner. It is more reasonable to assume that Viner believed the expression to be only accidental and therefore not worth mentioning.
30. In Lindgren (1967), p. 49, the expression is 'Heavenly antiquity'. According to Davis (1990), Smith is 'suggesting that Judaism evolved from a polytheistic animism' (nt 10, p. 343); this may be the reason why the essay was not published when Smith was still alive. We simply believe that this version is a typographical mistake. Smith wants to distinguish between 'savages' and the Greeks or Romans (*Heathen* antiquity) that he could not consider mere savages.
31. Macfie (1971), p. 598.
32. Ibid.
33. Macfie (1971), p. 596.
34. Persky, however, stresses the importance of *security* 'as much part of the invisible hand mechanism as the basic desire for profit' (p. 198). Hence this

paper can as well be classified among those in favour of the *market-plus-framework* hypothesis.

35. We employ 'signifier' according to the definition given by the linguist Ferdinand de Saussure, namely that of an acoustic image as opposed to the 'signified', which corresponds to the concept underlying the word.
36. *History of Ancient Physics*, p. 9.
37. As stressed above, the problem may be represented by a *functional* equation whose solution is a function, that is, a rule and not simply a value. Therefore, the *invisible hand* may stand for such an unknown rule.
38. *History of Astronomy*, IV, p. 76.
39. The paper is also included in this collection.
40. McCall (1991), pp. 2–3.

References

Ahmad. S. (1990) 'Adam Smith's Four Invisible Hands', *History of Political Economy*, vol. 22, no. 1, pp. 137–44.

Bonar, J. (1932) *A Catalogue of the Library of Adam Smith* (London: Macmillan).

Burke, E. (1990) *A Philosophical Enquiry into the Origin of our Ideas of the Sublime and Beautiful* (New York: Oxford University Press).

Coleridge, S. T. (1983) *Biographia Literaria* (Princeton, NJ: Princeton University Press).

Davis, J. R. (1990) 'Adam Smith on the Providential Reconciliation of Individual and Social Interests: Is Man Led by an Invisible Hand or Misled by a Sleight of Hand?', *History of Political Economy*, vol. 22, no. 2, pp. 341–52.

Dunn, W. C. (1941) 'Adam Smith and Edmund Burke: Complementary Contemporaries', *Southern Economic Journal*, vol. VII, January, pp. 330–46.

Fay, C. R. (1960) *The World of Adam Smith* (Cambridge: Heffer).

Leijonhufvud, A. (1986) 'Capitalism and the Factory System', in R. N. Langlois (ed.), *Economics as a Process* (Cambridge University Press), pp. 203–23.

Leijonhufvud, A. (1987) 'The Uses of the Past', paper written for the History of Economics Society Conference, Harvard University.

Leijonhufvud, A. (1989) 'Information Costs and the Division of Labour', *Economic Growth Policies*, ISSJ 120, pp. 165–176.

Lindgren, J. R. (ed.) (1967) *The Early Writings of Adam Smith* (New York: Augustus M. Kelley).

Macfie, A. (1971) 'The Invisible Hand of Jupiter', *Journal of the History of Ideas*, vol. 32, no. 4, pp. 595–99.

McCall, J. J. (1991) 'Vico and De Finetti: an Economic Perspective', unpublished, MS UCLA Working Paper No. 630.

Nicholson, S. (1910) *A Project of Empire* (London: Macmillan).

Oncken, A. (1897) 'The Consistency of Adam Smith', *The Economic Journal*, vol. 7, pp. 443–50.

Persky, J. (1989) 'Adam Smith's Invisible Hands', *Journal of Economic Perspectives*, vol. 3, no. 4, pp. 195–201.

Robbins, L. (1952) *The Theory of Economic Policy in English Classical Political Economy* (London: Macmillan).

Samuels, W. J. (1966) *The Classical Theory of Economic Policy* (Cleveland: World Publishing Company).

Smith, A. (1980) *The Principles which Lead and Direct Philosophical Enquiries; Illustrated by the History of Astronomy* (Oxford University Press).

Smith, A. (1976) *The Theory of Moral Sentiments* (Oxford University Press).

Smith, A. (1976) *An Inquiry into the Nature and Causes of the Wealth of Nations* (Chicago, University of Chicago Press).

Smith, A. (1983) *Lectures on Rhetoric and Belles Lettres* (Oxford University Press).

Viner, J. (1926) 'Adam Smith and Laissez Faire', *Adam Smith, 1776–1926. Lectures To Commemorate the Sesquicentennial of the Publication of The Wealth of Nations* (Chicago, University of Chicago Press).

Viner, J. (1977) 'Guide to John Rae's Life of Adam Smith' in J. Rae, *Life of Adam Smith* (Fairfield: Augustus M. Kelley).

Part III

9 Monetary Regimes, Inflation and Monetary Reform:

Michael Bordo and Lars Jonung*

INTRODUCTION

Since the late 1970s Axel Leijonhufvud has written extensively on monetary regimes and their connection with nominal and real economic performance. Monetary regimes are important because they determine whether countries follow stable or unstable monetary policies and hence have stable or inflationary price levels. Each monetary regime is associated with a given set of inflationary expectations of the private sector and a pattern of policy reactions to these expectations by the monetary authorities. The state of the private sector's expectations, specific to each regime, in turn greatly influences the response of real variables to monetary policy actions.

A number of useful taxonomies for the study of different monetary regimes in history have been created by Leijonhufvud, based on their ability to conform to constitutional rules; by their form of monetary control – monetary aggregate control versus the convertibility principle; and by their inflationary expectations responses.

Leijonhufvud has also contributed to our understanding of inflation. Traditional neoclassical models of the inflation tax with perfectly anticipated money growth and inflation rates imply minimal welfare losses associated with the loss of consumer surplus under the money demand curve. He argues that real world inflationary regimes are characterised by unreliable and unpredictable policy-making which engender price level uncertainty and substantial inflation forecast errors. The resultant uncertainty costs of inflationary monetary regimes are substantial, and for high inflation regimes such as those of Latin America they are disastrous. The

* For excellent research assistance we would like to thank Jakob Koenes, Joe Santos and Alex Hohmann. Helpful comments on an earlier draft were provided by conference participants, Tamin Bayoumi, Barry Eichengreen, Julio Santaella and Anna Schwartz.

costs elucidated include: excessive relative price variability with its attendant resource misallocation; the drying up of key asset markets; misallocation and reduction in capital formation; and the misallocation of both employment and economic activity towards inflation-guessing activity.

Leijonhufvud has made a number of proposals for monetary reform to restore inflationary monetary regimes to price stability. For the low to moderate inflation regimes that have characterised the industrialised countries since the mid-1960s, he recommends a Peel's Act–Friedman rule which would combine the stable nominal anchor characteristics of Friedman's (1960) constant money growth rule (the issue department) with flexibility to deal with emergencies (the banking department); a return to some form of convertibility regime based on a commodity basket; and bluebacking – issuing a parallel currency to appreciate in step with the expected depreciation of the existing currency. For high inflation countries, Leijonhufvud, along with Daniel Heymann, advocates orthodox stabilisation policies of fiscal reform and nominal anchoring, combined with incomes policy and bluebacking in the transition stages. Interestingly, Argentina followed the bluebacking proposal in its 1985 Austral Plan and the more orthodox programme in its 1991 stabilisation.

This paper discusses Leijonhufvud's writings on monetary regimes and inflation in the second section. The third section considers his various proposals for monetary reform and compares them to alternatives, including recent proposals for currency boards. A brief discussion of the recent stabilisation experiences of a number of countries highlights the main points. The fourth section examines evidence for fifteen countries (G-10 plus Switzerland and four high-inflation countries) since 1880, with respect to their experience of diverse monetary regimes. The performances of both nominal and real variables across regimes are compared to the predictions of Axel Leijonhufvud's theoretical writings. In conclusion, the last section presents a brief appraisal of Leijonhufvud's contribution to the analysis of monetary regimes and inflation.

MONETARY REGIMES AND INFLATION

In Leijonhufvud's work, the nature of the monetary regime determines the state of the private sector's price expectations, and hence the response of real activity to monetary policy. A monetary regime is defined as 'first, a system of expectations governing the behavior of the public. Second, it is a consistent pattern of behavior on the part of the monetary authorities such as will sustain these expectations' (Leijonhufvud, 1984a, p. 95).[1]

By incorporating expectations, a monetary regime differs from the earlier concept of a monetary standard, which referred simply to the institutions and arrangements governing the money supply. In the present lexicon, the private sector, with rational expectations, makes forecasts and decisions based on its understanding of the policy-makers' underlying model and likely policies consistent with that model. Policy-makers in turn make their forecasts and decisions based on how they expect the private sector to react. Under such a scheme, the private sector's response to, for example, expansionary monetary policy, will differ dramatically if price level expectations are anchored by the knowledge that the money supply process is constrained by adherence to a fixed exchange rate or to a constitutional rule restricting the growth rate of the money supply to the long-run growth rate of the economy, than if price expectations are based on guessing the monetary authorities' actions in a discretionary regime.

A distinction is made between a constitutional and a discretionary regime (Leijonhufvud, 1984a, 1987a and b). Under a constitutional regime, a set of constraints, often determined by force of law, puts limits on money growth. Under a discretionary regime, no such limits prevail – money growth is at the choice of the monetary authorities. This distinction is similar to the recent approach taken in the literature on the time inconsistency of optimal government policy (Kydland and Prescott (1977); Barro and Gordon (1983)). The absence of a credible commitment mechanism or monetary rule leads governments, in pursuing stabilisation policies, to produce an inflationary outcome. In the discretionary regime, once the monetary authority has announced a given rate of money growth, which the public expects it to validate, the authority then has an incentive to create a monetary surprise either to reduce unemployment or to reduce seigniorage revenue. The public, with rational expectations, will come to anticipate the authorities' perfidy, leading to an inflationary equilibrium. A credible precommitment mechanism or rule, by preventing the government from changing planned future policy, can preserve long-run price stability by anchoring the public's expectations.

Constitutional regimes or rules, according to Leijonhufvud (1984a, p. 95) should be contingent rules or have escape clauses.[2] The monetary authority follows the rule, for example, under the gold standard regime it keeps the price of the currency in terms of gold fixed – except in the event of a well understood emergency such as a major war, financial crisis or supply shock. Under these circumstances the monetary authority can suspend temporarily the rule (gold convertibility) and use seigniorage to finance its expenditure. The rule is contingent in the sense that the public understands that the suspension will only last for the duration of the emergency plus some period of adjustment. It assumes afterwards that the government will follow the deflationary policies necessary to restore the rule (resume gold payments at the original parity). Should the authorities misuse the escape clause, for example, not return to parity under the gold standard, the system of expectations consistent with following the original rule will change to that of a discretionary regime.

Leijonhufvud distinguishes between two types of monetary constitutions based on the quantity principle and the convertibility principle respectively (1984a, p. 99). According to the former, the price level is determined by the monetary authorities' control of some monetary aggregate. A leading example of such a regime is to have the monetary authority follow a Friedman (1960) constant money growth rule. Such a regime requires the monetary authority to have a monopoly over the issue of fiduciary money. It also requires a floating exchange rate. According to the latter, the monetary authority is committed to fixing the price of some commodity in terms of the national currency (historically gold, or silver, or both). Given the fixed commodity price, the private sector then determines the quantity of fiduciary money and the price level consistent with maintaining parity with the rest of the world.

Both types of constitutional regime anchor price level expectations. Under the Friedman rule, assuming no drift in long-run velocity, setting money growth equal to the long-run growth rate of the economy should produce stable prices (zero inflation) and similar expectations. Under the convertibility regime, assuming, for example, that it is based on gold, long-run price movements are determined by the fundamentals driving the demand and supply for monetary gold. Indeed, over the century in which the gold standard prevailed, the world price level underwent a series of alternating waves of deflation and inflation, but ultimately returned to its original level. Price level expectations under such a rule would be mean-reverting.

The two constitutional regimes differ in their possible short-run roles for monetary policy. The Friedman rule regime allows no short-run role for monetary policy to influence real variables – in the sense of a contingent rule – since fully anticipated money growth is completely neutral. By contrast, under the convertibility regime, because price expectations are anchored, the monetary authorities could influence temporarily *ex ante* real interest rates and hence affect investment expenditures and real activity (Leijonhufvud, 1984a, p. 101; 1990, p. 144).

In contrast to constitutional regimes, Leijonhufvud describes the operation of a discretionary fiat regime dubbed the random walk monetary standard (RWMS) (Leijonhufvud, 1986, 1988). Under the RWMS, the monetary authority decides one period at a time whether to change money growth. In making its decisions, only current economic and political considerations are employed. The authority is not concerned with the future. Under such a regime, the formation of price expectations is characterised by considerable uncertainty, which increases as the time horizon is lengthened. Moreover, in addition to the forecast errors of inflation increasing exponentially with the time horizon – because the nature of the policy-making process is so uncertain, individual price expectations tend to become increasingly dispersed. In addition, since policy-makers make judgements based on the anticipated responses of the private sector to their actions, as the private sector's forecasts became more unreliable, the likelihood increases that policy-makers will in time become more unpredictable in their actions.

A distinction is made between the RWMS regimes of low and moderate inflation and that of high inflation. Under low inflation, there need not be a direct connection between the fiscal deficit and the rate of growth of the monetary base. In the high inflation RWMS regimes, it is the inability of the fiscal authorities to raise revenues successfully because of an inefficient and often corrupt tax system, and the inability to stick to a 'hard' budget, that makes seigniorage the primary source of finance for government expenditure (Heymann and Leijonhufvud, 1995, ch. IV). In the high inflation regime, once the rate of inflation increases so that price change is measured in monthly rather than annual time, the degree of uncertainty reaches a level where economic activity is affected seriously. At this point the regime becomes unreliable. Only a drastic monetary reform returning the system to a more transparent, rules-based regime can prevent it from drifting towards the chaos of hyperinflation.

The History of Monetary Regimes

The history of monetary regimes since the 1880s can be described in terms of the taxonomy developed above. The gold standard evolved from a regime of 'tight convertibility' in its early stages to 'loose convertibility' at the end (Leijonhufvud, 1987b, pp. 60–2; 1990, pp. 142–4). Under a tight convertibility regime, for a closed economy with commercial banks, the monetary authority fixes the price of gold, banks issue notes, and the public prevents overissue by threatening to redeem their notes for coin. For the open economy, gold flows keep the domestic price level in line with the rest of the world. For the world as a whole, the price level is determined by the world's demand and supply of monetary gold.

The world moved away from 'tight convertibility' to 'loose convertibility' whereby the link between nations' monetary gold stocks and their money supplies became increasingly tenuous.[3] This was the consequence of a number of financial developments including the substitution of fiduciary money for coin by the private sector to economise on scarce resources; the substitution of fiat money for coin as base money by the banking system to lessen the risk of banking panics; the use of foreign exchange as a substitute for gold as international reserves. These developments led, by 1880, to the 'managed' gold standard whereby central banks were supposed to follow the 'rules of the game' to speed up adjustment to balance of payments disequilibrium. In fact, by the turn of the twentieth century many countries used their tools of monetary policy to shield their domestic economies from the full costs of adjustment (Bordo, 1984).

However, despite evidence of failures to conform to the rules of the game, the pre First World War gold standard worked (Schwartz, 1984). Under fixed exchange rates individual countries had some leeway to control their prices and interest rates in order to extend the period for adjustment. Ultimately they observed the restraints the standard imposed.

Further loosening of the gold constraints occurred with the development of the gold exchange standard in the interwar period, when most countries used the pound and the dollar extensively as international reserves. The system only lasted for six years, from 1925 to 1931. It foundered because holdings of reserve currencies had expanded relative to the stock of monetary gold, so the convertibility commitment lost credibility. Gold maldistribution and perverse sterilisation policies flouted the requirements that adherence to the gold standard imposed.

According to Leijonhufvud (1987a), the shift from 'tight' to 'loose' convertibility also meant a shift towards a regime requiring quality control. The Bretton Wood system (1946–71) he views as a regime based on control of the quantity of high-powered dollars, but one that the US monetary authorities operated for some twenty years to mimic a system following gold standard convertibility rules.

The Bretton Woods system, following the return in 1958 to current account convertibility by the Western European countries, evolved into a gold dollar exchange standard whereby the world increasingly created dollars as international reserves.[4] The USA maintained convertibility (for official transactions) of dollars into gold at $35 per ounce. The rest of the world pegged their currencies to the dollar. For the rest of the world, to the extent that they followed the 'rules of the game', the gold convertibility principle operated in full; however, the nominal anchor was the pegged exchange rate to the dollar.[5] For the USA, as key reserve centre, the balance of payments was not a binding constraint on money issue. According to Leijonhufvud, although the USA was not ultimately constrained by the convertibility principle, the monetary authorities acted as if they were, at least until the mid-1960s, by limiting monetary growth to the long-run growth rate of the economy.

The system broke down in the late 1960s and ended in August 1971 with the closing, by President Nixon, of the gold window. Leijonhufvud and others attribute its collapse in large part to the complete abandonment of the convertibility principle, both in the face of the growing demands of the Vietnam War-induced fiscal deficit and belief in the use of monetary policy to maintain full employment. The *de facto* convertibility regime, by anchoring price expectations, allowed the use of discretionary stabilisation policy. However, the shift to inflation finance and the vigorous manipulation of monetary policy led to a change in the public's expectations (Leijonhufvud, 1987a, p. 132). As the Federal Reserve switched to a discretionary monetary regime, the private sector tried to second-guess its actions, in turn attributing extrapolative rather than mean reverting expectations to short-run movements in the monetary base. Although in theory a constitutional regime could be based solely on the quantity control principle, it has not yet occurred.

The Real Effects of Alternative Monetary Regimes

Leijonhufvud (1987a and b, 1988); Heymann and Leijonhufvud (1995) criticises the standard neoclassical theory of expected inflation for trivalising the costs of unstable money.[6] According to that theory, if money growth and inflation are fully anticipated, then the only costs to society of maintaining, say, a 15 per cent inflation rate for ever is the loss of consumer surplus under the demand for money. Against this cost is weighed the much greater cost of disinflating, which would break existing expectations and lead to significant output losses. The prohibitive costs of disinflation are then used to make the case against reducing inflation.

Leijonhufvud argues that focus on the monetary regime and its expectation-generating mechanism can lead to a clearer understanding of the causes of inflation. Under the RWMS the cost of inflation is considerably higher than in the simple world of fully-anticipated steady money growth. Under the RWMS, price level uncertainty, manifest in forecast errors, increases with the time horizon, and this, in turn, has serious implications. The inability to forecast inflation will lead to mistakes in resource allocation. For example, if different agents have different inflationary expectations, *ex ante* real interest rates will differ, leading to misallocation of investment (Leijonhufvud, 1984a, p. 34). In addition, as the dispersion of inflation increases with time, both lenders and borrowers become increasingly reluctant to engage in long-term contracts. Hence long-term markets will thin out (ibid., p. 32). Moreover, with the increase in inflation uncertainty, resources and talent will be reallocated from production and distribution towards second-guessing the monetary authorities and shielding wealth from the effects of depreciating purchasing power. Increased inflation uncertainty is associated with increased relative price variability. This adds noise to the system, in turn increasing the likelihood of resource misallocation. Finally, as the increase in price uncertainty leads to the disappearance of markets, the political process emerges as an alternative and less efficient way to allocate and redistribute resources.

The losses associated with high inflation RWMS regimes are even greater. In the face of massive inflation uncertainty (for a case such as Argentina in 1989), a multiple currency standard emerges: spot markets for long-term assets are dollarised; rents and wages are indexed; and wholesale and retail markets for consumer goods use domestic currency (Leijonhufvud, 1992, p. 217; Heymann and Leijonhufvud, 1995, ch. 5). Multiple standards in turn create serious problems, such as, for example, in the real estate market, where prices are quoted in dollars yet no mortgage market exists. In the end, as inflation degenerates into hyperinflation, all transactions become speculative, shops begin to display signs reading 'closed for lack of prices' and monetary reform is the only alternative to anarchy.

PROPOSALS FOR MONETARY REFORMS

Axel Leijonhufvud has made a number of proposals to reform the monetary regimes of both low- and high-inflation countries. In addition, we compare his proposals with some alternatives and then examine actual experience with stabilisation programmes.

Low Inflation

For low to moderate inflation RWMS regimes, Leijonhufvud has proposed the Peel's Act–Friedman rule; a convertibility rule; and bluebacking. We describe each and then compare them to a number of alternative proposals.

The Peel's Act–Friedman rule

Leijonhufvud's suggestion for a constitutional amendment to move the USA from the present discretionary RWMS regime to a constitutional regime based on quantity control is a hybrid of the British Bank Charter Act of 1844, which created a rule for the managed gold standard, and Friedman's (1960) constant money growth rate rule (Leijonhufvud, 1984a; 1986). According to the Bank Charter Act, the Bank of England was to be divided into two departments: the issue department and the banking department. The issue department set the basic convertibility rule for the gold standard. It could issue up to £14 million in unbacked Bank of England notes; any extra issue had to be fully backed by gold reserves. Thus the monetary base would expand and contract with gold inflows and outflows, and the mixed-specie fiduciary system, by following the 'currency principle', would act like a pure gold standard. The banking department followed normal banking practices and also served as lender of last resort to the commercial banking system. However, it was constrained in its loan-making capacity by the note issue and gold reserves of the issue department.

Under Leijonhufvud's hybrid scheme, the Federal Reserve would be divided into an issue and a stabilisation department. The issue department would follow the currency principle and provide a nominal anchor by expanding the monetary base at the same rate as the long-run growth rate of the real economy, adjusted for drift in base velocity to reflect financial innovation. The stabilisation department would act like the banking department of the Bank of England. It would engage in short-run stabilisation policy. The reserves required for the stabilisation department to conduct its policies would be provided by setting the base initially some 10/15/20 per cent higher than the actual base before putting into practice the constant growth rate rule. The excess monetary base would then serve the same role as excess international reserves under the Bretton Woods system – to offset unexpected shocks to the economy. Because the long-run growth rate of the base would be anchored by the Friedman rule, market agents would follow mean-reverting price expectations and not view the policies of the stabilisation department as a form of RWMS.

A convertibility rule

Because financial innovation could make monetary control following the quantity principle no longer feasible, Leijonhufvud (1984a, p. 103) suggested that the monetary authorities peg the price of some basket of commodities (such as the scheme proposed by Hall (1982)) and then let the private sector determine the monetary aggregate endogenously.

Bluebacking

Leijonhufvud has proposed a method for abrupt disinflation that avoids both the traditional output and employment costs and the usual redistributions and bankruptcies resulting from cutting money growth and breaking the expectations of a well-established regime. His scheme is based on the issuance of a parallel currency called 'bluebacks'. These blueback notes would by law appreciate by the same amount as the existing currency – greenbacks – depreciates (Leijonhufvud, 1984b; Heymann and Leijonhufvud, 1995, ch. 3). For example, if the anticipated rate of inflation in greenbacks is 15 per cent per year and we start with the two types of currency trading at par, from then onwards 'bluebacks grow constantly in their legal capacity to extinguish debts contracted in greenbacks' (Leijonhufvud, 1984b, p. 21), and after one year 85 blueback cents would pay off a one greenback dollar debt; after two years it would take 71 blueback cents, and so on.

If expectations are rational, zero inflation would be achieved quickly and it would avoid the output losses and redistribution of wealth associated with conventional disinflation. In practice, a necessary condition for instituting such a scheme is that the original reason for inflation – to finance fiscal deficits – is reversed and that the public believes this. In that scenario, a bluebacking scheme would serve as a low-cost way of transiting from an inflationary to a stable price environment.

Alternative proposals

The rich literature on monetary reform contains a number of proposals for long-run price stability. The traditional approach to keeping inflation at bay for open economies is to maintain a fixed exchange rate. The basic idea is to create low-inflation credibility for the domestic currency by pegging it to a foreign currency issued by a highly credible central bank, such as the Federal Reserve or the Bundesbank. The pegged rate represents a method of 'importing' credibility and thus creating expectations of low rates of inflation. This was a major argument given in the late 1980s for the exchange rate mechanism (ERM) system under which several European central banks tried to create credibility for their domestic currencies by tying them, at a pegged rate, to the DM (Giavazzi and Pagano, 1988). Moreover, the success of the Bundesbank in turn was attributed to its independence from the fiscal authority. However, as history has shown, fixed-rate systems tend to break down sooner or later as the monetary authorities prove unable to maintain credibility in their commitments to the pegged rates in the face of massive shocks (Bordo, 1993b). This has been demonstrated most recently by the breakdown in 1992–3 of the ERM.

For economies with a *flexible exchange rate*, a number of well-known rules have been designed to maintain price stability, including Milton Friedman's (1960) *k*-per cent rule, and various proposals to target GNP growth as well as recommendations to stabilise the growth of the monetary base. In an earlier era, two proposals made were: Knut Wicksell's norm of price stabilisation of 1898, and David Davidson's norm where the price level should fall in proportion to the increase in productivity.[7]

Judging from the historical record, a fundamental weakness with all these plans to create low and stable inflation, including Leijonhufvud's proposals, is that they do not contain a mechanism that creates sufficient credibility. As long as money is produced by a government-controlled monopolist, the money supply process will be the subject of political pressure. Sooner or later the monetary system will be exposed to a disturbance that induces political manipulation of the money supply, creating inflation. Once inflation begins, it tends to become entrenched in the monetary regime.

One method to create strong credibility for the purchasing power of the currency is to take the money supply out of the control of the political system. Creating a truly independent central bank whose only responsibility is to maintain price stability is one way to do this (Cukierman, 1992), and this has seemingly been successful in New Zealand (Svensson (1993), Wood (1994)). An alternative is to establish a currency board, which is a monetary institution that issues domestic notes and coins fully backed by a reserve currency and fully convertible on demand into the reserve currency at a fixed exchange rate. The foreign reserve currency may be either a currency of a foreign central bank or specie. Under a currency board, there is no room for discretionary monetary policy and thus for inflationary policies – the money supply is completely isolated from the domestic political system.

Currency boards were common from the nineteenth century until the post-Second World War period. About 50–60 boards were operating in various European colonies in Africa, Asia and the Caribbean (Schwartz (1993), Hanke *et al.* (1993)). When these former colonies obtained independence, commonly one of their first steps was to set up a central bank. Eventually, most of these former colonies experienced high inflation, with the principal exception of those such as Singapore and Hong Kong, that maintained variants of currency boards.

The currency board solution, as we suggest below, may be practical for newly-independent countries facing the task of establishing credibility for their currencies, or for countries with a history of high and variable inflation, where the credibility of the existing central bank is low and likely to remain so in the foreseeable future. However, because currency boards require a country to give up its monetary sovereignty completely and because they do not have a provision for a lender of last resort, a country seeking an anti-inflation plan may not find this to be a palatable solution.

HIGH INFLATION

To extricate economies from high inflation RWMS regimes, Heymann and Leijonhufvud (1995, ch. 6) propose a very orthodox set of measures. The basic ingredients consist of fiscal reform leading to a long-run balanced budget and nominal exchange rate anchoring. Streamlining tax collection, raising taxes and rationalising the budgetary process will remove the imperative for monetisation of fiscal deficits. Fixing the nominal exchange rate will quickly reduce the prices of traded goods and act as a strong signal to price setters to reduce domestic goods prices. Other policies to ease the transition include incomes policies to offset the price inertia associated with backward looking indexation; and the bluebacking scheme.

Recent Stabilisation Plans

Two recent stabilisation plans in Argentina, the Austral Plan of 1985 and the Convertibility Law of 1991, illustrate key elements of the Heymann and Leijonhufvud proposals.

The Austral Plan

Argentina has instituted five plans since the 1970s to deal with an extremely serious high inflation problem – from 1980 to 1990 consumer prices increased on average by 300 per cent per year. The Austral Plan of June 1985 is of great interest because it incorporated Leijonhufvud's (1984b) bluebacking scheme. In mid-1985, inflation was running at 30 per cent per month and the fiscal deficit at above 15 per cent of GDP. The plan aimed at rapid disinflation to break inflationary expectations (Heymann, 1987 and 1991).

The plan contained three elements. First, the central bank would no longer finance the fiscal deficit. It was hoped that the reduction in inflation (through the Tanzi effect), in addition to other fiscal measures, would increase tax revenues sufficiently to relieve the inflationary pressure. Second, wages and prices were frozen. Third, a new currency – the austral – was created, worth 1000 pesos. According to the blueback scheme, payments denominated in pesos, resulting from previous contracts, would be made in australs at a conversion rate that changed daily, in such a way as to make the austral appreciate relative to the peso.

Initially, the plan was successful. Inflation fell to 3 per cent per month, as did the fiscal deficit. Output initially declined then recovered. However, the results were not permanent. Residual inflation led to further pressure on wages and prices, and the fiscal deficit expanded. By the end of 1986 inflation was up to 6–8 per cent per month and rising. The Austral Plan was shortlived because permanent fiscal balance was not achieved.

Thus Leijonhufvud's blueback scheme, when instituted under the Argentine Austral Plan of 1985, did not become credible because the monetary and fiscal processes of Argentina were not changed in a significant way at the same time as the plan was introduced. In other words, the monetary regime remained unchanged. Hence there was little reason for the Argentina public to revise its expectations about the future long-run behaviour of inflation. Inflation was expected to return to a high chronic level and eventually it did so.

The Convertibility Law of April 1991

The Menem Administration, which entered office in June 1989, faced a legacy of hyperinflation (greater than 20 per cent per month), a high fiscal deficit (11 per cent of GDP) and declining real output. In the following year, it enacted a number of important reforms to reduce the fiscal deficit. These included: simplification of the tax structure; reduced government expenditure; and privatisation of state-owned enterprises. These initiatives succeeded in reducing the deficit to 1.8 per cent by 1991, and converting it to a surplus in 1992 (Levin, 1993).

High inflation continued, leading the administration to implement an exchange-rate-based convertibility plan on 1 April 1991. The Convertibility Law established full convertibility of the Argentine currency at the rate of 10 000 australs per US dollar. It required the monetary base to be fully backed by international reserves. The indexation of austral-denominated contracts was outlawed. The Central bank immediately suspended all rediscounting and open market operations – effectively ending the use of discretionary monetary policy and the lender of last resort facility. Finally, on 1 January 1992 a new currency was instituted – the peso – equal to 10 000 australs.

The convertibility law transformed the Argentine central bank into a form of currency board pledged only to buy and sell pesos at the fixed exchange rate. Since Argentine international reserves in 1991 exceeded the dollar value of the monetary base (greatly eroded by hyperinflation), the new currency was fully backed.

Initially, the plan was highly successful. Within a year, inflation was reduced to less than 20 per cent. This plan is regarded by observers as more credible than earlier stabilisation plans because of the elimination of the fiscal deficit and the full backing of the currency (Levin, 1993). However, instituting a nominal exchange rate peg has created the potential for future difficulties. Pegging the exchange rate led to rapid declines in the prices of traded goods (determined in world markets) relative to the prices of non-traded goods (which are subject to considerable inflation inertia). This increase in the real exchange rate created a current account imbalance. At the same time, removal of capital controls combined with exchange rate stability, encouraged private capital inflows (largely repatriation of domestically-owned funds) sufficient to finance the deficit. The future impact of the high real exchange rate on aggregate demand and economic activity is likely to put increased pressure on the government to devalue – although a speculative attack on the peso was rebuffed successfully in November 1992. Given Argentina's history of hyperinflation, devaluation may harm seriously the government's credibility.

The Convertibility Law of 1991 incorporated many elements of proposals suggested by Heymann and Leijonhufvud. The costs of high inflation, which they elaborately describe, may have finally induced a sea change in a country with a long history of currency instability.

Other Stabilisation Experiences

A number of countries have suffered from extremely high inflation in recent decades and as a consequence experimented with various stabilisation programmes. Vegh (1992) surveys the experience of ten stabilisation programs in six high inflation countries (Argentina, Brazil, Chile, Israel, Mexico and Uruguay) since the 1960s, as well as stabilisation programs in hyperinflation countries.[8] Two types of programmes in high inflation countries are distinguished: heterodox stabilisations that rely on nominal exchange rate anchoring and incomes policies; and orthodox policies that do not include incomes policies. In all of the heterodox cases, inflation converged slowly to the rate of *devaluation*; real exchange rates appreciated; the current account deteriorated; and real activity followed a boom-and-bust cycle.

Of the ten cases, three were successful in permanently reducing the inflation rate to low levels: Chile (1978), Israel (1985) and Mexico (1987). The common element in the unsuccessful cases was the lack of a credible commitment to stick to the nominal anchor. Belief by agents that the peg would eventually be abandoned resulted in sustained inertia in domestic price inflation, in turn precipitating the inflationary consequences described above.

By comparison, stabilisation in a hyperinflation setting has had one outstanding success. In the case of Bolivian hyperinflation, the stabilisation of 1985 was successful in restoring price stability immediately. This may have been the case, according to Vegh, because under hyperinflation the need for seigniorage to finance the fiscal deficit is more apparent than under high inflation, so that a credible fiscal reform is viewed as sufficient to ensure price stability and because, under hyperinflation, conditions have deteriorated so badly that the imperative for reform is accepted by all. In contrast, under high inflation people seem to learn to live with it, thus keeping demand for anti-inflationary measures at bay.

Thus the Argentine stabilisation of 1991, successful so far, may, in addition to incorporating all the elements of an orthodox reform, have been successful because the economy had deteriorated into hyperinflation. This created the belief that there was no alternative to a fundamental change.

Currency Boards as a Method to Reduce High and Chronic Inflation

The fundamental obstacle facing the monetary and political authorities in countries with high and chronic inflation in planning a stabilisation programme to reduce the rate of inflation and keep it permanently at a low level, is to obtain *ex ante* credibility for the disinflationary policy package. In principle, if a programme of disinflation is fully credible, there would be no output loss caused by disinflation. Under these circumstances, the rate of inflation can be reduced quickly and painlessly. The history of various stabilisation programmes to reduce high and chronic inflation shows, however, that it is extremely difficult to get instantaneous as well as lasting credibility.[9]

The challenge to economists is to design a monetary regime that creates instant as well as lasting credibility in an economy with chronic inflation. A common method, often adopted, is to 'borrow' credibility from foreign sources. The orthodox approach is to tie the high inflation domestic currency to a low-inflation foreign currency through a fixed exchange rate. The weakness with this approach is that the fixed rate initially lacks credibility. This commonly forces the domestic central bank to maintain high domestic real interest rates to defend the fixed rate, in turn contracting the domestic economy and thus undermining the long run credibility of the fixed rate.

Another technique of borrowing credibility is to have an international organisation, such as the League of Nations or the International Monetary Fund (IMF), to enforce the commitment (see, for example, Santaella (1993)). Such foreign intervention, however, can lead to domestic political resentment which undermines credibility, in turn threatening the longevity of the regime.

One proposed method to stop high inflation, that avoids the weaknesses of the above-mentioned approaches, is for a currency board to introduce a new currency for domestic circulation that has high credibility immediately relative to the existing currency. The currency board can then function as a binding commitment, both in the transition to a low rate of inflation and to the subsequent maintenance of permanent stable rate. To create credibility it relies on an external arrangement, but here the commitment is much stronger than with a pegged exchange rate plus a central bank.

The transition to a low inflation monetary system based on a currency board may follow two strategies. The first one is to convert the existing inflationary central bank into a currency board – the strategy recently followed by Argentina.

The second strategy would be to set up a currency board that issues a currency parallel to that of the existing central bank. The advantage of this second solution is that during the stabilisation programme, existing contracts in the labour, product and capital markets made in terms of the inflationary central bank currency can be enforced, reducing the negative effects of inflation inertia on output. The parallel currency approach would in principle allow a transition during which budget constraints can gradually harden while the currency of the currency board enables the public to protect itself from high inflation. 'Soft' budgets will gradually turn into 'hard' ones, but the braking process will not be as abrupt. The process of currency competition and currency substitution will thus be gradual minimising output losses. This second strategy has much in common with Leijonhufvud's bluebacking scheme.

A currency board introduced in this way requires no 'preconditions' for monetary reform. Government finances, state enterprises or trade need not be reformed before the currency board can begin to issue its parallel currency. There are a few cases in history where a parallel currency has been a method to reduce high inflation (see, for example, Rostowski and Shapiro (1992), Hanke *et al.* (1993) and Siklos (1993) on two such cases in Russia in the 1920s).

If a high inflation Latin-American country were to establish a currency board, the currency board currency would presumably be tied to the US dollar, as is the case in Argentina. Such a currency would have a truly fixed exchange rate to the dollar and thus the same credibility as the US dollar. The Federal Reserve Board does not have a perfect inflationary record, but the relative credibility of the US dollar will in the foreseeable future be much stronger than that of the currency issued by the central bank of any Latin-American country.

To protect the currency board from political pressure and thus to foster the credibility of its currency, the main office of its board could be set up in a foreign country and the majority of the board's members be foreign citizens. Such an agreement, however, might engender resentment of foreign domination. To offset such concerns and to create public and political support for the new arrangement, part of the profits of the board from investment of its foreign currency reserves would be returned to the domestic government.[10]

A currency board reform for Russia, along these lines, has recently been proposed by Hanke *et al.* (1993). Russia in the 1990s is experiencing extreme inflation, and the Russian central bank has a history of high inflation, hence any attempt to create credibility for the rouble will be extremely costly. Russia may have to live with high real interest rates for a long time before credibility is established for the present central bank rouble. Thus Russia in the mid-1990s faces problems typical of many high-inflation economies in Latin America in the 1970s and 1980s.

In two countries, Estonia and Argentina, the central bank is presently mimicking currency boards. Estonia has tied its currency to the Deutschemark (DM) at the rate of eight Estonian kroons for one DM. As discussed above, one Argentine peso since January 1992 is equivalent to one US dollar.

Initially, these programmes have met with success. Their main weakness, however, is that a central bank that tries to mimic a currency board does not have the same high credibility as a traditional currency board. The political system may change the behaviour of the Estonian or Argentine central bank so that it will not continue to behave as a fully-fledged currency board, consequently starting the inflationary process anew. The future will show if these two central banks, ordered to operate as currency boards, represent a significant step towards less inflationary monetary regimes. If so, it will be a remarkable break with the past. If they fail, a possible next step would be to move towards traditional currency boards.

EMPIRICAL EVIDENCE

Axel Leijonhufvud's theoretical writings contain a number of predictions about the performance of key macro variables under alternative monetary regimes. His most important predictions concern the behaviour of the inflation rate under convertible and discretionary (RWMS) regimes; specifically, the inflation rate should be higher, more variable and more uncertain under discretionary monetary policy regimes than under convertibility regimes. Moreover, the forecast error of inflation should be increasing with the time horizon under the RWMS regime. Also, inflation variability and uncertainty should be greater under high RWMS regimes than low ones. Presumably other nominal variables such as the exchange rate (Leijonhufvud, 1984b) and nominal interest rates should be more variable under the RWMS-type regimes than under convertible regimes.

As a consequence of greater inflation uncertainty, he argues that, other things being equal, real output will be more variable under RWMS regimes than under convertible regimes. At the same time, real variables may be more subject to both real and nominal shocks under convertible regimes than under discretion (Leijonhufvud, 1990). Other real variables, such as the real exchange rate and real interest rates, may respond in a similar manner.

Leijonhufvud's insights into the behaviour of alternative monetary regimes pertain specifically to the performance of individual countries or, if a number of countries are part of a similar regime – of the whole system. His insights can also be applied to the relationship between countries under alternative monetary regimes or, to put it another way, to the issues raised by the debate over fixed versus flexible exchange rates.

According to the traditional view, adherence to a (commodity-based) fixed-exchange rate regime, such as the gold standard, ensured long-run price stability for the world as a whole because the fixed price of gold provided a nominal anchor to the world money supply. Individual nations, by pegging their currencies to gold, fixed their price levels to that of the world.[11] A fixed-rate system based on fiat money, however, may not provide a stable nominal anchor unless a credible commitment mechanism constrains the growth of the world's money supply (Giovannini, 1993). The disadvantage of fixed rates is that individual nations are exposed to both monetary and real shocks transmitted from the rest of the world via the balance of payments and other channels of transmission (Bordo and Schwartz, 1989). The advantage of floating exchange rates is to provide insulation from foreign shocks. The disadvantage is the absence of the discipline of the fixed-exchange-rate rule, since monetary authorities might adopt inflationary policies.

Theoretical developments in recent years have complicated the simple distinction between fixed and floating rates. In the presence of capital mobility, currency substitution, policy reactions and policy interdependence, floating rates no longer necessarily provide insulation from either real or monetary shocks (Bordo and Schwartz 1989). Moreover, according to recent real business cycle approaches, there may be no relationship between the international monetary regime and the transmission of real shocks (Baxter and Stockman, 1989). Nevertheless, empirical comparisons of regimes may shed light on these issues.

To make the case for one monetary regime over another, empirical and historical evidence on their performance is crucial. In this section we present some evidence on different aspects of the macro performance of alternative international monetary regimes since 1881.[12] The comparison for eleven (the G-10 plus Switzerland) industrialised countries (the USA, UK, Germany, France, Japan, Canada, Italy, Belgium, the Netherlands, Sweden and Switzerland) is based on annual data for the classical gold standard (1881–1913), the interwar period (1919–39), Bretton Woods (1946–70), and the regime of floating rates between the principal currencies (1971–89). The Bretton Woods period is divided into two subperiods: the preconvertible phase (1946–58) and the convertible phase (1959–70).[13] This classification of regimes corresponds roughly to Leijonhufvud's distinction between convertibility-based regimes (the gold standard and Bretton Woods convertible); loose convertibility (interwar, Bretton Woods preconvertible) and discretion or RWMS (the floating regime).

In addition, to reflect Leijonhufvud's interest in the performance of high inflation economies, we also examine the behaviour of a number of macro variables for four traditionally high inflation economies (Argentina, Brazil, Chile, Israel) for periods for which data is available. Although these countries did not strictly adhere to convertible regimes, there is considerable evidence that the prevalence of such regimes in the rest of the world influenced their macro performance.[14]

One important caveat is that the historical regimes presented here do not represent clear examples of fixed and floating rate regimes, or alternatively of convertibility rules versus discretion. The interwar period is composed of three regimes: general floating from 1919 to 1925; the gold exchange standard from 1926 to 1931; and a managed float to 1939.[15] The Bretton Woods regime cannot be characterised as a fixed exchange rate regime throughout its history: the preconvertibility period was close to the adjustable peg envisioned by its architects; the convertible period was close to a *de facto* fixed dollar standard.[16] Finally, although the period since 1973 has been characterised as a floating exchange rate regime, at certain times it has experienced varying degrees of management.

Stability and Convergence

Table 9.1 presents descriptive statistics on ten macro variables for each of the G-11 countries; the data for each variable are converted to a continuous annual series from 1880 to 1989. Table 9.2 presents similar data for a smaller set of data for the four high inflation countries. The ten variables are: the rate of inflation; real per capita growth; money growth; short-term and long-term nominal interest rates; short-term and long-term real interest rates; the absolute rates of change of nominal and real exchange rates; and the ratio of the government deficit (government expenditures less tax revenues) to GDP. The definition of the variable used, for example, M_1 versus M_2 was dictated by the availability of data over the entire period. For each variable and each country we present two summary statistics: the mean and standard deviation. For all the countries taken as a group, we show two summary statistics: the grand mean and a simple measure of convergence defined as the mean of the absolute differences between each country's summary statistic and the grand means of the group of countries.[17] We comment on the statistical results for each variable.

Inflation (Panel a)

The classical gold standard had the lowest rate of inflation of any monetary regime for all fifteen countries, and the interwar period displayed mild deflation for all except Brazil. The rate of inflation during the Bretton Woods period was on average (for every country except Japan) lower than during the subsequent floating exchange rate period. The average rate of inflation in the two Bretton Woods sub-periods was virtually the same for the G-11 countries. However, this comparison conceals the importance of two periods of rapid inflation: in the 1940s and 1950s, and in the late 1960s. For the high inflation countries in Table 9.2, inflation increased substantially between the preconvertible and Bretton Woods convertible periods – see Figure 9.1.[18] Thus the evidence based on country and period averages of very low inflation in the gold standard period and of a lower inflation rate during Bretton Woods than the subsequent floating period is consistent with the traditional view on price behaviour under fixed (commodity based) and flexible exchange rates.

In addition, the inflation rates show the highest degree of convergence between the G-11 countries during the classical gold standard and to a lesser extent during the Bretton Woods convertible sub-period, compared to the floating rate period and the mixed interwar regime. This evidence is also consistent with the traditional view of the operation of the classical price specie–flow mechanism and commodity arbitrage under fixed rates and

Table 9.1 Descriptive statistics of selected open economy macro variables, the G-11 countries, 1881–1990, annual data: mean, standard deviation

	Gold standard (1881–1913)		*Inter-War (1919–38)*		*Bretton Woods (total) (1946–70)*		*Bretton Woods (preconvertible) (1946–58)*		*Bretton Woods (convertible) (1959–70)*		*Floating exchange (1974–90)*	
a Inflation PGNP[a]												
United States	0.3	3.1	–1.8	7.6	2.4	2.6	2.8	3.5	2.6	1.5	5.6	2.4
United Kingdom	0.3	3.1	–1.5	7.8	3.7	2.2	4.6	2.5	3.4	1.5	9.4	6.1
Germany	0.6	2.6	–2.1	4.7	2.7	4.0	2.1	6.2	3.2	1.8	3.3	1.2
France	0.0	4.9	2.2	9.1	5.6	4.1	5.6	5.1	5.5	3.6	8.8	3.2
Japan	4.6	5.5	–1.7	7.3	4.5	4.6	4.7	7.3	5.1	1.3	2.6	2.4
Canada	0.4	1.4	–1.9	6.1	2.7	3.0	3.9	3.9	2.9	1.5	7.3	2.6
Italy	0.6	3.2	–1.1	11.7	3.8	11.5	5.8	16.0	3.8	2.1	12.9	4.6
Belgium	0.2	4.9	3.8	4.4	2.4	3.2	2.4	4.4	3.3	1.5	4.9	2.5
Netherlands	1.0	2.2	–3.1	5.6	4.1	3.1	4.5	3.2	4.9	3.1	3.6	2.9
Sweden	0.4	3.1	–2.3	6.1	3.9	4.0	4.1	5.4	4.2	1.7	8.4	3.0
Switzerland	n.a.		–2.2	4.3	2.8	1.9	1.8	1.9	4.2	1.0	3.5	1.8
Mean	0.8	3.4	–1.1	6.8	3.5	4.0	3.8	5.4	3.9	1.9	6.4	3.0
Convergence	0.8	1.0	1.4	1.6	0.8	1.4	1.1	2.2	0.9	0.6	2.4	0.9
b Real per capita growth[a]												
United States	1.8	5.1	0.0	8.1	2.0	2.8	1.8	3.4	2.9	1.9	2.1	2.7
United Kingdom	1.1	2.4	1.2	4.5	2.1	1.8	2.1	2.2	2.3	1.4	1.5	4.2
Germany	1.7	2.9	2.6	8.5	5.0	3.3	7.3	3.9	3.5	2.6	2.1	1.9
France	1.5	4.6	1.3	7.2	3.9	2.1	4.6	2.7	3.9	1.3	1.7	1.5
Japan	1.4	3.8	2.0	6.1	8.1	2.7	5.7	1.1	8.9	2.5	3.5	1.1

Table 9.1 Continued

	Gold standard (1881–1913)		Inter-War (1919–38)		Bretton Woods (total) (1946–70)		Bretton Woods (preconvertible) (1946–58)		Bretton Woods (convertible) (1959–70)		Floating exchange (1974–90)	
Canada	2.3	2.8	0.2	8.8	2.5	2.6	2.4	3.3	3.5	1.7	1.3	2.4
Italy	1.0	4.0	0.9	4.7	5.6	3.3	5.2	4.4	5.8	1.9	2.5	2.2
Belgium	n.a.		3.8	6.4	3.1	2.2	2.4	2.4	4.1	1.3	1.8	2.0
Netherlands	0.9	2.4	0.9	3.8	3.6	3.3	4.1	4.0	4.2	2.7	1.1	1.8
Sweden	2.4	4.3	2.6	6.8	3.3	1.5	3.1	1.7	3.6	1.5	1.6	2.3
Switzerland	n.a.		0.1	2.6	3.0	3.1	2.7	4.1	2.9	1.7	1.5	2.5
Mean	1.6	3.6	1.4	6.1	3.8	2.6	3.8	3.0	4.1	1.9	1.9	2.2
Convergence	0.4	0.9	0.9	1.4	1.3	0.5	1.4	0.8	1.1	0.5	0.5	0.6
c Money growth[a]												
United States	6.1	4.1	0.6	8.6	6.3	5.8	6.4	8.3	7.0	1.5	8.6	2.4
United Kingdom	2.1	1.7	0.8	4.7	3.2	3.2	1.7	2.9	5.5	2.9	13.5	5.5
Germany	5.7	4.7	1.3	10.1	12.8	5.9	17.6	5.6	10.9	4.7	5.7	4.5
France	2.1	4.7	6.4	8.5	11.5	7.5	14.7	7.2	8.6	6.5	8.8	3.4
Japan	7.2	14.5	0.5	9.7	16.2	16.2	20.1	21.8	14.9	7.0	5.8	6.3
Canada	7.4	5.3	1.1	4.7	6.0	4.0	5.1	3.2	8.3	3.9	10.6	3.9
Italy	3.2	3.1	3.6	6.2	13.3	7.8	15.9	10.5	12.4	2.0	13.4	4.9
Belgium	n.a.		6.3	9.8	4.5	3.2	3.8	2.2	5.5	4.2	4.4	3.7
Netherlands	4.2	3.5	–1.0	8.1	5.5	4.9	3.2	5.8	8.1	1.7	6.6	4.6
Sweden	5.9	3.7	–0.8	4.9	7.1	3.5	6.7	3.0	8.0	3.8	8.6	4.8
Switzerland	n.a.		8.1	7.9	5.8	4.2	3.3	3.7	7.9	3.3	3.4	7.0
Mean	4.9	5.0	2.4	7.6	8.4	6.0	9.0	6.7	8.8	3.8	8.1	4.6
Convergence	1.8	2.2	2.0	1.6	3.5	2.2	5.3	3.3	2.1	1.4	2.3	0.9

Table 9.1 *Continued*

	Gold standard (1881–1913)		*Inter-War (1919–38)*		*Bretton Woods (total) (1946–70)*		*Bretton Woods (preconvertible) (1946–58)*		*Bretton Woods (convertible) (1959–70)*		*Floating exchange (1974–90)*	
d Short-term interest rate												
United States	4.8	0.9	3.5	2.0	3.4	1.9	2.0	0.9	4.8	1.6	8.9	2.6
United Kingdom	2.8	0.8	3.0	1.8	4.0	2.5	2.3	1.8	5.8	1.6	11.2	2.1
Germany	3.2	0.9	4.6	1.6	4.0	1.5	4.1	1.1	4.0	1.7	5.9	2.4
France	2.5	0.6	3.1	1.4	4.1	1.9	3.2	1.5	5.1	1.9	10.3	2.6
Japan	2.4	0.5	2.0	0.5	6.5	0.8	6.9	1.0	6.3	0.6	5.2	2.0
Canada	n.a.		0.9	0.4	2.9	2.0	1.4	1.1	4.6	1.3	10.3	2.8
Italy	n.a.		n.a.		n.a.		n.a.		n.a.		n.a.	
Belgium	2.8	0.7	3.4	1.4	3.0	2.0	1.5	0.6	4.6	1.6	9.8	2.8
Netherlands	2.8	0.8	2.6	1.5	2.5	1.6	1.6	1.0	3.5	1.6	6.7	1.9
Sweden	n.a.		n.a.		5.9	1.4	n.a.		5.9	1.4	10.2	2.7
Switzerland	3.6	0.6	2.5	1.1	1.9	0.7	1.6	0.4	2.3	0.9	2.5	2.1
Mean	3.1	0.7	2.8	1.3	3.8	1.6	2.7	1.0	4.7	1.4	8.1	2.4
Convergence	0.6	0.1	0.8	0.4	0.9	0.4	1.3	0.3	0.7	0.3	1.9	0.3
e Long-term interest rest												
United States	3.8	0.3	4.2	0.6	3.9	1.3	3.0	0.4	5.0	1.1	10.4	2.1
United Kingdom	2.9	0.2	4.1	0.7	5.2	1.8	3.9	0.8	6.6	1.3	12.1	2.8
Germany	3.7	0.2	6.8	1.6	6.3	0.7	5.9	0.5	6.7	0.7	7.8	1.5
France	3.2	0.3	4.6	0.8	5.7	0.8	5.8	0.5	5.7	1.0	10.9	2.4
Japan	n.a.		n.a.		7.0	0.1	n.a.		7.0	0.1	7.1	1.8
Canada	3.5	0.4	4.7	0.8	4.5	1.5	3.3	0.5	5.8	1.1	11.0	2.0
Italy	4.2	0.5	5.9	0.6	6.0	0.7	6.3	0.4	5.7	0.7	13.7	3.3

Table 9.1 *Continued*

	Gold standard (1881–1913)		*Inter-War (1919–38)*		*Bretton Woods (total) (1946–70)*		*Bretton Woods (preconvertible) (1946–58)*		*Bretton Woods (convertible) (1959–70)*		*Floating exchange (1974–90)*	
Belgium	3.2	0.3	4.7	0.8	5.2	1.1	4.4	0.2	6.1	0.9	9.9	2.0
Netherlands	3.3	0.3	4.1	0.7	4.3	1.3	3.3	0.4	5.3	1.3	8.4	1.5
Sweden	3.8	0.3	4.5	0.4	4.3	1.6	2.9	0.3	5.7	1.2	11.1	1.7
Switzerland	3.8	0.2	4.7	0.9	3.5	0.8	3.0	0.3	3.9	0.9	4.9	1.1
Mean	3.5	0.3	4.8	0.8	5.1	1.1	4.2	0.4	5.8	0.9	9.8	2.0
Convergence	0.3	0.1	0.6	0.2	0.8	0.4	1.1	0.1	0.5	0.3	1.5	0.5
f Real short-term interest rate[b]												
United States	4.8	2.0	3.8	6.7	0.3	3.9	–1.6	4.7	2.4	0.4	2.5	2.8
United Kingdom	2.9	2.3	4.2	7.1	–0.1	3.4	–2.4	3.2	2.3	1.1	1.3	5.1
Germany	2.4	2.3	5.1	5.2	2.2	2.6	3.0	3.6	1.6	1.5	2.5	1.9
France	2.8	6.4	1.2	14.7	–0.9	5.2	–3.3	6.9	1.2	1.4	2.1	2.8
Japan	–1.5	5.5	1.4	8.8	1.9	2.5	3.4	3.5	1.1	1.6	1.4	3.5
Canada	n.a.		–0.8	1.3	–0.3	4.2	–2.6	4.8	2.2	0.8	3.1	3.2
Italy	n.a.		n.a.		n.a.		n.a.		n.a.		n.a.	
Belgium	2.8	4.8	0.6	9.6	0.4	3.8	–1.1	4.8	2.0	1.4	3.9	3.5
Netherlands	2.2	2.8	4.3	5.8	–1.7	3.0	–3.1	3.5	–0.3	1.4	2.5	3.3
Sweden	n.a.		n.a.		1.8	1.5	n.a.		1.8	1.5	1.9	3.6
Switzerland	n.a.		4.4	5.3	–0.1	1.8	0.3	2.0	–0.5	1.5	–1.1	2.7
Mean	2.3	3.7	2.7	7.2	0.4	3.2	–0.8	4.1	1.4	1.3	2.0	3.2
Convergence	1.1	1.6	1.9	2.6	1.0	0.8	2.0	1.1	0.7	0.3	0.6	0.5

Table 9.1 *Continued*

	Gold standard (1881–1913)		*Inter-War (1919–38)*		*Bretton Woods (total) (1946–70)*		*Bretton Woods (preconvertible) (1946–58)*		*Bretton Woods (convertible) (1959–70)*		*Floating exchange (1974–90)*	
g Real long-term interest rate[b]												
United States	3.7	2.2	4.6	6.8	0.8	3.6	–0.7	4.4	2.5	0.7	3.9	3.8
United Kingdom	3.0	2.5	5.4	7.1	1.1	2.8	–0.8	2.6	3.2	1.0	2.2	3.7
Germany	2.9	2.4	6.9	6.0	4.3	2.8	4.3	4.4	4.3	0.9	4.4	0.9
France	3.5	6.5	1.0	15.1	0.4	4.4	–1.2	6.2	1.8	1.0	2.7	3.1
Japan	n.a.		n.a.		1.7	1.3	n.a.		1.7	1.3	2.0	4.2
Canada	2.9	4.1	5.4	6.5	1.3	3.8	–0.7	4.4	3.4	0.7	3.8	3.1
Italy	4.2	2.4	3.4	9.3	0.3	9.4	–1.5	12.9	2.3	2.2	1.4	4.8
Belgium	3.2	5.0	1.8	9.8	2.7	3.4	1.8	4.6	3.5	1.1	4.1	3.3
Netherlands	2.4	2.6	5.8	5.6	0.1	3.1	–1.3	3.5	1.6	1.7	4.2	2.4
Sweden	3.4	3.3	6.1	7.0	0.5	3.3	–0.9	3.9	2.0	1.4	2.8	3.0
Switzerland	n.a.		6.6	5.5	1.4	1.7	1.7	1.8	1.1	1.4	1.3	1.6
Mean	3.2	3.4	4.7	7.9	1.3	3.6	0.1	4.9	2.5	1.2	3.0	3.1
Convergence	0.4	1.2	1.5	1.9	0.9	1.2	1.4	1.6	0.8	0.4	0.9	0.8
h Nominal exchange rate[d]												
United States[e]					0.7	0.8			0.7	0.8	5.0	4.7
United Kingdom	0.2	0.2	6.8	7.9	2.4	6.2	3.3	7.9	1.5	4.0	10.0	4.7
Germany	0.2	0.1	3.9	9.5	1.8	3.8	2.4	5.3	1.3	2.1	9.3	8.2
France	0.0	0.0	18.4	15.6	2.5	7.7	4.4	11.3	1.1	3.3	10.7	7.8
Japan	2.9	4.5	6.8	9.0	15.9	37.2	30.5	47.9	0.2	0.2	8.8	9.5
Canada	0.0	0.0	2.7	3.4	1.6	1.9	1.8	1.9	1.4	2.0	3.7	2.4

Table 9.1 *Continued*

	Gold standard (1881–1913)		*Inter-War (1919–38)*		*Bretton Woods (total) (1946–70)*		*Bretton Woods (preconvertible) (1946–58)*		*Bretton Woods (convertible) (1959–70)*		*Floating exchange (1974–90)*	
Italy	1.4	1.5	13.6	20.1	7.4	20.6	14.1	27.4	0.2	0.2	10.9	9.0
Belgium	2.6	3.1	13.2	16.5	0.8	2.0	1.3	2.8	0.3	0.2	10.6	8.5
Netherlands	0.2	0.3	6.5	8.8	1.8	5.6	3.0	7.7	0.6	1.0	9.9	7.9
Sweden	0.2	0.2	7.7	10.0	2.2	7.3	4.0	9.9	0.2	0.2	8.4	7.0
Switzerland	0.3	0.2	7.7	9.6	0.2	0.3	0.2	0.3	0.2	0.2	10.8	9.3
Mean	0.8	1.0	8.7	11.0	3.4	8.5	6.5	12.2	0.7	1.3	8.9	7.2
Convergence	0.9	1.2	3.7	3.7	2.7	6.7	5.7	9.0	0.5	1.1	1.7	1.7
i Real exchange rate[cd]												
United States					1.7	1.0			1.7	1.0	7.4	6.5
United Kingdom	1.7	1.5	8.5	10.0	4.2	6.5	5.9	8.2	2.5	3.9	12.3	6.1
Germany	2.4	1.2	5.8	9.2	2.8	5.1	3.9	7.3	1.9	1.8	8.8	8.2
France	4.3	5.0	9.0	6.9	4.1	5.6	6.2	7.7	2.5	2.9	9.2	7.7
Japan	6.6	5.5	7.8	7.2	3.5	2.9	4.8	4.1	2.7	1.5	9.6	8.9
Canada	2.6	2.2	3.2	2.8	2.4	2.3	3.3	2.5	1.5	1.7	3.8	2.0
Italy	2.1	1.7	13.3	16.9	8.0	18.7	13.1	25.2	2.4	1.6	8.6	7.8
Belgium	5.4	3.9	9.0	9.8	2.3	2.9	3.4	3.8	1.3	0.9	10.0	8.4
Netherlands	3.6	2.6	8.6	8.9	2.9	4.0	3.7	5.4	2.1	1.4	9.4	8.0
Sweden	2.8	1.5	7.9	10.1	4.1	6.6	6.0	8.8	2.0	1.1	7.5	7.2
Switzerland	n.a.		6.7	8.5	2.3	2.3	2.8	3.0	1.8	0.8	10.2	9.0
Mean	3.5	2.8	8.0	9.0	3.5	3.6	5.3	7.6	0.2	1.7	8.8	7.3
Convergence	1.3	1.3	1.6	2.2	1.1	1.4	1.8	3.5	0.5	0.6	1.3	1.3

Table 9.1 *Continued*

	Gold standard (1881–1913)		*Inter-War (1919–38)*		*Bretton Woods (total) (1946–70)*		*Bretton Woods (preconvertible) (1946–70)*		*Bretton Woods (convertible) (1959–70)*		*Floating exchange (1974–90)*	
j Government budget deficit as a percentage of nominal GNP												
United States	–0.3	0.6	1.9	4.4	–0.1	2.1	–0.5	2.9	0.4	0.6	3.0	1.5
United Kingdom	0.1	0.8	–0.7	1.7	–2.3	2.8	–2.2	3.6	–2.4	1.6	2.2	1.6
Germany	2.8	1.0	1.4	1.1	0.1	1.8	–0.2	2.6	0.4	0.8	1.5	0.8
France	0.6	1.1	3.8	3.8	2.2	2.3	3.5	2.8	1.0	0.9	1.5	1.3
Japan	–3.1	3.3	–5.1	3.3	–1.0	3.2	–2.7	3.7	0.8	0.8	4.5	1.9
Canada	0.5	1.1	1.5	2.5	–1.1	1.7	–1.7	1.9	–0.4	1.1	2.9	2.0
Italy	0.9	1.1	8.6	9.7	3.2	4.0	5.0	4.9	1.2	1.0	9.5	2.2
Belgium	n.a.		3.9	1.2	4.8	4.8	6.8	6.1	2.7	0.9	7.3	2.7
Netherlands	2.1	0.3	5.6	3.2	0.6	5.2	1.7	7.1	–0.6	0.7	3.8	2.4
Sweden	0.0	0.6	1.6	1.2	–0.2	2.6	1.8	1.6	–2.4	1.6	1.1	3.8
Switzerland	n.a.		–0.0	0.5	–0.2	1.1	0.2	1.4	–0.5	0.6	0.1	0.7
Mean	0.4	1.1	2.1	3.0	0.6	2.9	1.1	3.5	0.0	1.0	3.4	1.9
Convergence	1.1	0.5	2.4	1.5	1.6	1.0	2.4	1.2	1.0	0.3	1.7	0.5

Notes:
[a]Mean growth rate calculated as the time coefficient from a regression of the natural logarithm of the variable on a constant and a time trend.
[b]Calculated as the nominal interest rate minus the annual rate of change of the CPI.
[c]Calculated as the nominal exchange rate divided by the ratio of foreign to the US CPI.
[d]Absolute rate of change.
[e]Trade-weighted nominal and real exchange rate starting in 1960.
Date sources: See data appendix.

Table 9.2 Descriptive statistics of selected open economy macro variables, four high inflation countries, 1881–1989, annual data: mean, standard deviation

	Gold standard (1881–1913)		Inter-War (1919–38)		Bretton Woods (total) (1946–70)		Bretton Woods (preconvertible) (1946–58		Bretton Woods (convertible) (1959–70)		Floating exchange (1974–89	
a Inflation PGN[a]												
Argentina	2.6	14.2	–1.7	6.3	22.9	14.7	18.4	9.8	19.9	7.2	122.9	77.0
Brazil	4.3	17.5	3.0	6.3	24.3	14.8	13.2	6.1	35.7	15.1	94.5	87.0
Chile	5.2	9.1	5.5	7.3	27.7	13.0	29.0	15.7	26.6	10.5	30.9	41.2
Israel					7.9	9.3	14.9	12.3	5.7	3.2	69.3	42.9
Mean	4.0	13.6	2.3	6.6	20.7	13.0	18.9	11.0	2.2	9.0	79.4	62.0
Convergence	1.0	3.0	2.6	0.4	6.4	1.8	5.1	3.0	9.2	3.8	29.3	20.0
b Real per capita growth[a]												
Argentina	1.4	6.3	0.6	4.6	1.5	4.7	0.8	4.3	2.7	4.2	–1.4	4.2
Brazil	1.2	8.0	1.7	8.7	4.8	4.2	3.5	2.1	6.4	5.6	0.9	4.3
Chile	3.3	4.7	1.2	12.0	0.7	5.9	1.5	7.0	1.6	5.0	1.6	7.4
Israel					5.1	5.1	4.3	7.2	5.0	3.3	1.8	8.9
Mean	2.0	6.3	1.2	8.4	3.0	5.0	2.5	5.2	3.9	4.5	0.7	6.2
Convergence	0.9	1.1	0.4	2.6	1.9	0.5	1.4	2.0	1.8	0.8	1.1	2.0
c Money growth[a]												
Argentina	6.7	12.1	1.8	8.9	22.2	10.1	19.0	7.4	25.1	12.3	124.6	82.6
Brazil	6.9	25.8	9.2	12.4	32.6	21.5	17.9	8.9	48.4	19.9	91.5	79.8
Chile	–3.7	9.7	5.1	15.1	29.7	12.8	28.3	11.3	32.5	14.8	32.2	39.3
Israel					15.4	8.9	17.9	9.3	13.9	8.4	66.1	39.3
Mean	3.3	15.9	5.4	12.1	25.0	13.3	20.8	9.2	30.0	13.9	78.6	60.3
Convergence	4.7	6.6	2.6	2.2	6.2	4.1	3.8	1.1	10.5	3.5	29.5	21.0

Table 9.2 Continued

	Gold standard (1881–1913)		Inter-War (1919–38)		Bretton Woods (total) (1946–70)		Bretton Woods (preconvertible) (1946–58)		Bretton Woods (convertible) (1959–70)		Floating exchange (1974–89)	
d Nominal exchange rate[d]												
Argentina	7.6	10.9	9.7	7.6	18.8	16.7	20.2	16.6	17.3	17.5	130.5	85.4
Brazil	9.6	12.6	13.8	14.7	22.5	22.3	11.6	17.8	34.3	21.0	95.5	84.6
Chile	9.5	8.9	13.3	19.4	24.3	17.9	27.4	21.6	21.0	12.9	40.3	49.0
Israel	n.a.		n.a.		12.0	26.3	19.7	35.5	5.5	13.8	46.9	47.4
Mean	8.9	10.8	12.3	13.9	19.4	20.8	19.7	22.9	19.5	16.3	78.3	66.6
Convergence	0.9	1.3	1.7	4.2	4.0	3.5	4.1	6.3	8.1	3.0	34.7	18.4
e Real exchange rate[cd]												
Argentina	6.0	5.0	12.6	10.9	13.1	10.2	13.2	11.7	12.9	8.7	32.6	23.3
Brazil	6.4	5.7	15.1	10.3	10.7	8.7	11.2	8.9	10.1	8.9	9.8	8.5
Chile	13.9	6.6	14.6	16.3	15.7	13.9	19.4	15.3	11.7	11.5	12.1	14.1
Israel	n.a.		n.a.		14.2	20.3	22.6	26.0	7.2	10.8	8.2	5.5
Mean	8.8	5.8	14.1	12.5	13.4	13.3	16.6	15.5	10.5	10.0	15.7	12.9
Convergence	3.4	0.6	1.0	2.5	1.6	3.8	4.4	5.3	1.8	1.2	8.5	5.9
f Government budget deficit as a percentage of nominal GNP												
Argentina	2.5	2.4	1.7	1.8	6.0	3.7	8.7	3.0	3.0	1.1	9.3	6.8
Brazil	2.0	2.4	1.9	1.6	1.5	1.3	1.1	1.3	2.0	1.3	5.5	5.8
Chile	2.4	4.4	1.8	1.7	1.9	1.6	1.3	1.4	2.5	1.7	1.1	2.9
Israel	n.a.		n.a.		5.1	6.5	0.8	1.3	8.3	6.9	17.6	7.1
Mean	2.3	3.1	1.8	1.7	3.6	3.3	3.0	1.8	4.0	2.8	8.4	5.7
Convergence	0.2	0.9	0.1	0.1	1.9	1.8	2.9	0.6	2.2	2.1	5.1	1.4

Notes: As Table 9.1.
Date sources: See data appendix.

Figure 9.1 Inflation rates, 1880–1989

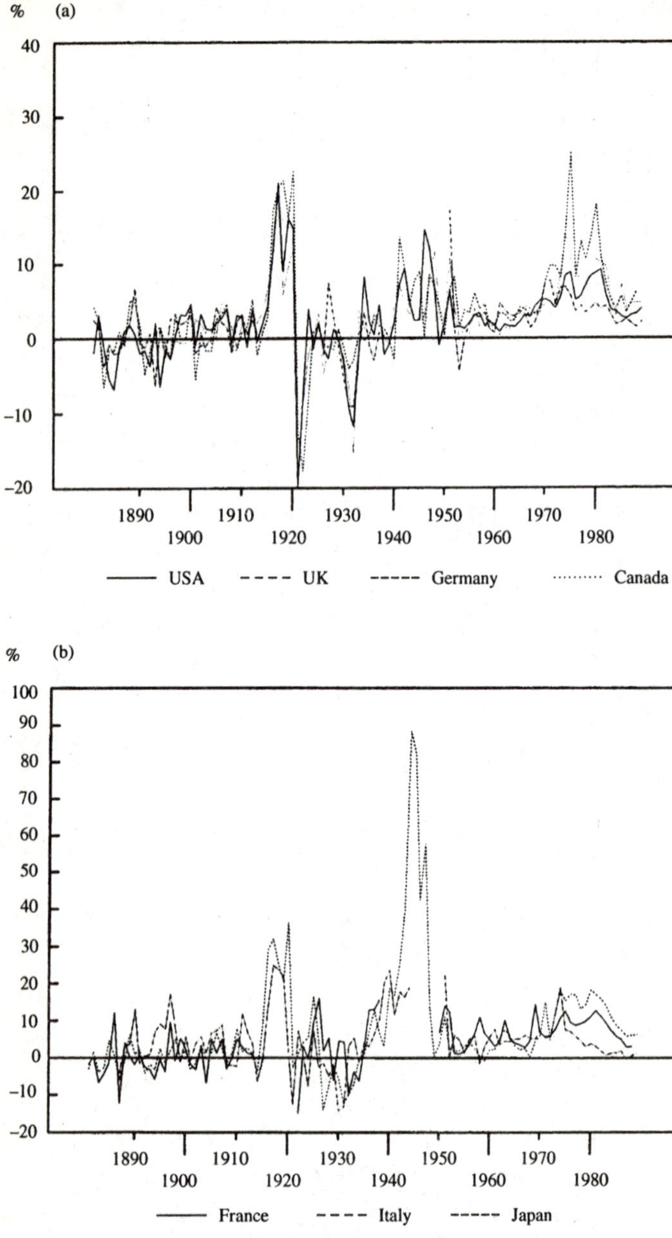

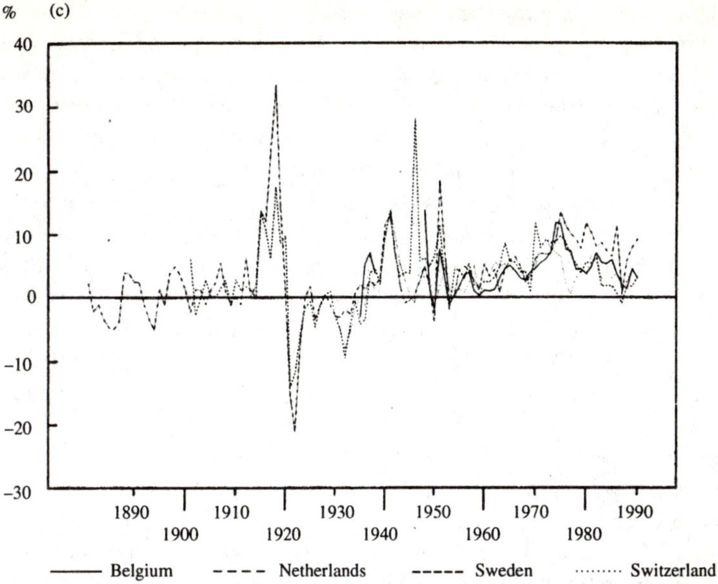
%
(c)
40
30
20
10
0
–10
–20
–30
1890
1900
1910
1920
1930
1940
1950
1960
1970
1980
1990
Belgium
Netherlands
Sweden
Switzerland

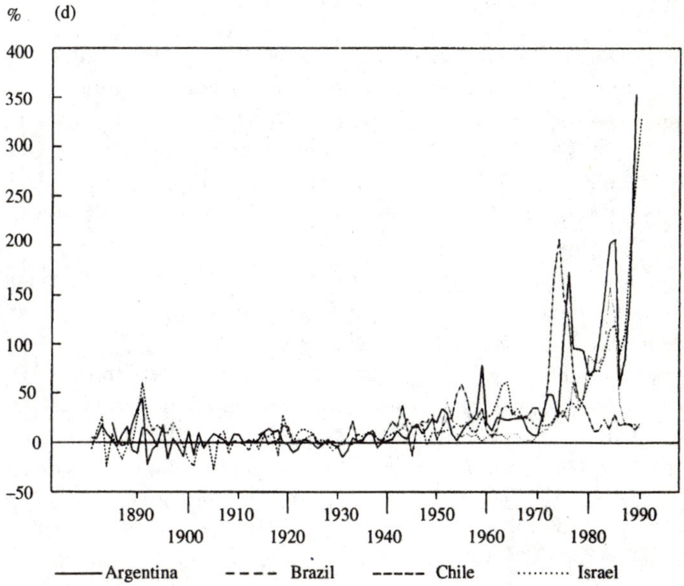
%
(d)
400
350
300
250
200
150
100
50
0
–50
1890
1900
1910
1920
1930
1940
1950
1960
1970
1980
1990
Argentina
Brazil
Chile
Israel

insulation and greater monetary independence under floating rates.[19] For the high inflation countries for the post-war period, although the inflation rates are considerably higher than for the G-11, we also observe greater convergence under Bretton Woods than during the float. This suggests that although these countries did not follow convertibility rules directly, the fact that the G-11 *did*, may have influenced regime performance.[20]

The Bretton Woods convertible sub-period had the most stable inflation rate of any regime for the G-11 judged by the standard deviation. By contrast, for the G-11 the preconvertible Bretton Woods period exhibited greater inflation variability than either the gold standard or the recent float. The evidence of a high degree of price stability in the convertible phase of Bretton Woods is also consistent with Leijonhufvud's writing and the traditional view that convertible fixed-rate (commodity-based) regimes provide a stable nominal anchor; however, the remarkable price stability during this period may also reflect the absence of major shocks.

For the high inflation countries, inflation variability does not differ much between the gold standard and Bretton Woods Regimes. It is considerably higher than in the G-11 for those periods. As with the G-11, inflation variability was much greater in the floating rate period than under Bretton Woods. The dramatic increase in inflation variability probably reflects the operation of a RWMS regime.

Real per capita GNP (Panel b)

Generally, the Bretton Woods period, especially the convertible period, exhibited the most rapid output growth of any monetary regime for the G-11 countries, and, not surprisingly, the interwar period the lowest (see Figure 9.2). Like the G-11, for the high inflation countries, the Bretton Woods convertible period displayed the most rapid growth, but for Argentina, Brazil and Israel the floating period was the lowest. For these countries, as Heymann and Leijonhufvud suggest, the shift from a loosely convertible to a purely discretionary regime may have been related to deteriorating economic performance. Output variability was also lowest in the convertibility sub-period of Bretton Woods for the G-11, but because of higher variability in the preconvertible period, the Bretton Woods system as a whole was more variable than the floating period. Both pre-Second World War regimes exhibit higher variability than their post Second World War counterparts.[21] For the high inflation countries for which we have data, the convertible Bretton Woods period also emerges as the least variable.

For the G-11, the Bretton Woods regime also exhibited the lowest divergence of output variability between countries of any regime, with the

Figure 9.2 Per capita income growth rates, 1880–1989

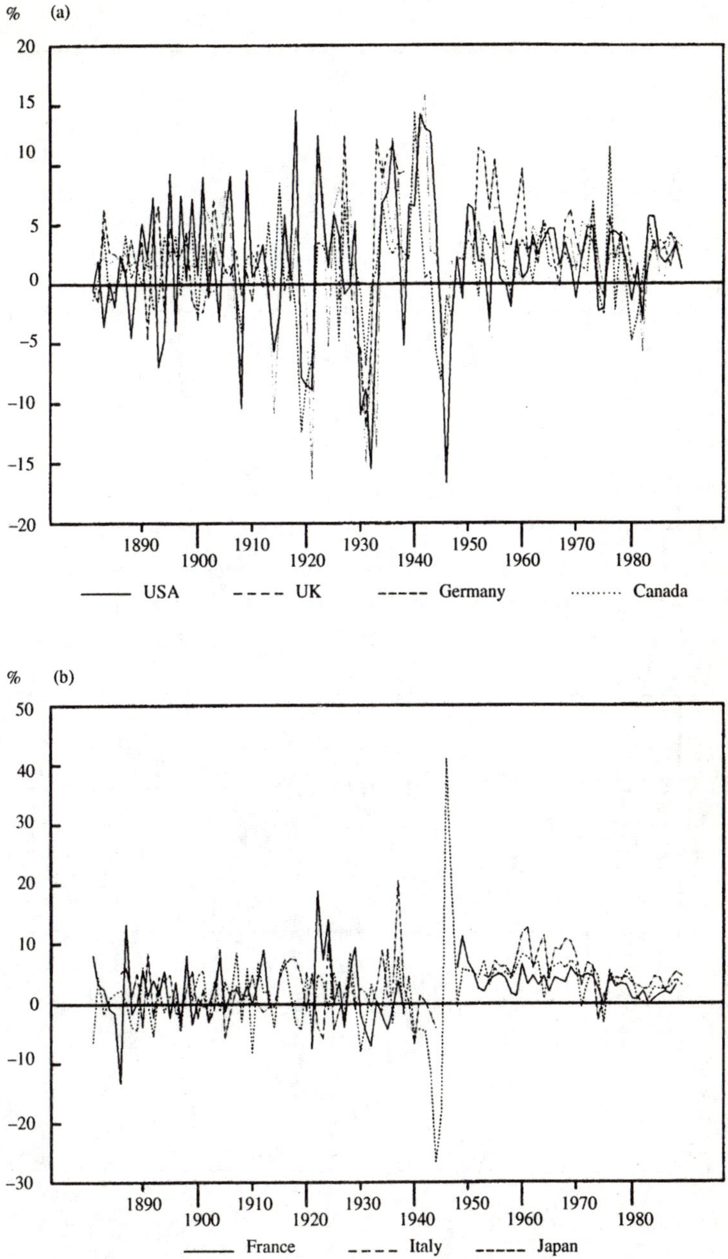

Figure 9.2 *Continued*

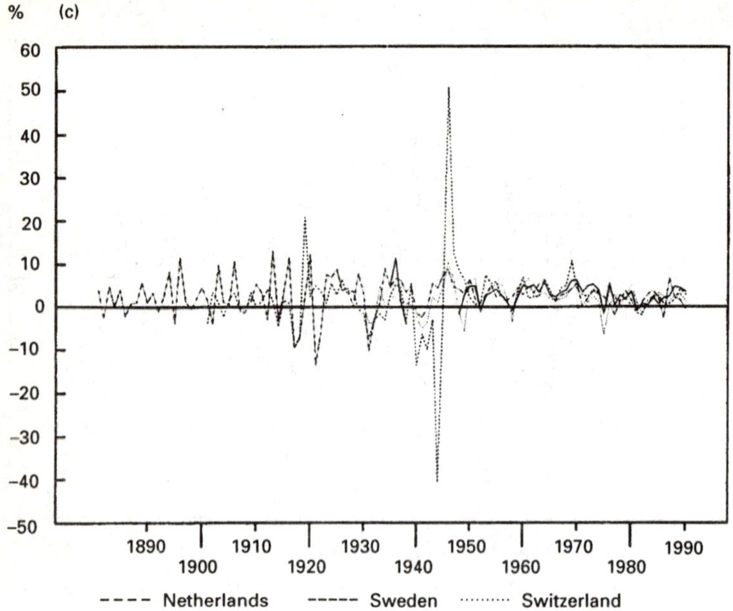

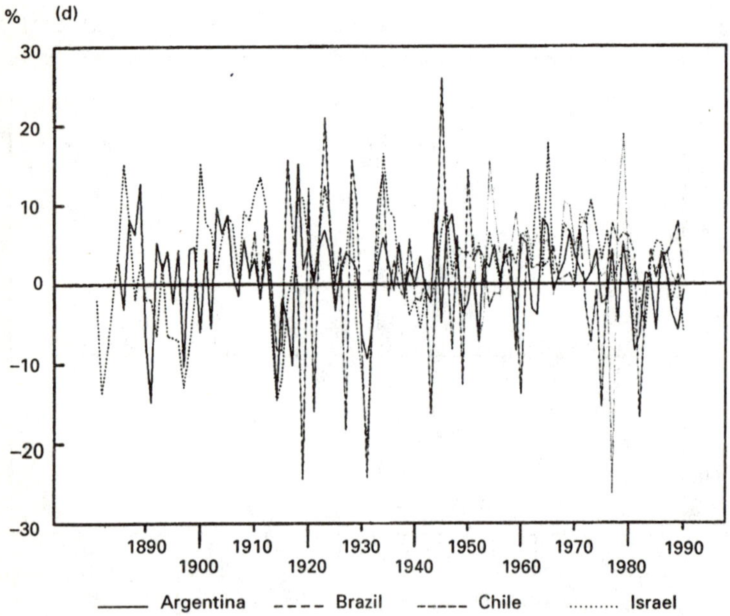

inter-war regime the highest.[22] For the high inflation countries, such as the G-11, the Bretton Woods convertible regime exhibited the lowest degree of divergence, along with the gold standard. The inter-war showed the highest. The greater convergence of output variability under Bretton Woods may reflect conformity between countries' business fluctuations, created by the operation of the fixed-exchange-rate regime (Bordo and Schwartz, (1989) and Darby and Lothian (1989)).

Money growth (M_2) (Panel c)

It was considerably more rapid across all countries post-Second World War than before the war (see Figure 9.3). For the G-11 there is not much difference between Bretton Woods and the subsequent floating regime. Within the Bretton Woods regime, money growth was more rapid in the preconvertibility period than in the convertibility period. For the high inflation countries, money growth rates accelerated over the entire post-war period, reaching their highest levels under the float. To the extent that the world adhered to convertibility rules they seem to have been more effective in constraining money growth in these countries before 1959. However, the comparison between the total Bretton Woods period and the subsequent float (for all except Chile) suggests that the convertibility regime in the rest of the world still may have had some influence.

Money growth rates showed the least divergence between G-11 countries during the fixed-exchange-rate gold standard and the convertible Bretton Woods regime, with the greatest divergence in the preconvertible Bretton Woods period and the inter-war period. For the high inflation countries, divergence in money growth rates increased as the twentieth century progressed, reaching its peak with the recent float.

Like inflation and real output variability, for the G-11, money growth variability was lowest in the convertible Bretton Woods period. This, however, was not the case for the preconvertible period, which was the most variable of any regime. It also exhibited the great divergence in variability between countries. To the extent that one of the properties of adherence to a fixed-exchange-rate regime is conformity of monetary growth rates between countries, these results are sympathetic to the view that the Bretton Woods system really began in 1959. For the high inflation countries, money growth variability did not increase significantly between the earlier periods and the Bretton Woods period, but did so dramatically between Bretton Woods and the float. This could be viewed as strong evidence for the operation of a RWMS monetary regime. Divergence of money growth variability also increased dramatically for those countries between the Bretton Woods and floating regimes.

Figure 9.3 Money growth rates, 1880–1989

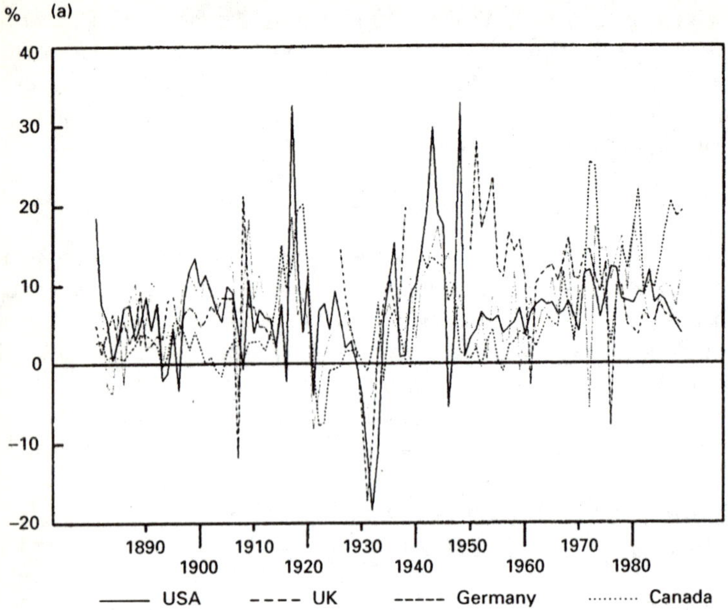

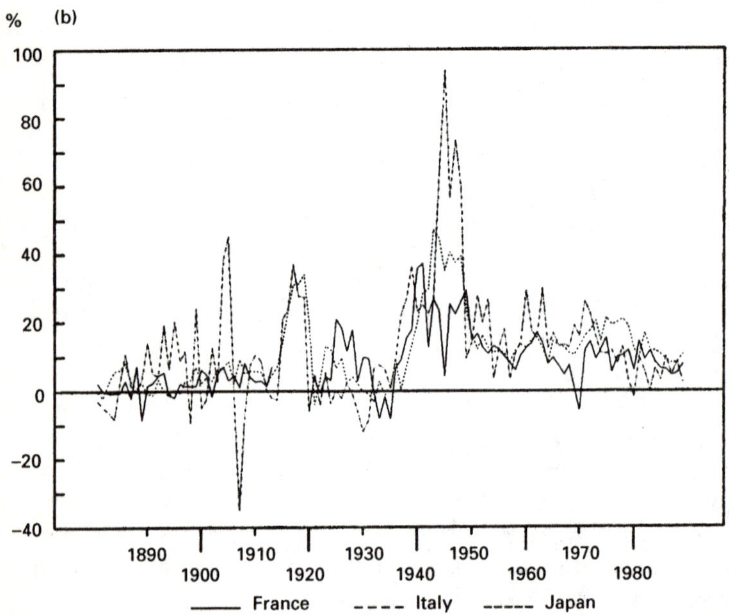

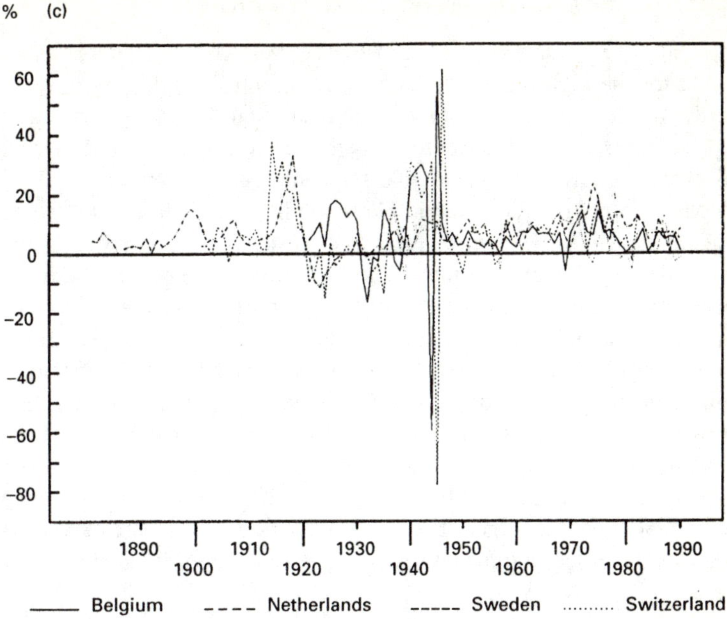
% (c)
60
40
20
0
-20
-40
-60
-80
1890
1900
1910
1920
1930
1940
1950
1960
1970
1980
1990
Belgium
Netherlands
Sweden
Switzerland

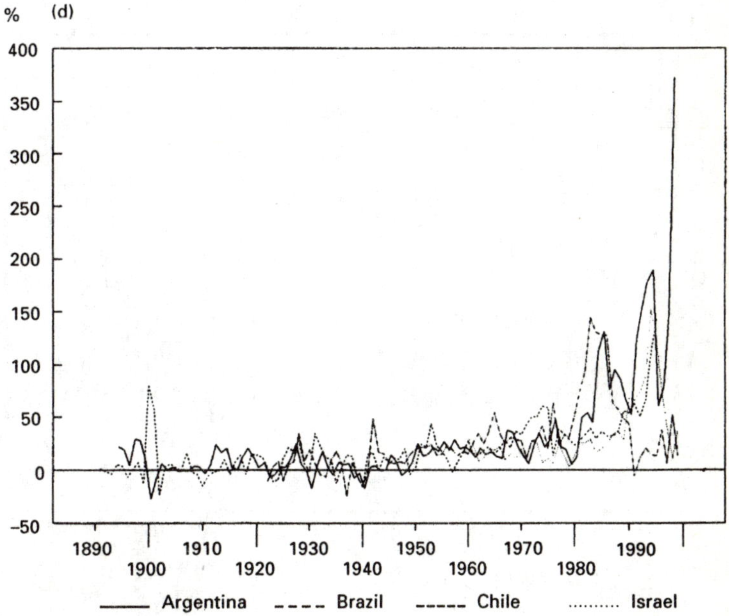
% (d)
400
350
300
250
200
150
100
50
0
-50
1890
1900
1910
1920
1930
1940
1950
1960
1970
1980
1990
Argentina
Brazil
Chile
Israel

Short-term and long-term interest rates (Panels d and e)

The underlying data can be seen in Figures 9.4 and 9.5. Reliable data are absent for the high inflation countries. As in other nominal series for the G-11, the degree of convergence of mean short-term interest rates is highest in the convertible Bretton Woods period. Long-term rates are most closely related in the classical gold standard, with the convertible Bretton Woods period not far behind. These findings are similar to these of McKinnon (1988), who views them as evidence of capital market integration under fixed exchange rates. The lack of convergence in the preconvertibility Bretton Woods period reflects the presence of pervasive capital controls. Convergence of nominal interest rates would not be expected under floating exchange rates. Convergence of standard deviations is also highest in the gold standard period, followed by Bretton Woods. Long-term rates were most stable and least divergent under the classical gold standard, followed by the two Bretton Woods sub-periods, with floating exchange rates the least stable. The evidence that nominal interest rates are more stable and convergent between countries under fixed exchange rate (commodity-based) regimes is consistent with the traditional view.

Figure 9.4 Short-term interest rates, 1880–1989

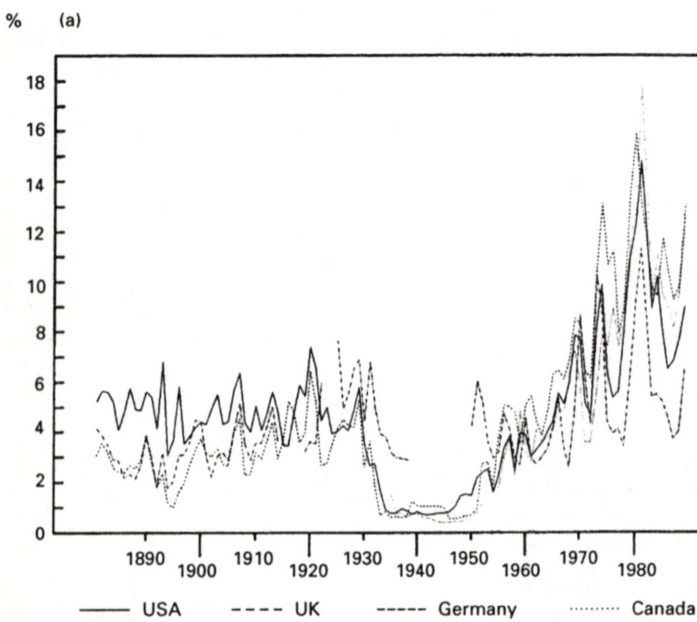

% (b)

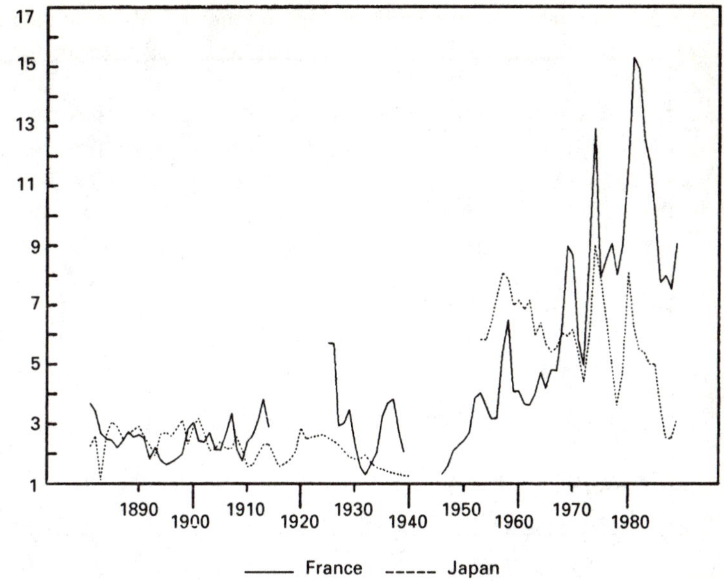
17
15
13
11
9
7
5
3
1
1890
1900
1910
1920
1930
1940
1950
1960
1970
1980
France
Japan

% (c)

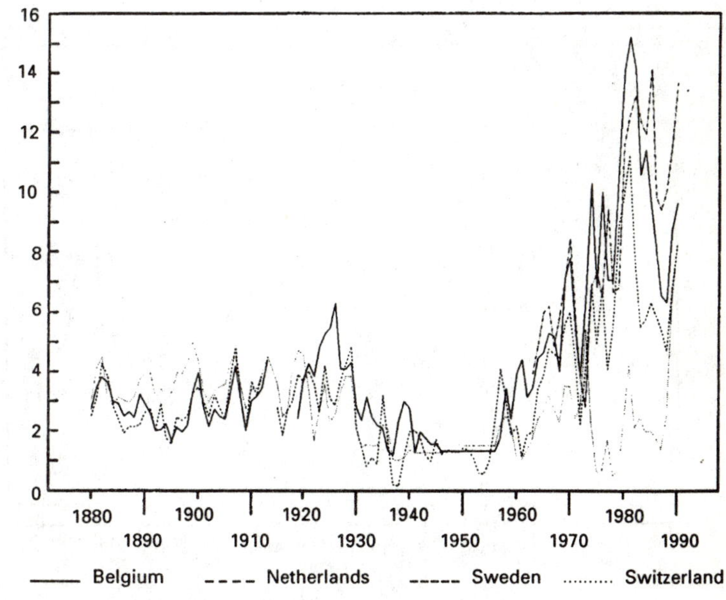
16
14
12
10
8
6
4
2
0
1880
1890
1900
1910
1920
1930
1940
1950
1960
1970
1980
1990
Belgium
Netherlands
Sweden
Switzerland

Figure 9.5 Long-term interest rates, 1880–1989

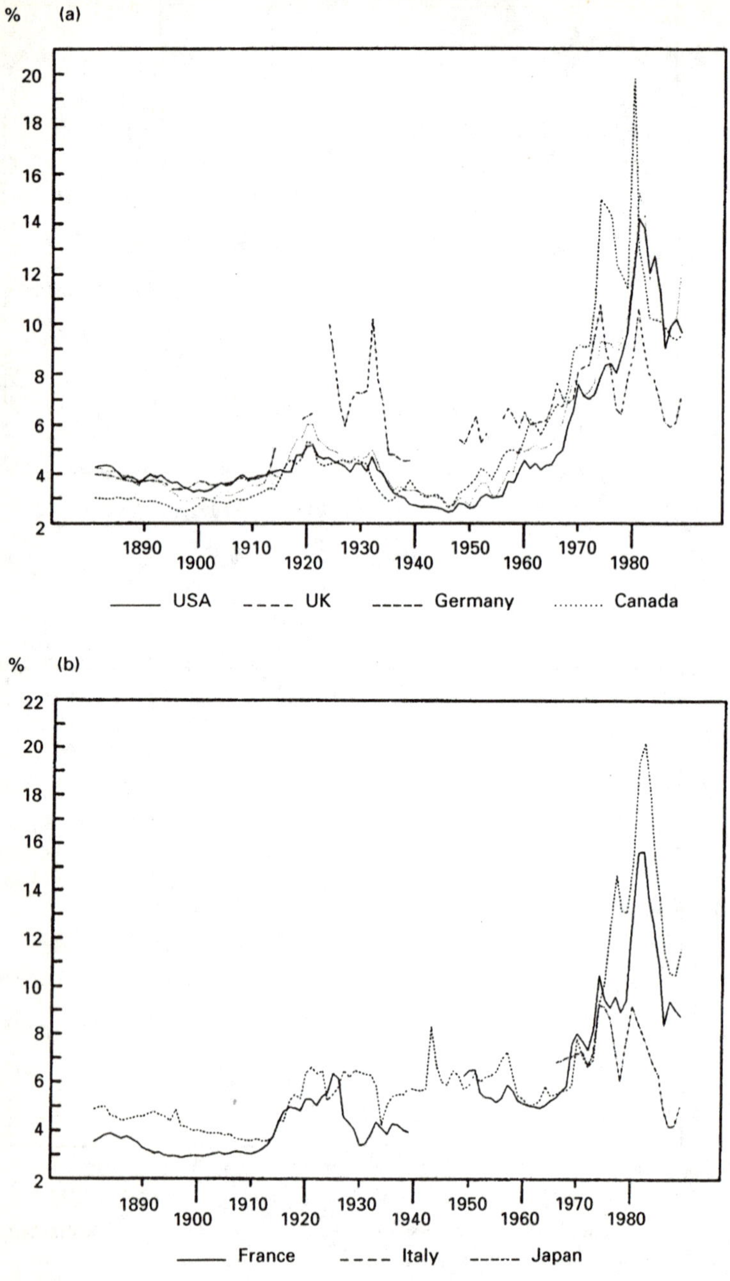

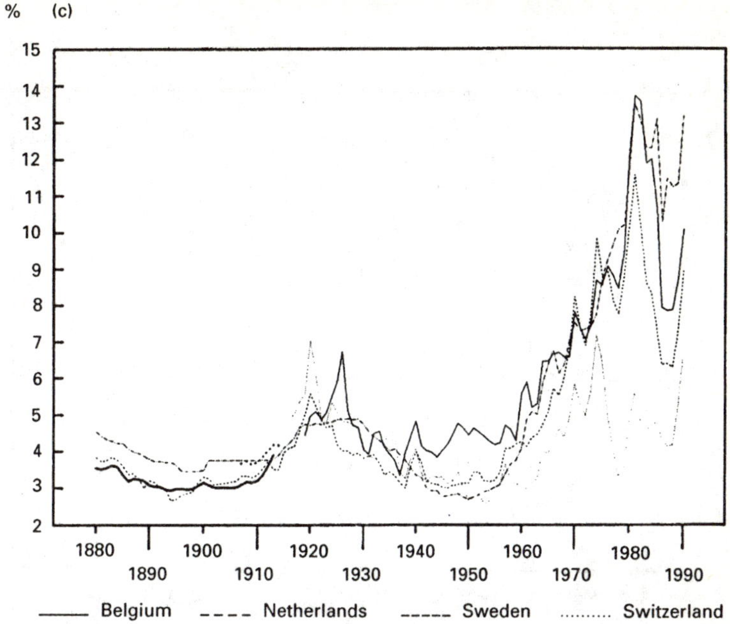

Real short-term and real long-term interest rates (Panels f and g)

For the underlying data see Figures 9.6 and 9.7. The real interest rates are *ex-post* rates calculated using the rate of change of a consumer price index.[23] Unlike the nominal series for the G-11, the degree of convergence in means between real short-term interest rates is lowest in the floating exchange rate period, next lowest in the Bretton Woods convertible period, and highest in the preconvertible period. For the long-term real rates, as in the case of nominal rates, convergence is highest under the gold standard, followed by the Bretton Woods convertible regime; it is lowest under preconvertible Bretton Woods. The real short-term interest rate is most stable across countries during the Bretton Woods convertible period. It also shows the least amount of divergence in standard deviations. The same holds for real long-term interest rates.

The behaviour of real interest rates across regimes is consistent with McKinnon's (1988) explanation. He argued that fixed exchange rates encourage capital market integration by eliminating devaluation risk. This reduces variability in short-term real interest rates. Similarly, real long-term interest rates are stabilised by pooling across markets, which reduces capital market risk.

Figure 9.6 Real short-term interest rates, 1880–1989

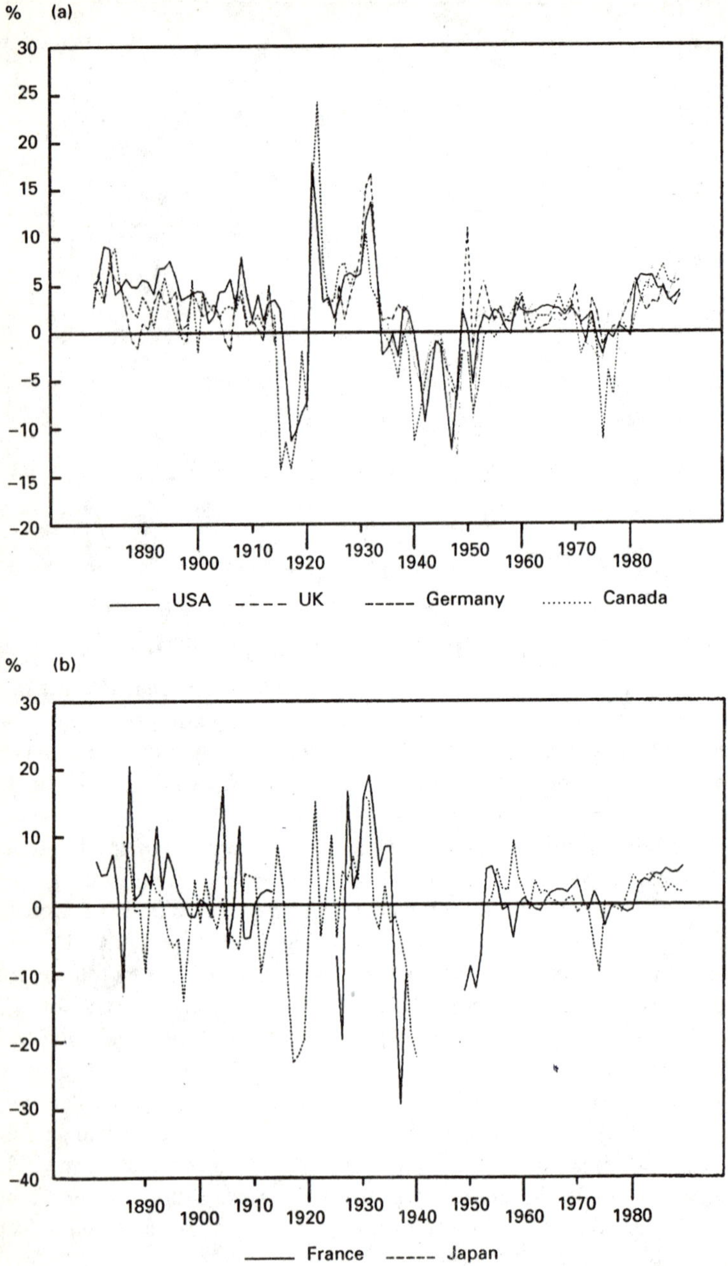

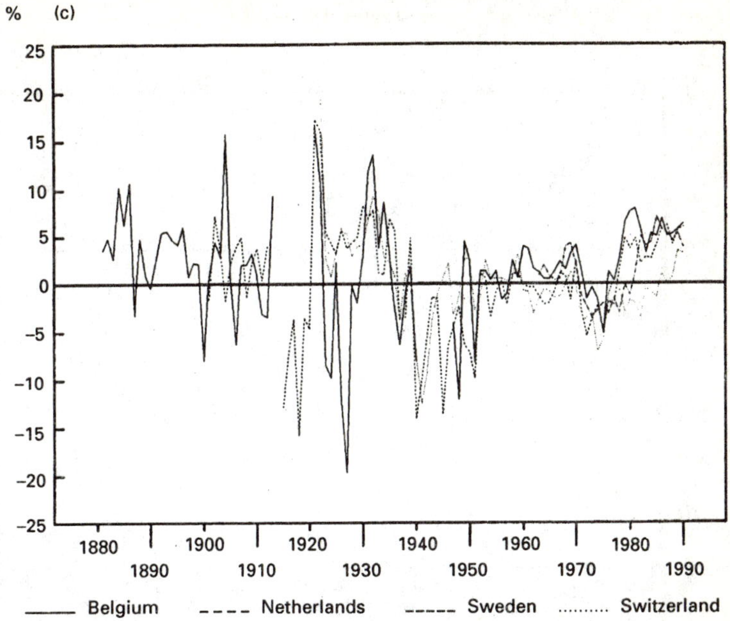

Figure 9.7 Real long-term interest rates, 1880–1989

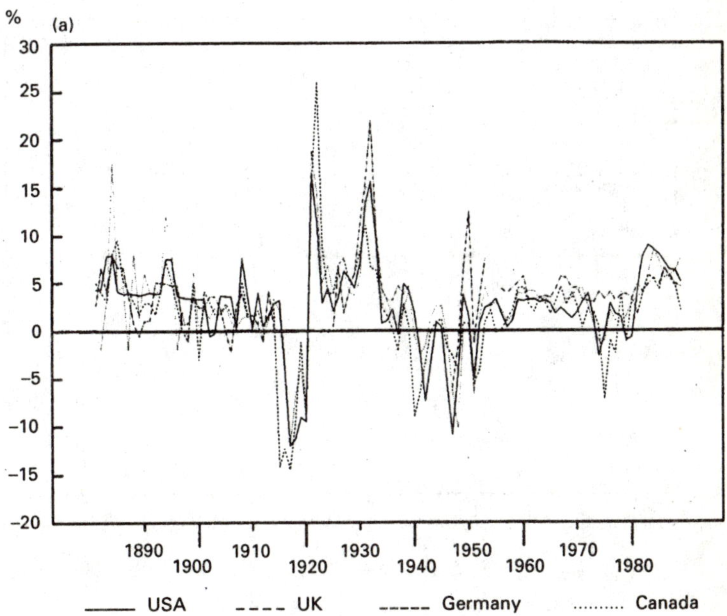

Figure 9.7 Continued

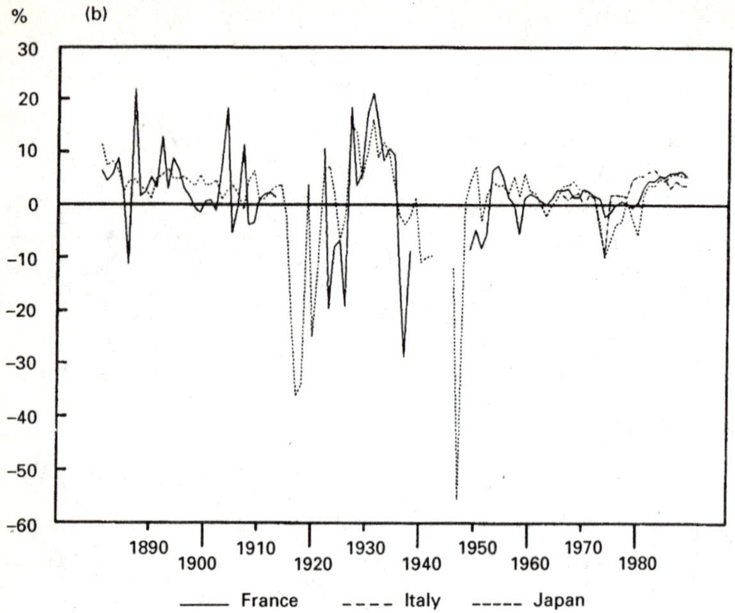

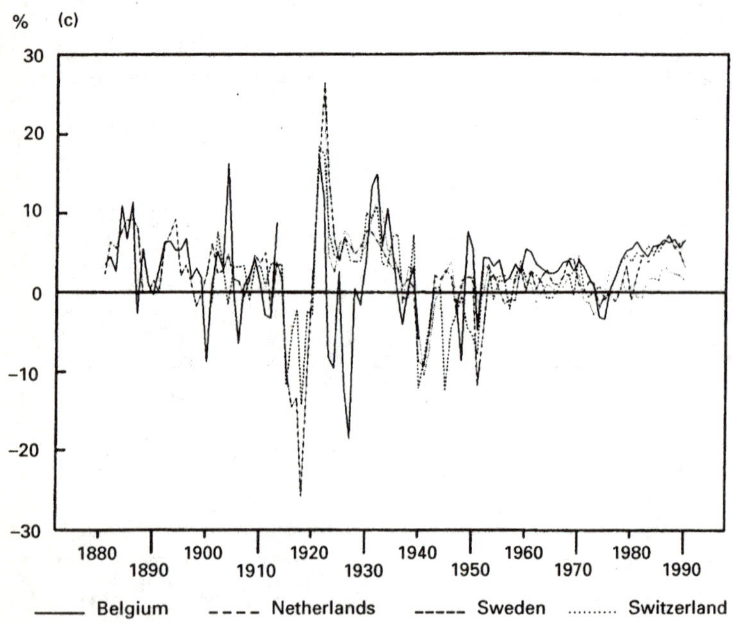

Nominal and real exchange rates (Panels h and i)

For the underlying data see Figures 9.8 and 9.9. For the G-11 countries, the lowest mean rate of change of the nominal exchange rate and the least divergence between rates of change occurred during the Bretton Woods convertible and gold standard periods, with the former exhibiting the lowest degree of divergence. Exchange rates during the preconvertibility Bretton Woods regime changed almost as much as during the floating period. This mainly reflected the major devaluations of 1949 (see Figure 9.8 and Bordo (1993a) table 2). Nominal exchange rates were least variable in the gold standard and convertible Bretton Woods periods, and the most variable and most divergent in the Bretton Woods preconvertible period. For the high inflation countries, both the average rate of change of the nominal exchange rate and its standard deviation were a multiple of the G-11 for all regimes. Both measures increased between the pre- and post-Second World War periods, reaching a peak with the recent float. Divergence of both measures also increased as the century progressed.

As with the nominal exchange rate, the lowest mean rate of change in the real exchange rate across G-11 countries, and the least divergence between countries, was in the Bretton Woods convertible period, with the

Figure 9.8 Absolute change in nominal exchange rates, 1880–1989

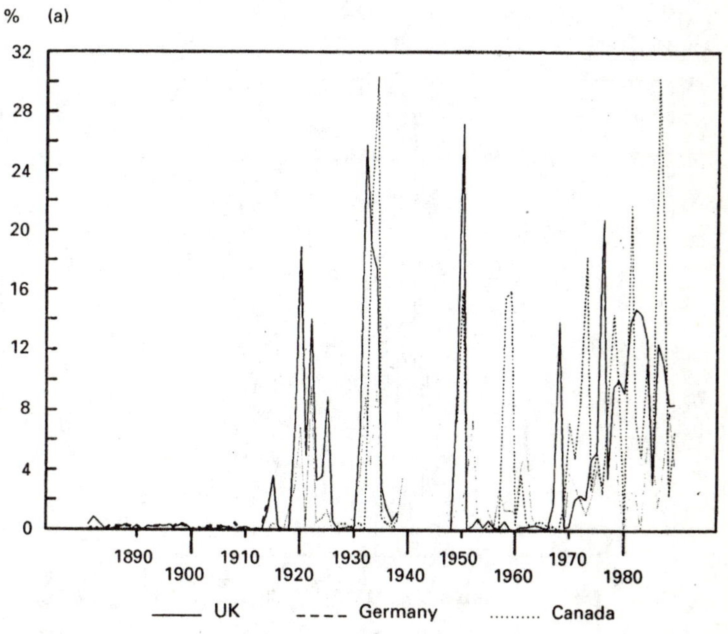

Figure 9.8 *Continued*

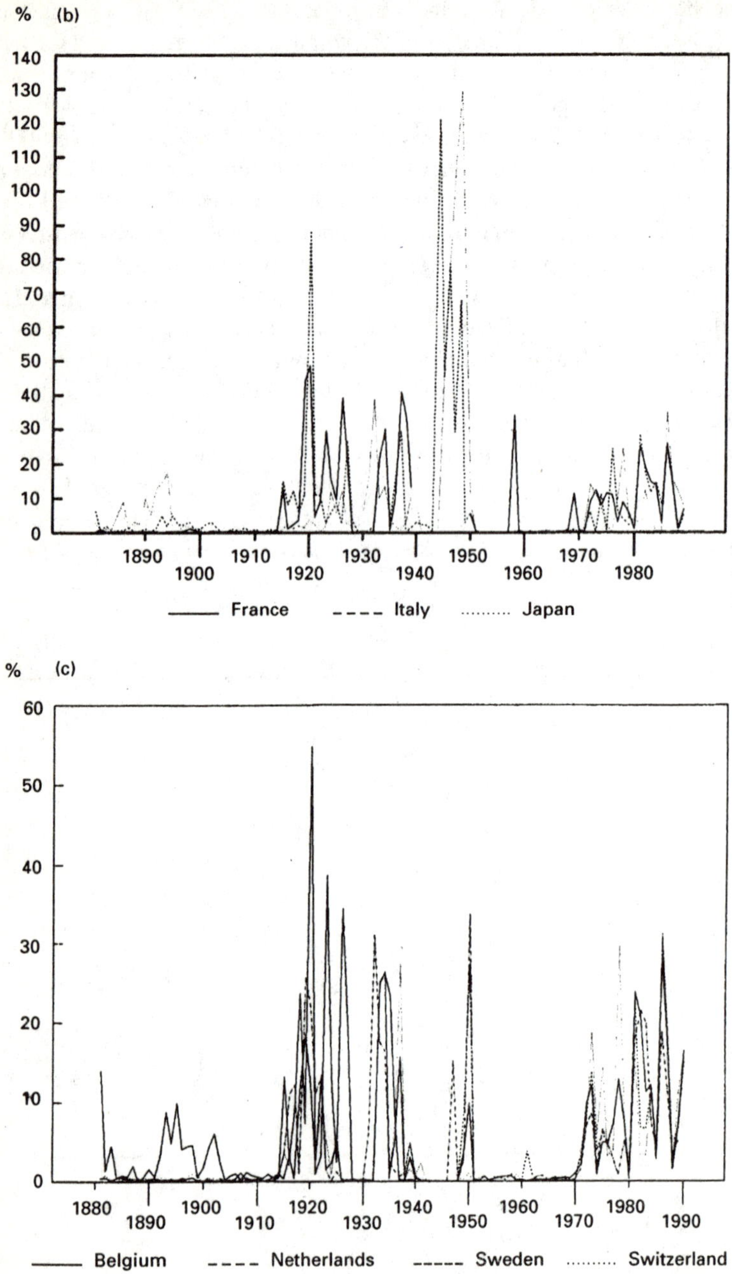

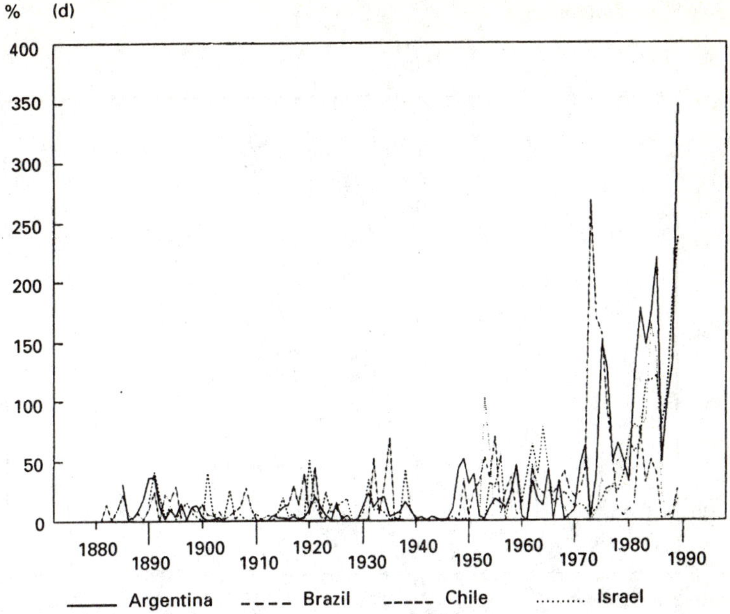

Figure 9.9 Absolute change in real exchange rates, 1880–1989

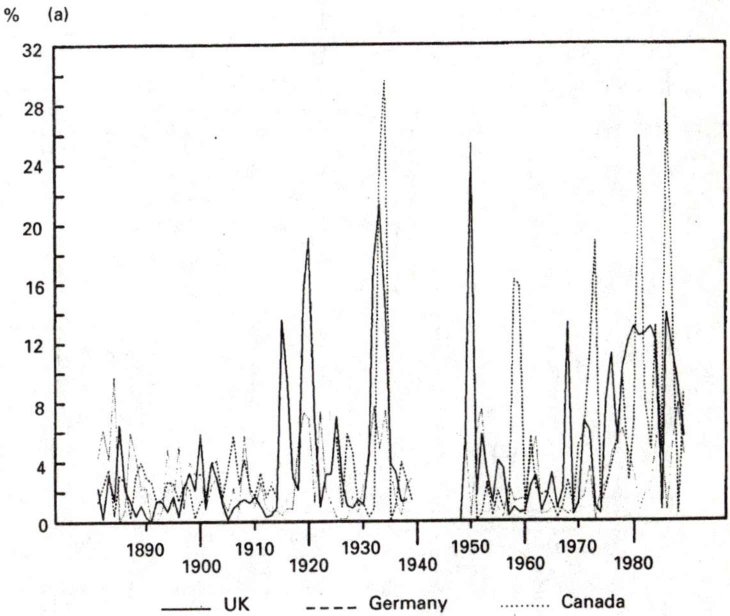

Figure 9.9 Continued

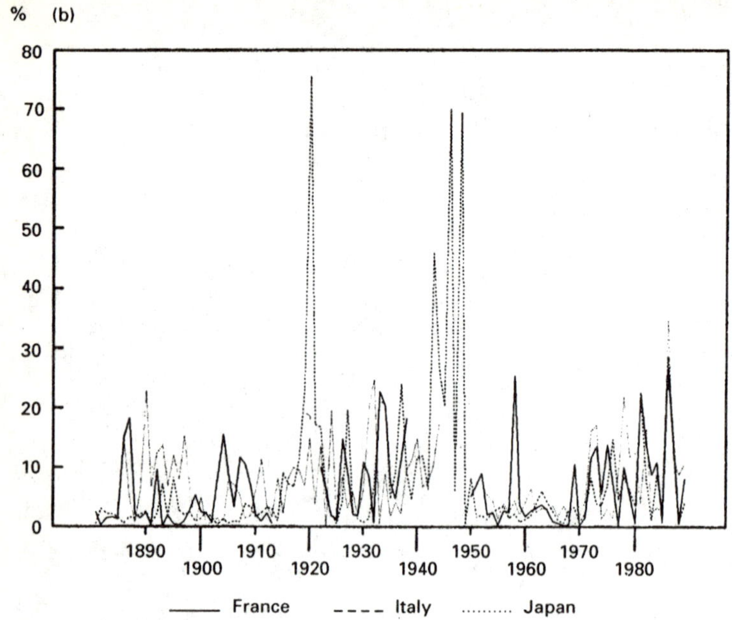

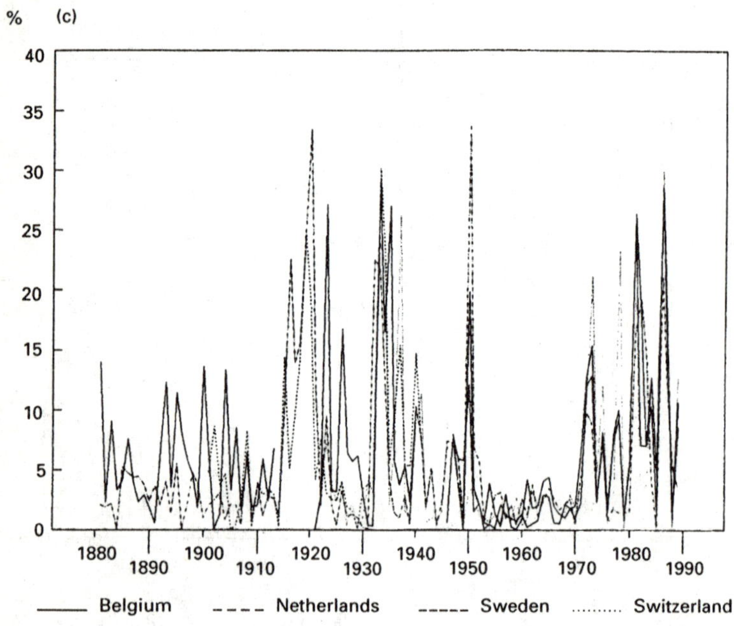

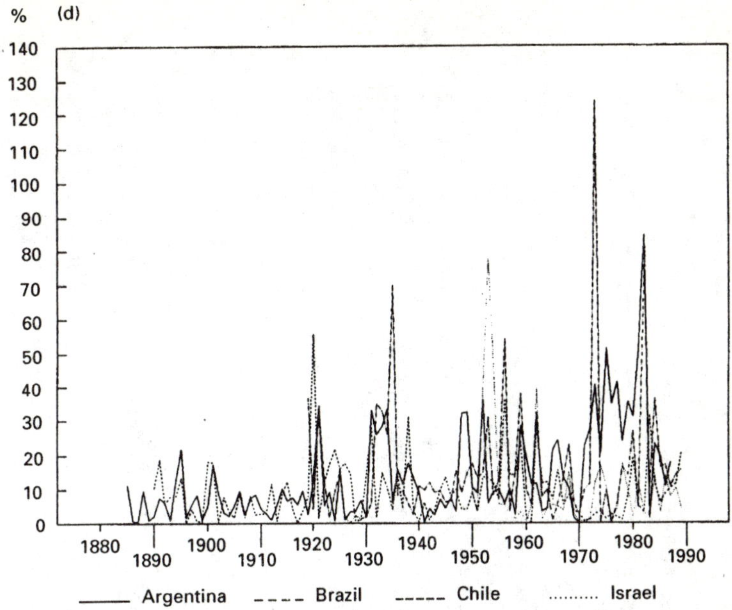

gold standard period next in size of these measures (see Figure 9.9). The highest rate of change was in the floating period. Similarly, the lowest standard deviation across countries, and the least divergence between standard deviations, was the Bretton Woods convertible period, with the gold standard again next in these rankings. The other regimes were characterised by much greater variability and divergence. For the high-inflation countries, although the orders of magnitude are somewhat greater than the G-11, the same pattern emerges across regimes.

These results shed light on the relationship between the nominal exchange rate regime and the behaviour of real exchange rates. Mussa (1986) presented evidence for ten industrialised countries in the post-Second World War period, showing the similarity between nominal and real exchange rate variability under floating rate. His explanation for greater real exchange rate variability under floating rates than under fixed rates is nominal price rigidity.[24] The explanation may be questioned: for example, fixed nominal exchange rates may produce greater trade stability that will be reflected in the real exchange rate, as is evident for both the Bretton Woods and gold standard periods. Yet, as Eichengreen (1992a) points out and as can be seen in Table 9.5 here, these results could be explained by the fact that both periods were characterised by few shocks.[25]

Table 9.3 Persistence of CPI inflation: G-11 countries and four high inflation countries, 1880–1990, annual data: coefficient of AR1 regression; (standard error); t-statistic for unit root test

	United States			*United Kingdom*			*Germany*			*France*		
Gold standard	0.27	(0.18)	4.05	0.30	(0.17)	4.03	0.51	(0.16)	3.06	- 0.22	(0.18)	6.78
Interwar	0.45	(0.17)	3.18	0.35	(0.19)	3.37	0.51	(0.21)	2.33	0.42	(0.24)	2.42
Bretton Woods (total)	0.49	(0.19)	2.68	0.33	(0.20)	3.38	- 0.03	(0.21)	4.90	0.56	(0.16)	2.75
Bretton Woods (preconvertible)	0.41	(0.27)	2.15	0.15	(0.29)	2.96	- 0.07	(0.31)	3.45	0.60	(0.27)	1.48
Bretton Woods (convertible)	1.07	(0.20)	- 0.33	0.57	(0.34)	1.27	0.44	(0.31)	1.81	0.12	(0.14)	6.29
Floating exchange	0.68	(0.18)	1.76	0.69	(0.19)	1.67	0.83	(0.14)	1.21	0.85	(0.16)	0.94
Post-Second World War	0.65	(0.12)	2.92	0.75	(0.10)	2.50	0.31	(0.15)	4.60	0.69	(0.11)	2.82
	Japan			*Canada*			*Italy*					
Gold standard	0.22	(0.18)	4.33[a]	0.08	(0.18)	5.11	0.28	(0.14)	5.14			
Interwar	0.70	(0.25)	1.20	0.35	(0.20)	3.25	0.28	(0.18)	4.00			
Bretton Woods (total)	0.52	(0.18)	2.67	0.39	(0.19)	3.21	0.21	(0.12)	6.58			
Bretton Woods (preconvertible)	0.47	(0.07)	1.96	0.32	(0.28)	2.43	0.18	(0.18)	4.56			
Bretton Woods (convertible)	0.18	(0.31)	2.64	0.81	(0.20)	0.95	0.38	(0.29)	2.14			
Floating exchange	0.70	(0.19)	1.58	0.75	(0.17)	1.47	0.75	(0.17)	1.47			
Post-Second World War	0.54	(0.13)	3.54	0.65	(0.11)	3.18	0.28	(0.10)	7.20			

Table 9.3 Continued

	Belgium			*Netherlands*			*Sweden*			*Switzerland*		
Gold standard	0.11	(0.18)	4.83	–0.36	(0.26)	5.14	0.42	(0.17)	3.53	n.a.		
Interwar	0.49	(0.20)	2.57	0.34	(0.16)	4.14	0.50	(0.11)	4.44	0.16	(0.15)	5.77
Bretton Woods (total)	–0.09	(0.22)	5.07	0.31	(0.15)	4.58	0.13	(0.21)	4.26	0.24	(0.20)	3.75
Bretton Woods (preconvertible)	–0.16	(0.33)	3.58	0.26	(0.22)	3.33	0.17	(0.29)	2.89	–0.07	(0.30)	3.56
Bretton Woods (convertible)	0.72	(0.21)	1.32	0.50	(0.26)	1.91	–0.19	(0.35)	3.38	0.40	(0.29)	2.07
Floating exchange	0.78	(0.16)	1.40	0.88	(0.11)	1.05	0.53	(0.21)	2.25	0.69	(0.18)	1.80
Post-Second World War	0.41	(0.14)	4.13	0.52	(0.11)	4.41	0.54	(0.13)	3.62	0.60	(0.12)	3.28
	Argentina			*Brazil*			*Chile*			*Israel*		
Gold standard	0.29	(0.30)	2.37	(0.47)	(0.17)	3.11	0.23	(0.31)	2.48	n.a.		
Interwar	–0.17	(0.19)	6.14	0.53	(0.19)	2.53	0.16	(0.23)	3.68	n.a.		
Bretton Woods (total)	0.27	(0.20)	3.61	0.81	(0.12)	1.58	0.56	(0.16)	2.66	0.46	(0.20)	2.66
Bretton Woods (preconvertible)	0.41	(0.29)	2.04	0.49	(0.26)	1.95	0.63	(0.21)	1.77	0.40	(0.35)	1.74
Bretton Woods (convertible)	0.16	(0.32)	2.63	0.68	(0.22)	1.49	0.33	(0.30)	2.20	0.31	(0.30)	2.31
Floating exchange	0.74	(0.22)	1.18	1.29	(0.08)	–3.62	0.85	(0.14)	1.11	0.72	(0.17)	1.63
Post-Second World War	0.93	(0.10)	0.70	1.28	(0.05)	–5.91	0.83	(0.08)	2.04	0.82	(0.09)	2.00

Notes: 5% significant level for unit root test with 25 observations is 3.00.
[a] GNP deflator was used because of unavailability of CPI data.
Data sources: As Table 9.1.

Finally, based on monthly data for 1880–1986 for the UK, and the USA, Grilli and Kaminsky (1991) show that, with the exception of the post-Second World War period, no clear connection exists between the nominal exchange rate regime and the variability of real exchange rates. Our results for the G-11 show a clear correlation between nominal exchange rate rigidity and lower real exchange rate variability for the gold standard and Bretton Woods convertible regime. For the preconvertible Bretton Woods period – *de jure* a type of fixed exchange rate regime – the correlation is not evident. We do not distinguish between fixed and flexible periods in the inter-war segment (as do Grilli and Kaminsky), hence that period cannot be used in the comparison.[26]

Government deficit (Panel j)

For the underlying data see Figure 9.10. For the G-11 countries, the average ratio of the government deficit to GNP is lowest on average during the Bretton Woods convertible period, followed by the gold standard. The highest ratio is for the recent float. The ratios did not differ much between most of the countries across the regimes, with the key exception

Figure 9.10 Government deficit as a percentage of nominal GNP, 1880–1990

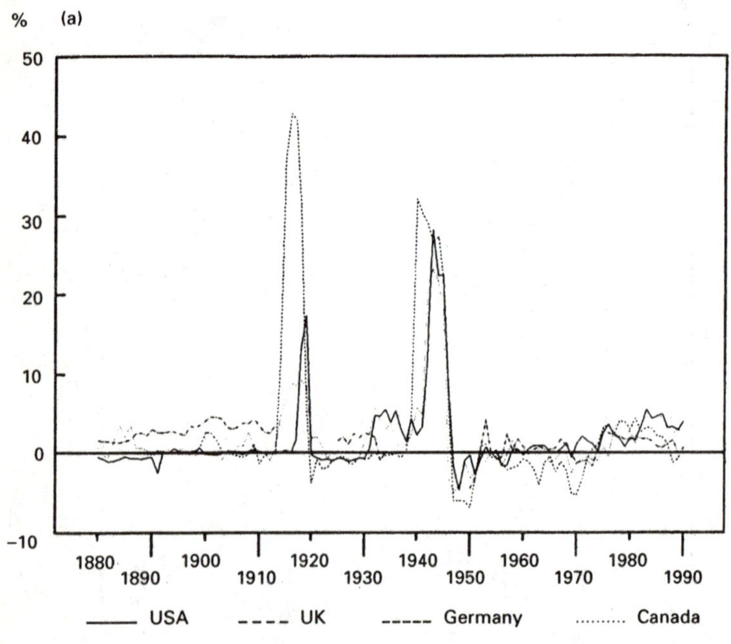

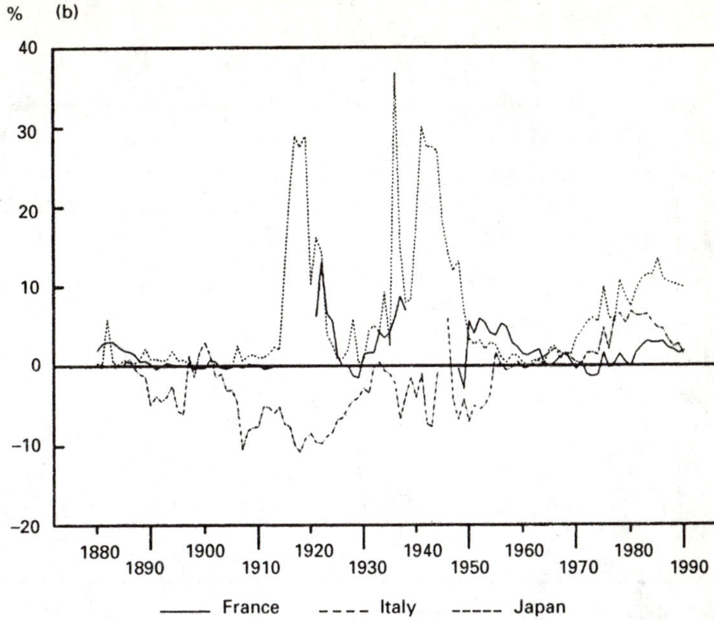

%
(b)
40
30
20
10
0
−10
−20
1880
1890
1900
1910
1920
1930
1940
1950
1960
1970
1980
1990
France
Italy
Japan

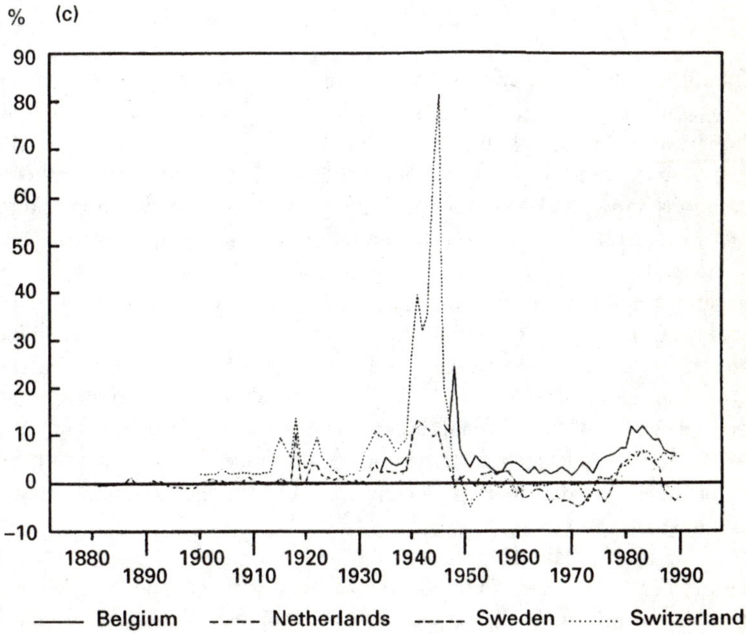

%
(c)
90
80
70
60
50
40
30
20
10
0
−10
1880
1890
1900
1910
1920
1930
1940
1950
1960
1970
1980
1990
Belgium
Netherlands
Sweden
Switzerland

Figure 9.10 *Continued*

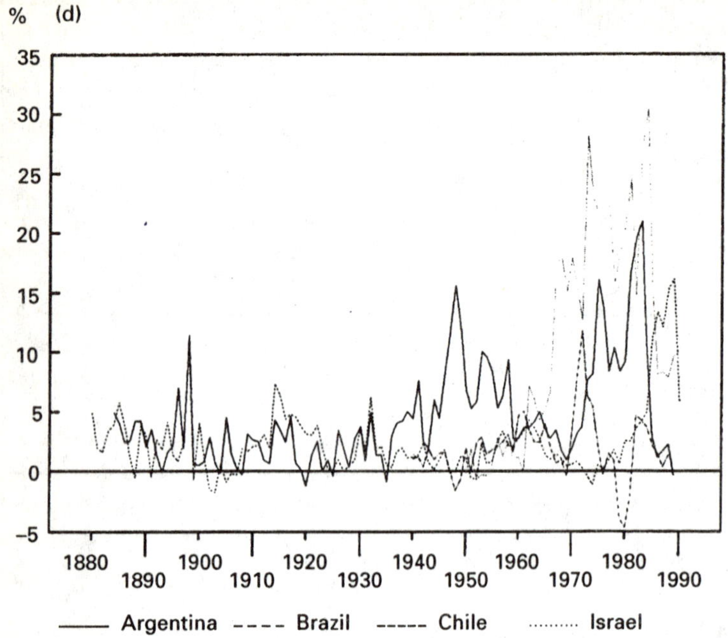

of Italy and Belgium, which ran considerably higher deficits than the others under the preconvertible Bretton Woods period and the recent float. The standard deviations in fiscal policy followed a pattern similar to the means. As in the case of monetary policy, as both means and standard deviations show, fiscal policy converged most closely between countries during the Bretton Woods convertible regime and the gold standard. Thus adherence to constitutional regimes may have constrained fiscal policy in the same way as it did monetary policy. On the other hand, more limited fiscal needs during those regimes may have made it easier to adhere to the convertibility rule. Finally, except for the recent float, the correlation between the fiscal deficit and inflation is weak. As Heymann and Leijonhufvud (1995) point out, under convertible as well as low inflation RWMS regimes, fiscal authorities possess sufficient credibility to have liberal access to bond markets – hence fiscal and monetary policy can be quite independent.

For the high inflation countries (see Table 9.2, Panel f) the deficit as a share of GNP increases dramatically between the pre- and post-Second World War periods. In the post-war period it increases between regimes, reaching a peak with the float. In contrast to the G-11, the fiscal deficit is quite significant in the Bretton Woods convertible period. A closer correlation between the fiscal deficit, money growth and inflation can be observed across regimes for the high inflation countries that is the case for the G-11. As Heymann and Leijonhufvud, and others, have argued, a close connection between fiscal deficits and inflationary finance is a hallmark of the absence of credibility of high inflation random walk monetary regimes.

In summary, the gold standard and convertible Bretton Woods regime exhibited the most stability and closest convergences of financial variable for the G-11 countries. In addition, the Bretton Woods convertible regime exhibited the best overall macro performance of any regime. As the summary statistics in Table 9.1 show, both nominal and real variables were most stable in this period. However, the floating rate regime on most criteria, was not far behind the Bretton Woods convertible regime. On all criteria, the inter-war period performed the worst.

These results agree with the views of Leijonhufvud and others that convertibility rules in the past have been associated with superior performance of nominal variables. However, there is little evidence that adherence to such rules has been associated with better real performance, as can be seen in a comparison for the G-11 between the recent float and the gold standard.

For the high-inflation countries, there is strong evidence linking poor nominal performance with the absence of a link to a convertible regime. Also, for at least three countries, real performance deteriorated markedly after the world shifted from Bretton Woods to a floating regime. This suggests, in agreement with Leijonhufvud, that a high inflation RWMS regime has serious real consequences.

Inflation Persistence

A second piece of evidence on regime performance is the persistence of inflation. Evidence of persistence in the inflation rate suggests that market agents expect the monetary authorities to follow an inflationary policy continually, or alternatively a RWMS regime; its absence would be consistent with the belief that the authorities are following a stable monetary rule such as the gold standard's convertibility rule. Barsky (1987) presented evidence for the UK and USA, based on both autocorrelations and time series models that inflation under the gold standard was very nearly a white-noise process, whereas in the post-Second World War period, the inflation rate exhibited considerable persistence. Alogoskoufis and Smith (1991) also show, based on AR(1) regressions of the inflation rate, that inflation persistence in the two countries increased between the classical gold standard period and the inter-war period and between the inter-war period and the post-Second World War period.[27]

Table 9.3 presented the inflation-rate coefficient (from the type of AR(1) regressions on CPI inflation estimated by Alogoskoufis and Smith (1991), for all fifteen countries over successive regimes since 1880,[28] as well as the standard errors and the Dickey–Fuller tests for a unit root.[29] The results, as in Alogoskoufis and Smith, show an increase in inflation persistence for most G-11 countries between the classical gold standard and the inter-war period, and also between the inter-war period and the post-Second World War period as a whole. Within the post-Second World War period, inflation persistence is generally lower (with the exceptions of France, Japan and Sweden) in the preconvertible Bretton Woods than the convertible period. This suggests that though the immediate post-Second World War period was characterised by rapid inflation, market agents may have expected a return to a stable price regime. The higher degree of persistence in the convertible regime suggests that this expectation lost credence. Finally, the evidence that persistence was generally highest during the float may imply the public's realisation that there was no longer a stable nominal anchor.

For the high inflation countries, the pattern is quite similar to the G-11, with inflation persistence increasing over time and reaching its peak under the post-1973 float. In sum, the evidence on inflation persistence in Table 9.3 is consistent with Leijonhufvud's (1987a) description of the gradual evolution away from convertible and towards RWMS regimes.

Forecast Errors in Inflation and Growth

A third piece of evidence relates to the forecast errors of inflation and real output growth. According to Meltzer and Robinson (1989), 'a welfare maximizing monetary rule would reduce variability to the minimum inherent in nature and institutional arrangements' (p. 179). They measure variability by the mean absolute error (MAE) of a one period forecast based on the univariate multi-state Kalman filter (MSKF). Following their approach, Table 9.4 presents the MAEs for inflation and real growth for the G-11 countries over successive regimes. The MSKF forecasts incorporate both transitory and permanent shocks to the rate of change series.[30] According to Leijonhufvud (1984a), inflation forecast errors should increase with time under a RWMS regime. Accordingly, we present the forecast errors for both variables at one-, five- and ten-year horizons.

Table 9.4 Forecast errors in inflation and real growth: G-11 and high inflation countries, 1880–1989, annual data: mean absolute errors using the multi-state Kalman filter

	Gold Standard 1880–1913		*Interwar 1919–39*		*Bretton Woods (total) 1946–70*		*Bretton Woods (preconvertible) 1946–58*		*Bretton Woods (convertible) 1959–70*		*Floating exchange 1973–89*	
	Growth	*Inflation*	*Growth*	*Inflation*	*Growth*	*Inflation*	*Growth*	*Inflation*	*Growth*	*Inflation*	*Growth*	*Inflation*
United States												
1	2.00	1.48	4.90	4.55	3.22	2.33	5.01	3.47	1.28	1.11	1.67	2.06
5	7.16	6.31	22.26	30.82	20.54	13.51	33.24	22.23	6.79	4.08	5.46	13.03
10	12.86	11.87	19.00	48.54	30.55	27.48	45.08	44.15	14.80	9.42	9.23	26.64
United Kingdom												
1	1.42	2.10	4.11	4.41	1.41	1.47	1.82	1.76	0.97	1.15	2.71	4.61
5	4.85	6.80	22.01	37.20	10.80	5.95	17.54	6.73	3.50	5.11	8.24	28.59
10	9.39	10.28	29.66	70.02	20.52	23.39	31.70	31.21	8.40	14.92	7.54	61.10
Germany												
1	1.69	1.69	8.07	3.97	2.77	2.59	3.24	5.00	2.49	1.18	1.45	1.30
5	5.73	7.39	37.00	22.65	12.99	12.15	18.40	45.80	11.63	3.74	7.02	7.85
10	7.45	9.44	31.21	40.05	29.44	31.92	n.a.		29.44	31.92	19.80	16.31
France												
1	2.25	2.48	5.54	7.08	1.51	3.13	2.28	4.62	0.87	2.14	1.44	2.82
5	9.48	8.55	31.40	34.08	7.35	9.47	15.76	25.78	3.15	4.03	7.93	13.61
10	16.58	11.29	79.95	59.54	19.11	20.29	n.a.	n.a.	19.56	20.29	14.42	22.30
Japan												
1	1.69	3.95	2.69	6.83	1.59	2.63	0.85	5.48	1.89	0.96	1.54	2.95
5	5.31	16.95	12.57	39.21	8.03	14.53	n.a.	48.57	8.37	6.02	10.62	16.82
10	12.71	37.64	20.40	61.50	26.50	37.10	n.a.	n.a.	26.50	37.10	26.01	30.06

Table 9.4 Continued

	1880–1913		Interwar 1919–39		Bretton Woods (total) 1946–70		Bretton Woods (preconvertible) 1946–58		Bretton Woods (convertible) 1959–70		Floating exchange 1973–89	
	Growth	Inflation	Growth	Inflation	Growth	Inflation	Growth	Inflation	Growth	Inflation	Growth	Inflation
Canada												
1	1.58	0.80	7.16	3.99	2.19	1.92	3.06	2.96	1.24	0.79	1.72	2.29
5	8.11	3.44	29.66	25.54	15.16	8.73	21.01	12.14	8.83	5.03	7.75	14.04
10	11.70	6.35	27.85	45.63	25.21	18.56	34.21	23.91	15.46	12.78	17.09	33.67
Italy												
1	2.08	2.01	2.08	7.58	4.21	6.35	7.04	10.54	1.15	1.81	1.14	3.83
5	6.72	6.62	8.44	56.11	21.00	74.57	36.60	135.02	4.10	9.10	7.62	26.55
10	14.26	11.39	9.06	123.71	40.53	194.17	70.48	352.43	8.07	22.73	17.99	57.08
Belgium												
1	n.a.	1.22	1.36	1.93	1.48	2.06	1.71	2.95	1.29	1.30	1.41	2.16
5	n.a.	6.74	7.41	8.21	5.34	10.60	6.38	20.26	4.82	5.77	7.53	12.75
10	n.a.	8.80	9.37	21.02	11.80	21.64	n.a.	n.a.	9.58	13.53	13.91	28.74
Netherlands												
1	1.59	1.36	3.23	3.09	4.48	2.87	7.34	3.84	1.38	1.82	1.48	1.93
5	5.42	5.66	14.44	26.35	27.64	10.54	47.44	13.73	6.19	7.08	9.29	14.68
10	17.20	21.07	24.23	43.99	53.94	33.80	83.14	49.54	22.31	16.74	19.48	29.18
Sweden												
1	1.86	2.24	3.79	5.80	1.33	2.28	1.63	3.08	1.01	1.40	1.58	2.72
5	4.48	9.57	16.70	43.53	7.84	10.20	11.55	15.70	3.81	4.24	6.14	14.41
10	8.84	14.08	17.31	83.53	12.42	21.90	15.28	29.35	9.32	13.83	14.66	33.46

Table 9.4 Continued

	1880–1913		*Interwar 1919–39*		*Bretton Woods (total) 1946–70*		*Bretton Woods (preconvertible) 1946–58*		*Bretton Woods (convertible) 1959–70*		*Floating exchange 1973–89*	
	Growth	*Inflation*	*Growth*	*Inflation*	*Growth*	*Inflation*	*Growth*	*Inflation*	*Growth*	*Inflation*	*Growth*	*Inflation*
Switzerland												
1	n.a.	n.a.	1.74	3.50	2.30	1.41	3.20	1.44	1.32	1.38	1.73	1.77
5	n.a.	n.a.	8.75	18.88	10.33	7.94	14.84	8.85	5.45	6.95	8.70	7.56
10	n.a.	n.a.	n.a.	n.a.	18.91	26.37	25.09	38.98	12.22	12.69	19.79	13.42
Average G-11												
1	1.80	1.93	4.06	4.79	2.41	2.64	3.38	4.10	1.35	1.37	1.62	2.59
5	6.36	7.80	19.15	31.14	13.37	16.20	22.28	32.26	6.06	5.56	7.85	15.44
10	12.33	14.22	26.80	59.75	26.27	41.51	43.57	81.37	15.97	18.72	16.36	32.00
Argentina												
1	3.52	8.16	3.26	5.40	3.27	11.29	2.94	12.19	3.63	10.32	3.28	53.99
5	11.09	33.69	15.98	25.04	7.31	54.82	9.42	56.19	5.03	53.34	10.84	209.39
10	23.82	65.62	34.14	36.36	7.53	126.59	8.12	125.98	6.89	127.25	21.75	405.75
Brazil												
1	5.66	10.64	6.41	6.15	2.89	11.27	2.32	5.48	3.50	17.54	3.57	27.21
5	30.91	56.59	26.33	30.51	11.85	62.39	10.73	31.51	13.06	95.83	14.72	139.19
10	60.82	124.21	47.78	52.63	28.03	120.78	30.35	70.68	25.52	175.05	24.70	284.69
Chile												
1	3.44	8.14	6.53	6.60	3.19	9.09	3.71	10.69	2.62	7.36	5.00	32.03
5	16.79	26.29	21.32	29.62	12.55	47.34	11.67	49.22	13.50	45.31	18.45	255.08
10	23.71	53.70	25.81	55.24	18.26	90.72	17.44	88.18	19.15	92.41	17.27	581.52

Table 9.4 Continued

	1880–1913		*Interwar 1919–39*		*Bretton Woods (total) 1946–70*		*Bretton Woods (preconvertible) 1946–58*		*Bretton Woods (convertible) 1959–70*		*Floating exchange 1973–89*	
	Growth	*Inflation*	*Growth*	*Inflation*	*Growth*	*Inflation*	*Growth*	*Inflation*	*Growth*	*Inflation*	*Growth*	*Inflation*
Israel												
1	n.a.	n.a.	n.a.	n.a.	3.94	6.87	7.13	12.15	2.09	3.78	4.47	24.67
5	n.a.	n.a.	n.a.	n.a.	12.48	42.49	42.24	67.76	5.04	36.17	16.98	169.49
10	n.a.	n.a.	n.a.	n.a.	29.75	119.62	n.a.	n.a.	29.75	119.62	32.97	289.91
Average high inflation countries												
1	4.21	8.98	5.40	6.05	3.32	9.63	4.03	10.13	2.96	9.75	4.08	34.48
5	19.60	38.86	21.21	28.39	11.05	51.76	18.52	51.17	9.16	57.66	15.25	193.29
10	36.12	81.18	35.91	48.08	20.89	114.43	18.64	94.95	20.33	128.58	24.17	390.47

The lowest forecast errors at the one-year horizon for inflation on average, for the G-11 countries, were for the Bretton Woods convertible period, followed by the gold standard and the floating rate period. The highest were for the inter-war period, followed by the preconvertible Bretton Woods period. The most notable exception to the pattern was for the UK, where the floating rate period exhibited the largest variability. For the high-inflation countries, inflation forecast errors were considerably lower in all periods compared to the recent float.

For the G-11, the inflation forecast error increases with time across all regimes, but much more so under the recent float – a pattern predicted by Leijonhufvud's RWMS regime. For the high-inflation countries, forecast errors increased dramatically with time in the post-war period, especially since 1973.

For real growth, as for the inflation rate, the lowest MAE, at the one year horizon on average, for the G-11 occurred in the convertible Bretton Woods period. An exception to this pattern was Japan. The highest MAE was again in the inter-war and the preconvertible Bretton Woods period. The floating period, though more variable than the convertible Bretton Woods period, was slightly less variable than the gold standard. Like inflation, forecast errors increase with time across all regimes, but the most variable regime was the preconvertible Bretton woods period. For the high-inflation countries, such as the G-11, real growth forecast errors at the one-year horizon were lowest in the Bretton Woods convertible period. At longer time horizons, the errors increased, but not dramatically as in the case of inflation.

The results for inflation forecast errors are highly consistent with Leijonhufvud's theory. They increase with time at a much greater rate for non-convertible monetary regimes. This pattern is most dramatic for high-inflation countries. Although forecast errors in real growth tend to be lowest at the one-year horizon during the Bretton Woods convertible regime, there does not appear to be a clear pattern linking real growth uncertainty to the monetary regime.

Demand and Supply Disturbances

An important issue is the extent to which the performance of alternative monetary regimes, as revealed by the data in the preceeding tables, reflects the operation of the monetary regime in constraining policy actions or the presence or absence of shocks to the underlying environment. One way to shed light on this issue, following earlier work by Bayoumi and Eichengreen (1992, 1993, 1994a and b) is to identify underlying shocks to aggregate supply and demand. According to them, aggregate supply shocks reflect shocks to the environment and are independent of the regime, but aggregate demand shocks probably reflect policy actions and are specific to the regime.

The approach used to calculate aggregate supply and demand shocks is an extension of the bivariate structural vector autoregression (VAR) methodology developed by Blanchard and Quah (1989). Following Bayoumi and Eichengreen (1994a), we estimated a two-variable VAR in the rate of change of the price level and output.[31] Restrictions on the VAR identify an aggregate demand disturbance, which is assumed to have only a temporary impact on output and a permanent impact on the price level, and an aggregate supply disturbance, which is assumed to have a permanent impact on both prices and output.[32] Overidentifying restrictions: namely, that demand shocks are positively correlated and supply shocks are negatively correlated with prices, can be tested by examining the impulse response functions to the shocks.

The methodology has important limitations, which suggest that the results should be viewed with caution. The key limitation is that one can easily imagine frameworks in which demand shocks have permanent effects on output, while supply shocks have only temporary effects.[33]

We estimated supply (permanent) and demand (temporary) shocks, using annual data for each of the fifteen countries, over alternative regimes, in the period 1880–1989. The VARs are based on three separate sets of data (to the extent available): 1880–1913, 1919–39 and 1946–89, omitting the war years because complete data on them was available for only a few of the countries. The VAR have two lags. We also did the estimation for aggregated price and output data for the G-11 and the high-inflation countries.

Figures 9.11 and 9.12 show the impulse responses to a one standard deviation shock in aggregate demand and aggregate supply, on output and prices for the G-11 aggregate and the high inflation countries aggregate, by regime.[34] The overidentifying restrictions that demand shocks be positively correlated and supply shocks negatively correlated with the price level are satisfied for the two country grouping aggregates for the post-Second World War regimes. But for the period before the Second World War, for most countries, as reflected in the aggregates shown in Figures 9.11 and 9.12, they are not. Supply shocks were positively correlated with prices.

Keating and Nye (1991) attempted to explain this result for the earlier periods by possible hysteresis effects. Bayoumi and Eichengreen (1994a) argue that the positive response to favourable aggregate supply shocks

Figure 9.11 Impulse response functions of demand and supply shocks on prices (dotted lines) and output (solid lines), G-11 aggregate by regimes, 1881–1990, annual data.

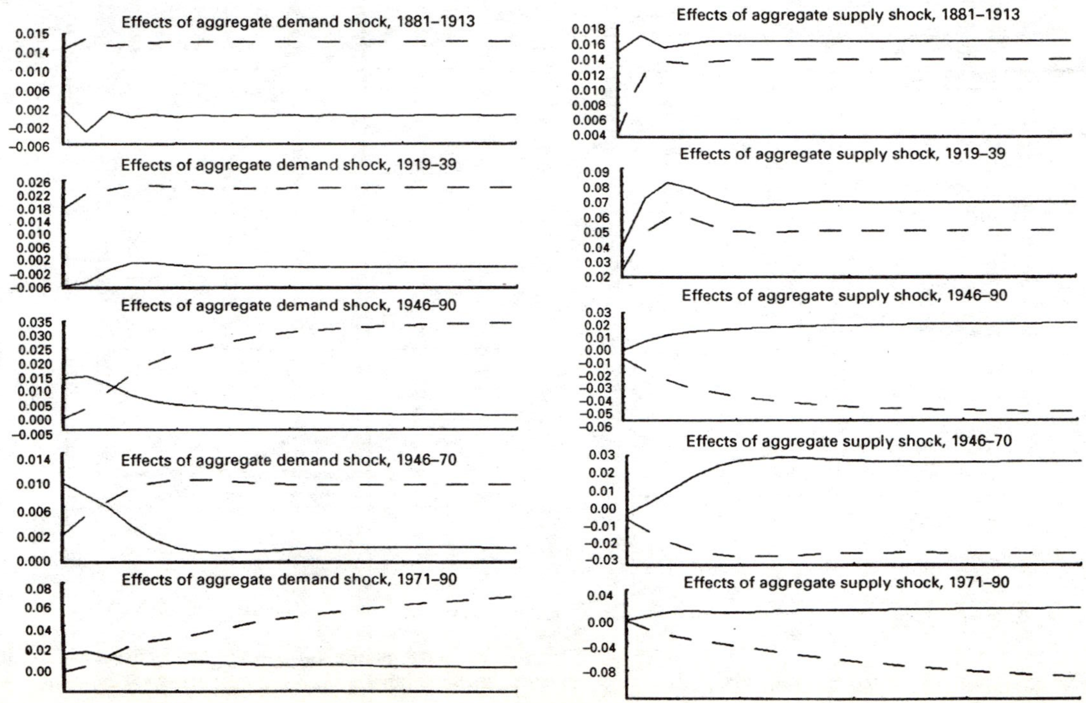

Figure 9.12 Impulse response functions of demand and supply shocks on prices (dotted lines) and output (solid lines), high inflation countries aggregate by regimes, 1881–1990, annual data

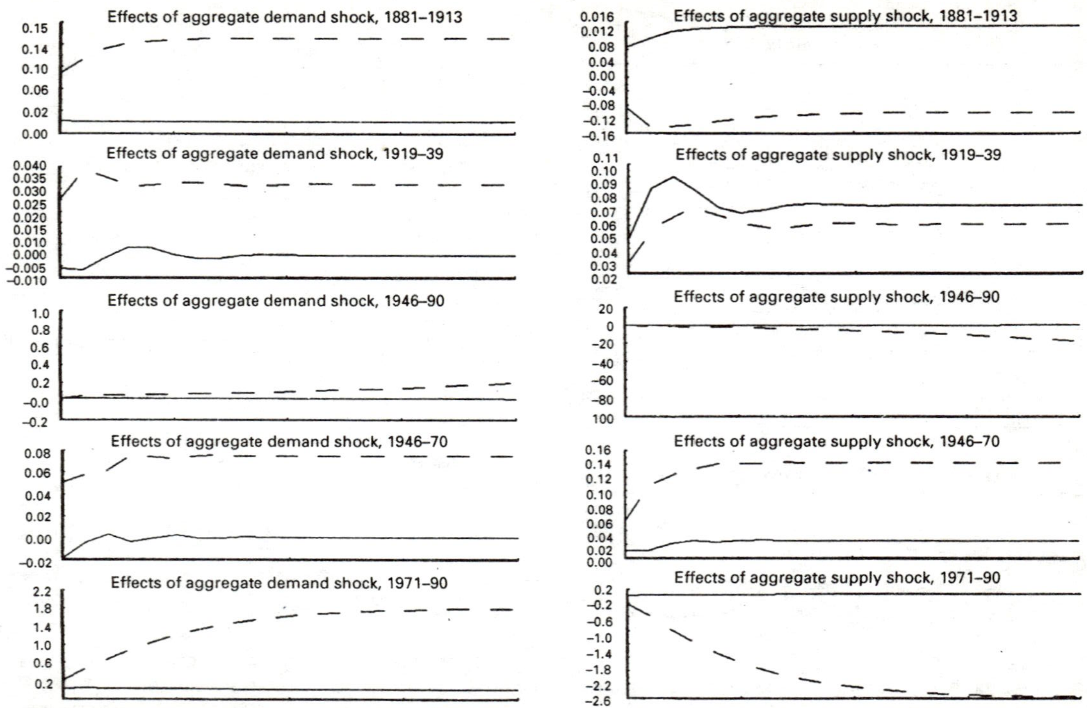

may be, in part, caused by the operation of the classical gold standard – as the price level rises, it reduces the relative price of gold and its flow supply. It may also reflect the temporary nature of agricultural supply shocks being treated as demand shocks. These results may also reflect a limitation of the Blanchard–Quah methodology.

Table 9.5 presents the standard deviations of supply and demand shocks for the fifteen countries by regimes. We also present the aggregate of shocks for the G-11 countries, and for the four high-inflation countries. In addition, we show, following Bayoumi and Eichengreen (1994a), the weighted average of the individual country shocks. Figures 9.13 and 9.14 show the shocks for the G-11 aggregate, for the high inflation aggregate and for each of the 15 countries.

Table 9.5 shows for the G-11 aggregate that the convertible Bretton Woods regime was the most tranquil of all the regimes – neither supply nor demand shocks dominated. However, it was not that much less turbulent than the succeeding float. The inter-war period, unsurprisingly, shows the largest supply and demand shocks.[35] Sizeable supply and demand shocks which are two or three times greater than the post- Second World War period also characterise the classical gold standard.[36]

For individual G-11 countries, the Bretton Woods convertible period was the most stable in eight countries, and flexible rates in three. However, the difference between the convertible Bretton Woods period and the float was not great in any country. The inter-war period, as expected, was the most volatile. Both types of shock were the largest in every country except the UK. Finally, in the majority of countries, with the principal exceptions of the UK, Germany and the Netherlands, both supply – and to a lesser extent demand – shocks were greater in the gold standard period than in the post-Second World War period.

For the high inflation countries, in marked contrast to the G-11, demand shocks exceeded supply shocks in all regimes except the inter-war. In the post-Second World War period, demand shocks considerably exceeded supply shocks, especially under the float. This evidence, to the extent that demand shocks capture discretionary policy actions, provides strong support for Leijonhufvud's views on the RWMS regime.

The dispersion of demand shocks across countries, as measured by the GNP weighted standard deviation of the individual country shocks around the G-11 aggregate, reveals very little difference between the gold standard and the post-Second World War regimes, with floating regime displaying the highest degree of convergence. Dispersion is much greater in the inter-war period. The dispersion of supply shocks is considerably greater during the gold standard and the inter-war periods than in any of

Table 9.5 Supply (permanent) and demand (temporary) shocks, 1880–1990, annual data: standard deviations of shocks (%): dispersion of shocks across countries (%)

	Gold standard 1883–1913		*Inter-war 1921–39*		*Bretton Woods (total) 1948–70*		*Bretton Woods (preconvertible) 1948–58*		*Bretton Woods (convertible) 1959–70*		*Floating exchange 1973–89*		*Post Second World War 1948–89*	
	D	*S*	*D*	*S*	*D*	*S*	*D*	*S*	*D*	*S*	*D*	*S*	*D*	*S*
United States	2.03	3.81	4.46	6.74	2.33	1.54	3.11	1.99	1.37	1.07	1.72	1.94	2.08	1.69
United Kingdom	2.66	2.16	1.93	3.52	2.62	1.95	3.06	2.61	2.26	1.14	3.57	4.31	3.03	3.10
Germany	2.36	2.32	4.47	3.13	2.88	2.65	1.85	2.87	3.34	2.66	1.66	1.39	2.36	2.20
France	4.58	3.75	7.17	5.19	3.50	1.75	3.23	1.61	3.77	1.84	1.93	1.52	2.84	1.70
Japan	4.85	3.39	6.28	5.36	3.18	2.56	4.05	1.97	2.83	2.78	2.39	2.50	2.75	2.69
Canada	0.93	2.75	4.01	8.61	2.42	2.60	3.08	3.70	1.74	0.98	2.66	2.10	2.48	2.47
Italy	3.16	3.12	7.40	4.14	2.76	1.75	2.93	1.85	2.56	1.73	3.58	1.91	3.30	2.47
Belgium	n.a.		5.82	3.37	2.14	1.67	2.97	2.00	1.16	1.15	1.55	1.99	1.91	1.81
Netherlands	0.81	0.81	2.79	3.42	2.97	2.67	3.13	2.46	2.77	2.86	1.94	1.85	2.50	2.39
Sweden	2.57	3.03	3.08	5.10	3.89	2.96	5.37	4.18	1.97	1.26	2.32	2.40	3.44	2.86
Switzerland	n.a.		3.54	2.08	2.50	1.72	2.84	2.24	1.46	1.04	2.24	1.44	2.50	1.73

Table 9.5 Continued

	Gold standard 1883–1913		*Inter-war 1921–39*		*Bretton Woods (total) 1948–70*		*Bretton Woods (preconvertible) 1948–58*		*Bretton Woods (convertible) 1959–70*		*Floating exchange 1973–89*		*Post Second World War 1948–89*	
	D	*S*	*D*	*S*	*D*	*S*	*D*	*S*	*D*	*S*	*D*	*S*	*D*	*S*
G-11	1.55	2.27	3.61	3.82	1.10	0.85	1.79	1.19	0.65	0.77	1.63	0.89	1.41	0.86
G-11*	1.52	2.25	2.42	4.43	1.08	0.82	1.61	1.05	0.69	0.73	1.36	1.62	1.24	1.33
Dispersion	4.38	5.70	7.53	8.71	4.83	3.17	4.29	2.54	5.01	3.38	4.14	3.32	4.43	3.30

Notes:
G-11: G-11 aggregate data.
G-11: Weighted average of individual country shocks; the weights are calculated as the share of each country's national income in the total income in the G-11 countries in 1970, where the GNP/GDP data are converted to dollars using the actual exchange rate.
1880–1913: United States 53%, United Kingdom 6%, Germany 10%, France 8%, Japan 11%, Canada 4%, Italy 6%, Belgium 0%, Netherlands 0%, Sweden 2%, Switzerland 0%.
1919–38: United States 58%, United Kingdom 7%, Germany 0%, France 8%, Japan 12%, Canada 5%, Italy 6%, Belgium 0%, Netherlands 2%, Sweden 2%, Switzerland 0%.
1946–90: United States 51%, United Kingdom 7%, Germany 9%, France 8%, Japan 10%, Canada 4%, Italy 5%, Belgium 1%, Netherlands 2%, Sweden 2%, Switzerland 1%.
Dispersion: mean value of $\Sigma(\text{weight}_i\ (\text{shock}_i - \Sigma\ \text{weight}_I{}^* \ \text{shock}_I)^{2/05}$, for i = United States, United Kingdom, Germany, France, Japan, Canada, Italy, Belgium, Netherlands, Sweden, Switzerland.

Table 9.5 Continued

	Gold standard 1883–1913		*Inter-war 1921–39*		*Bretton Woods (total) 1948–70*		*Bretton Woods (preconvertible) 1948–58*		*Bretton Woods (convertible) 1959–70*		*Floating exchange 1973–89*		*Post Second World War 1948–89*	
	D	*S*	*D*	*S*	*D*	*S*	*D*	*S*	*D*	*S*	*D*	*S*	*D*	*S*
Argentina	9.62	6.77	5.48	3.67	17.10	4.58	11.97	3.94	21.32	5.25	98.32	8.42	68.05	6.62
Brazil	15.91	6.69	5.76	6.15	16.67	5.54	12.46	4.00	17.97	5.42	18.16	7.50	18.47	5.78
Chile	8.55	6.90	9.98	9.41	12.73	5.62	14.46	6.91	11.56	4.20	27.00	8.06	20.86	6.79
Israel	n.a.		n.a.		5.94	3.16	6.47	3.50	5.95	2.99	28.06	12.29	20.12	9.28
High Inflation	11.32	5.19	5.93	5.98	10.36	5.99	12.44	4.01	17.13	6.01	26.89	10.02	20.11	9.04
High Inflation	9.32	5.81	4.86	4.72	11.91	5.23	10.76	4.52	19.67	6.37	30.72	9.86	23.23	8.69
Dispersion	7.40	4.95	4.27	4.23	20.01	6.14	18.12	5.20	21.31	6.79	50.28	13.51	32.68	9.51

High Inflation: High inflation countries – aggregate data.
High Inflation: Weighted average of individual country shocks; the weights are calculated as the share of each country's national income in the total income in the G-11 countries in 1970, where the GNP/GDP data are converted to dollars using the actual exchange rate.
1946–90: Argentina 31%, Brazil 53%, Chile 9%, Israel 7%.
1880–1939: Argentina 34%, Brazil 56%, Chile 10%.

Figure 9.13 Supply and demand shocks: G-11 aggregate, 1880–1989, annual data

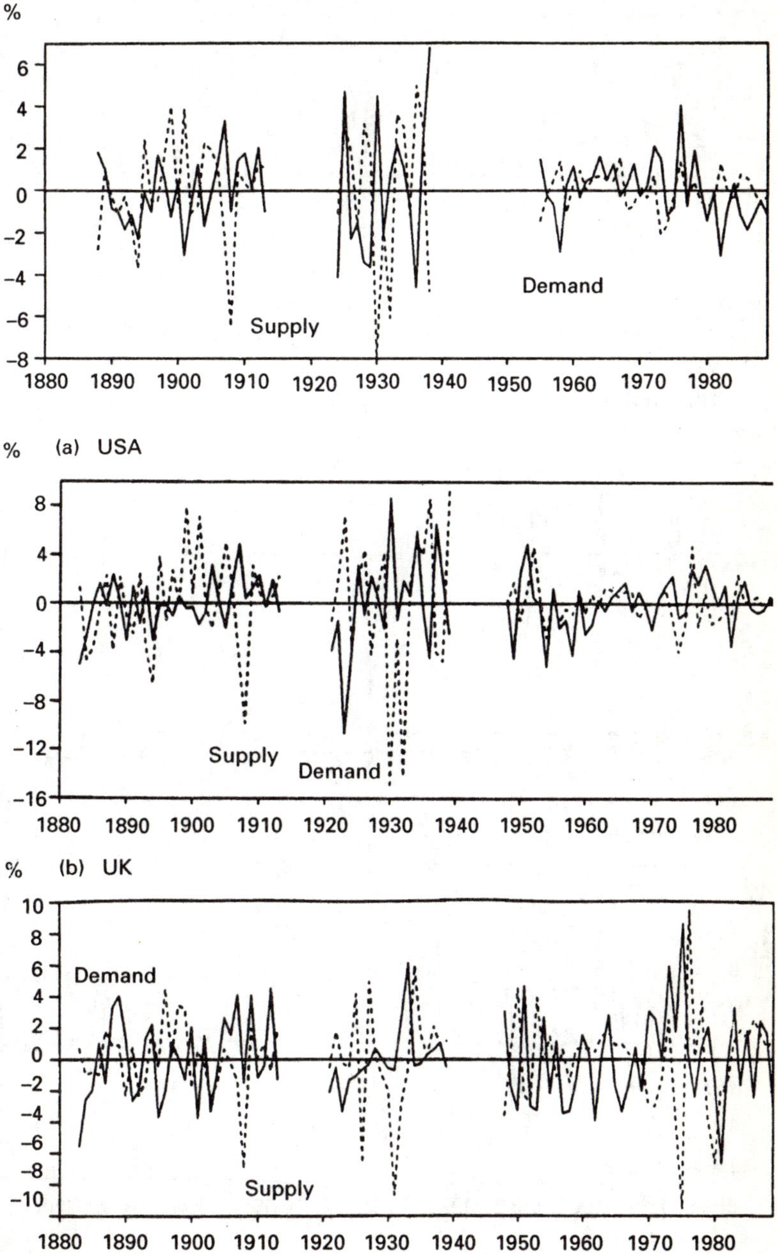

Figure 9.13 *Continued*

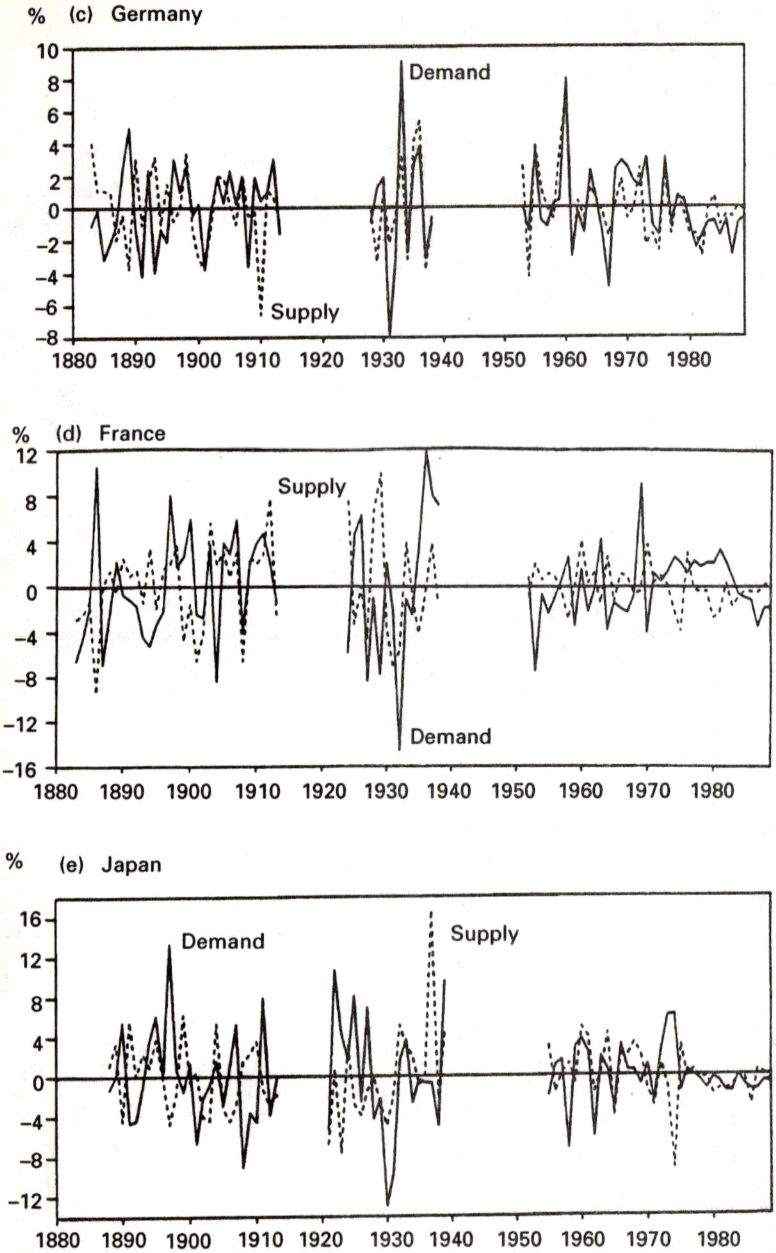

Figure 9.13 *Continued*

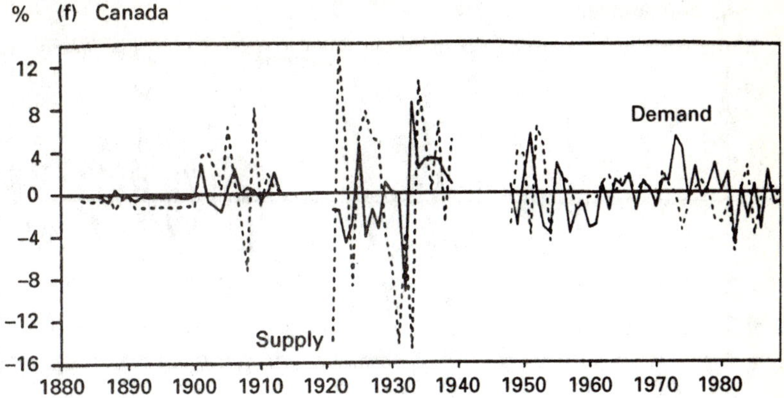

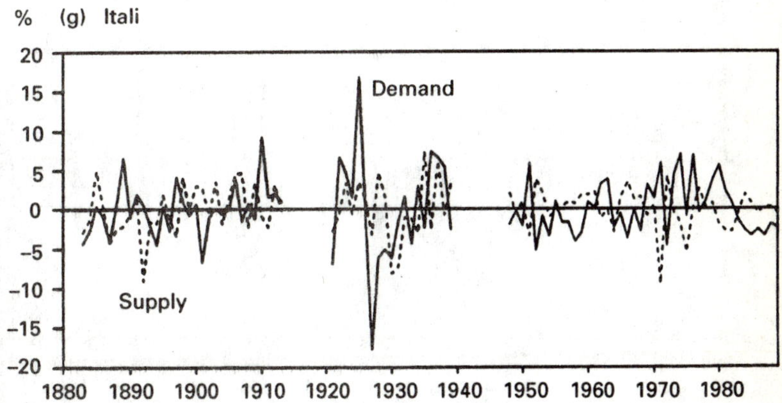

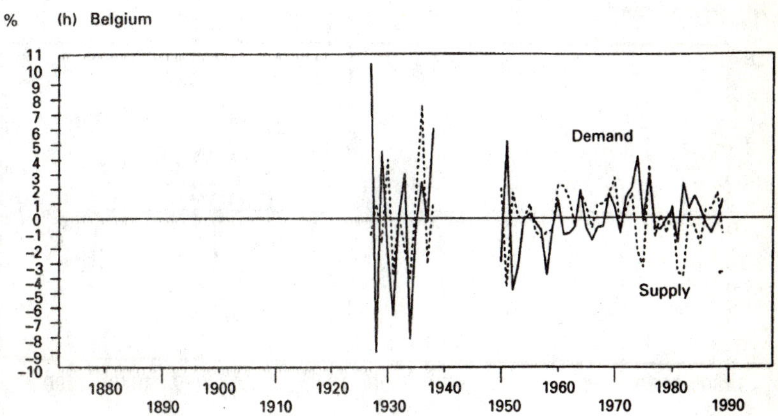

Figure 9.13 *Continued*

% (i) Netherlands

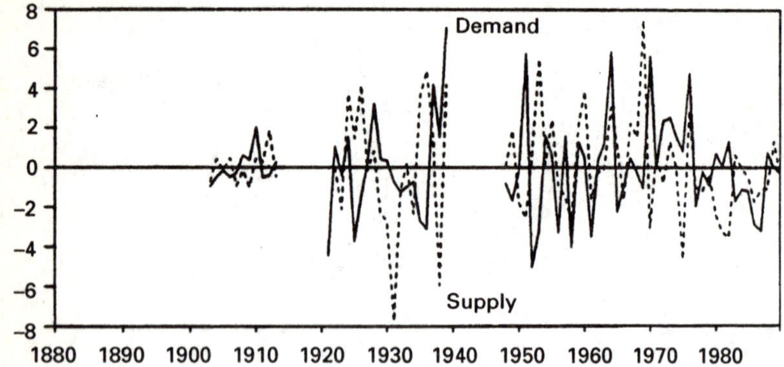

% (j) Sweden

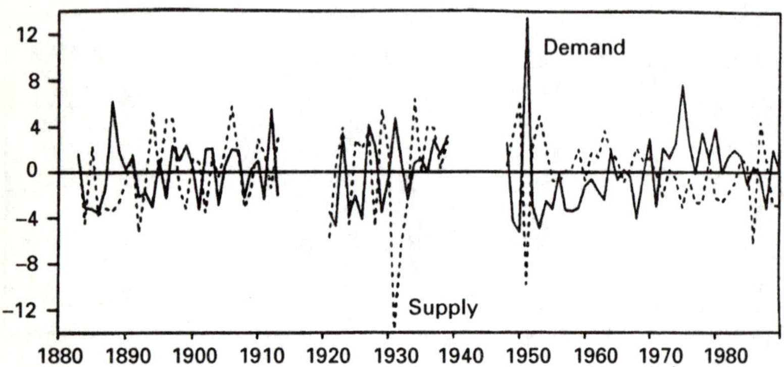

% (k) Switzerland

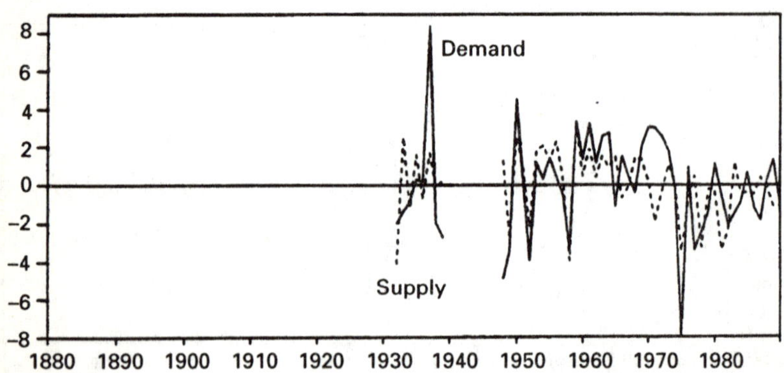

Figure 9.14 Supply and demand shocks: high inflation countries aggregate, 1880–1989, annual data

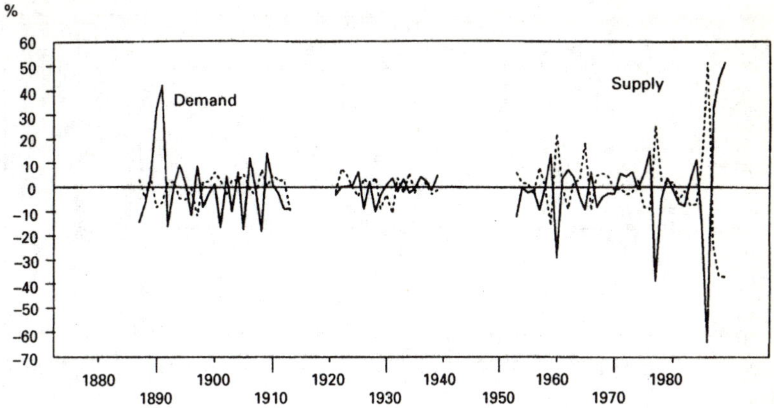

(a) Argentina

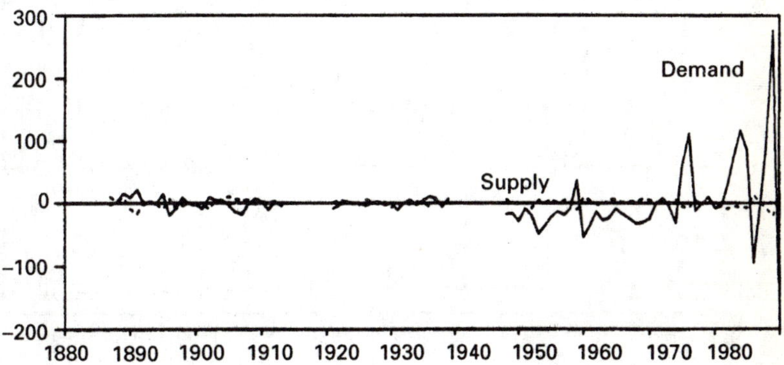

(b) Brazil

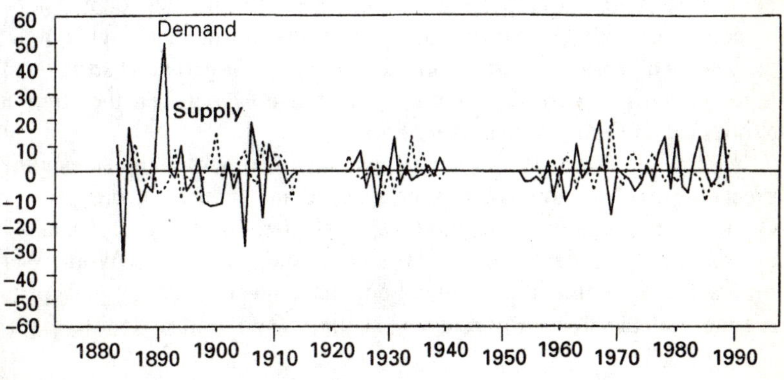

Figure 9.14 *Continued*

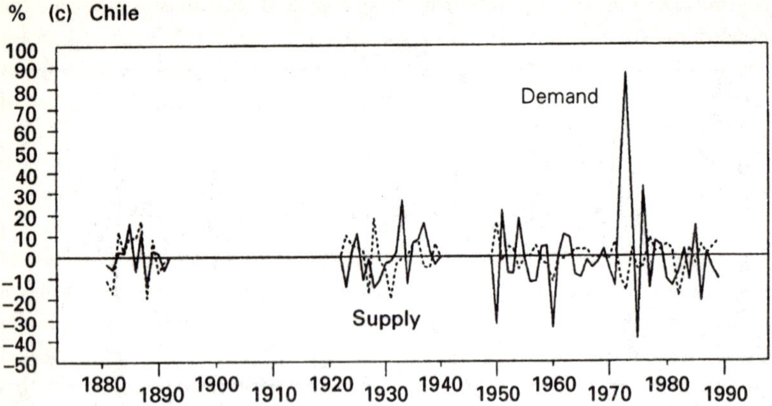

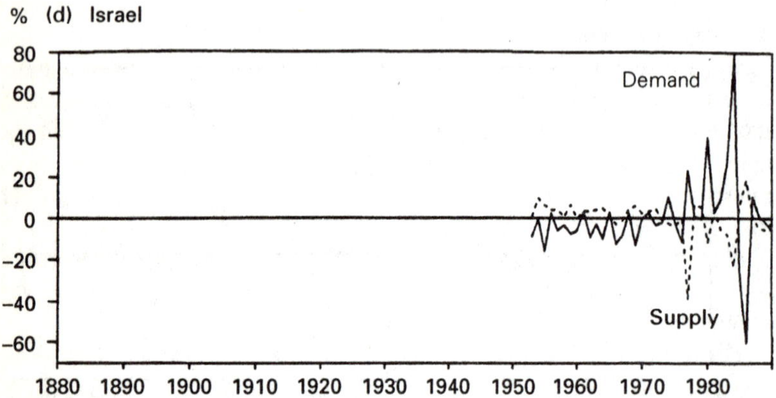

the post-Second World War regimes. For the high-inflation countries, the dispersion of both demand and supply shocks is considerably higher in the post-Second World War era compared to the pre-war period. Dispersion of both types of shocks also increases considerably between the Bretton Woods regime and the subsequent float.

In sum, the evidence on demand and supply shocks complements the preceding evidence on stability, persistence and forecast errors. For the G-11 countries, the gold standard was characterised by higher demand – and especially higher supply – shocks than the post-Second World War regimes, and within that period both the convertible Bretton Woods regime and the float are relatively stable.[37] By contrast, for the high-inflation countries, demand shocks exceeded supply shocks across all regimes, and especially under the post-war float. For these countries, the constraints of the convertible regime appear to be much weaker than for the G-11, although one could argue that the much greater incidence of supply shocks in those countries may in part account for the greater use of discretionary policy.

Summary

Leijonhufvud emphasises important difference between convertible and discretionary monetary regimes. Under convertible regimes, the existence of a credible nominal anchor constrains the money supply process; under discretion, no such constraint exists. Under a discretionary regime, inflation uncertainly has serious effects on economic welfare. In this section we presented some empirical evidence on the performance of alternative monetary regimes in history. In many aspects, the evidence is consistent with Leijonhufvud's views.

First, under two convertible regimes, the classical gold standard and the convertible Bretton Woods regime, we observe, for both industrialised and less-developed high-inflation countries, greater stability of nominal variables – the inflation rate, nominal interest rates and nominal exchange rates – than under regimes characterised by loose convertibility or outright discretion.

Second, we observe lower inflation persistence under convertible regimes – evidence for the credibility of commitment to maintain the nominal anchor.

Third, the forecast errors of inflation and of real growth tend to be lowest during the convertible Bretton Woods regime. Also, forecast errors of inflation increase with time at a much greater rate under discretionary than under convertible regimes, as Leijonhufvud predicted. This result is most evident for the high inflation countries since the Second World War.

Fourth, the evidence on real variables is less transparent. There is no clear pattern for the G-11 countries showing that real variables perform worse under discretion or loose convertibility than under tight convertibility. For the high-inflation countries, however, this has not been the case since 1973, (see Heymann and Leijonhufvud, 1995).

Fifth, the evidence on demand and supply shocks for the G-11 countries suggests that it is unlikely that the convertible regimes prevailed because of the absence supply shocks, since the size of demand and supply shocks was quite similar across both types of regime. The durability or fragility of past convertible regimes probably had more to do with regime design

(Bordo, 1993b). For the high-inflation countries, demand shocks far exceeded supply shocks across all regimes – evidence for the operation of a discretionary regime, although supply shocks for these countries also were much larger than for the G-11. Greater supply shocks, in addition to traditionally higher fiscal deficits, may account for the greater prevalence in those countries of RWMS regimes.

CONCLUSION

Axel Leijonhufvud has made important contributions to our understanding of the relationship between the money supply process and inflation. These include his emphasis on the concept of a monetary regime as incorporating both monetary institutions and their link to the public's expectations; a reassessment of the costs of inflation; and new proposals for monetary reform.

Understanding the monetary regime for Leijonhufvud is crucial to understanding how alternative monetary strategies will impinge on both nominal and real activity. His analysis, by emphasising expectations, expands on the earlier concept of a monetary standard, which referred simply to the institutions and arrangements governing the money supply. The distinction made between constitutional and discretionary regimes represents a useful way of dichotomising monetary regimes in history. The world evolved over the century from 1881 from a tight, convertibility-based constitutional regime to a looser regime and then ultimately to a regime of pure discretion – the RWMS. The distinction between constitutional regimes and RWMS operationalises the earlier distinction made by Kydland and Prescott (1977) between rules and discretion, which emphasises the importance of credible rules in preventing monetary authorities from following time-inconsistent policies.

The concept of RWMS also adds insights to our understanding of the costs of inflation. The essence of the RWMS regime, because it is based on purely discretionary actions, is that it creates inflation uncertainty. Inflation uncertainty – which under the RWMS increases with time – in turn has the serious real consequence of distorting resource allocation. These costs, under conditions of high inflation, far outweigh the traditional neoclassical welfare losses of expected inflation.

Finally, to restore price stability and eliminate the RWMS, Leijonhufvud has advocated a number of proposals for reform. For low-to-moderate inflation he has suggested two schemes: a possible return to a commodity based convertibility rule; and the Peel-Friedman rule. The latter would combine the rule characteristics of a constant monetary growth rate with some flexibility to allow monetary authorities to deal with unexpected contingencies, without disturbing the expectations of the constitutional regime. For high-inflation countries, he proposed the blue-back scheme – issuing a parallel currency to appreciate in step with the expected depreciation of the existing currency – as a way of easing the transition to price stability. This scheme would operate in combination with an orthodox reform package of restoring budget balance and nominal exchange rate anchoring. Argentina incorporate bluebacking in its short-lived 1985 Austral Plan. Had it also followed the other prescriptions for reform and been a true regime shift as may have occurred with the convertibility plan of 1991, Leijonhufvud's proposal would have probably been successful. Historical experience suggests that the use of monetary institutions that are sharply isolated from domestic political pressure, such as a currency board, would be more successful in countries with inflationary histories than a central bank in reducing high inflation in the short run and maintaining a low rate in the long run. Inflation is a political phenomenon. To eliminate inflation after long periods of rapidly rising prices it is most promising to make fundamental changes in the monetary institutional framework: that is to create a new monetary regime.

In the fourth section of this chapter we presented a body of empirical evidence on regime performance in history which is very sympathetic to the distinctions made in Leijonhufvud's writings between types of regimes and their characteristics. We examine annual data for over a century for fifteen countries divided into industrialised and high-inflation economies. Our classification of historical regimes were the gold standard; inter-war; Bretton Woods preconvertible; Bretton Woods convertible; and floating. These correspond roughly to Leijonhufvud's distinction between convertibility-based regimes (the gold standard and Bretton Woods convertible); loose convertibility (inter-war, Bretton Woods preconvertible) and discretion or RWMS (the floating regime).

We find, as Leijonhufvud predicted, that inflation and other nominal variable (interest rates and nominal exchange rates) are more variable in loose convertible and discretionary regimes than under tight convertibility regimes. Also, they are more tightly integrated across countries, as predicted by traditional open-economy macro theory. In addition, inflation persistence is lower under the gold standard than in subsequent looser regimes – evidence consistent with Leijonhufvud's classification of regimes. Moreover, in strong support of Leijonhufvud, inflation forecast errors are greater under loose convertibility and discretion than under tight convertibility, and they increase most rapidly with the time horizon under the recent float. Also, in line with Leijonhufvud's work, we find little correlation between the budget deficit and inflation for the G-11 countries, but a marked one for the high-inflation countries. Finally, for the industrialised countries, since little difference between demand (policy induced) and supply (environment induced) shocks was detected, regime performance does not appear to be accounted for by the presence or absence of supply shocks. However, for the high-inflation countries, the response to larger supply shocks than in the G-11 may, along with higher fiscal deficits, explain the very high demand shocks which were detected in all regimes.

Finally, with respect to real performance for the G-11, little evidence was detected linking it to the nature of the regime. But for high inflation countries, as Leijonhufvud predicted, the shift from a convertibility influenced regime to high inflation in the 1970s is associated with a decline in economic growth.

In summary, Leijonhufvud's writings on monetary regimes provide a highly useful perspective for students of economics and economic history; his views on inflation suggest fruitful topics for empirical research, and his proposals for monetary reform are of great value to policy-makers.

Notes

1. See also Leijonhufvud (1986, 1987a, 1987b).
2. See also Grossman and Van Huyck (1988); Flood and Isard (1989); DeKock and Grilli (1992); Bordo and Kydland (1995).
3. See Redish (1993) for a discussion of the tension between maintaining the credibility of the gold convertibility nominal anchor and national monetary sovereignty.
4. For a discussion of the history and the problems of the Bretton Woods system, see Bordo (1993a).
5. For a discussion of the 'rules of the game' of the Bretton Woods system, see McKinnon (1993).
6. For a recent defence of the neoclassical theory, see Lucas (1993).
7. Both of these proposals were infact adopted as guidelines for Swedish monetary policy. See Jonung (1979).
8. See also Bruno (1993).
9. See, for example, the surveys by Bruno (1993), Paldam (1993) and Vegh (1992). As a rule, the output loss of a stabilisation programme to eliminate hyperinflation is smaller than the loss of reducing chronic or even rapid inflation.
10. See Hanke *et al.* (1993) and Hanke and Schuler (1991) on various techniques to insulate a currency board from external pressure.

11. However, the common world price level under the gold standard exhibited secular periods of deflation and inflation, reflecting shocks to the demand for and supply of gold (Bordo (1981), Rockoff (1984)). A well-designed monetary rule, it is argued, could have avoided the long-run swings that characterised the price level under the gold standard (Cagan, 1984).
12. The approach taken below follows that taken in Bordo (1993a and b), which focuses on regime performance for the G-7 countries.
13. We also examined the period (1946–73) which includes the three years of transition from the Bretton Woods adjustable peg to the present floating regime. The results are similar to those of the 1946–70 period.
14. For the performance of Latin-American countries during the classical standard, see Fishlow (1987) and Eichengreen (1992a); and for the performance of all four countries during Bretton Woods, see Edwards and Santaella (1993).
15. To be more exact, the USA stayed on gold until 1933 and France until 1936. For a detailed comparison of the performances of these three regimes in the inter-war period, see Eichengreen (1992a).
16. Within the G-11, Canada floated from 1950 to 1961.
17. This is a very crude measure of covergence or divergence between the different countries' summary statistics. Because it is based on the average for the whole period, it suppresses unusual movements within particular subperiods. Bayoumi and Eichengreen (1994b) presented an alternative measure of convergence GDP-weighted standard deviation of the individual country series around the G-7 aggregate. Bordo (1993a) calculated this alternative measure of convergence for the G-7 countries. The results are very close to those reported here for virtually every variable.
18. The data sources for Figure 9.1 and all subsequent figures are listed in the data appendix.
19. For similar evidence, see Bordo (1981); Darby *et al.* (1983); and Darby and Lothian (1989).
20. According to Edwards and Santaella (1993), the IMF had limited success during the Bretton Woods period in using access to its reserves as a weapon to constrain inflationary excesses in a number of developing countries.
21. Baxter and Stockman (1989) and Eichengreen (1993) use residuals from a linear trend to the logarithm of real output as a detrending filter rather than the logarithmic first difference used here. According to their results, real output variability is not greater in the floating than in the fixed period.
22. However, using their alternative measure of convergence – the GDP-weighted standard deviation of the individual country series around the G-7 aggregate – Bayoumi and Eichengreen (1994a) report that the lowest degree of dispersion of real GDP growth was in the floating rate period, followed by the Bretton Woods convertible period. Similar results hold for the real GNP per capita data in Table 9.1. For Bayoumi and Eichengreen (1994a) the decline in the dispersion of real growth and the rise in the dispersion of inflation rates between the Bretton Woods convertible period and the float have the following explanations: the move to flexible rates allowed countries to stabilise their relative growth rates in the face of asymmetric supply shocks at the expense of their relative inflation rates. They also report that,

when they apply the linear trend filter of Baxter and Stockman (1989), evidence of a rise in the cross-country correlation between output movements after 1970 is considerably reduced.

23. Define the real interest rate as $r_t = i_t - \Delta\log P_t$; where i_t is the nominal interest rate and $\Delta\log P_t = \log P_t - \log P_{t-1}$ is the percentage change in the consumer price index.
24. Also see Dornbusch (1976).
25. Stockman (1983, 1988) argues that greater variability in real exchange rates under floating rates than under fixed rates reflects the response of real exchange rates to productivity shocks, with changes in the real exchange rate producing nominal exchange rate volatility. This volatility is offset under fixed rates by exchange market intervention.
26. Meltzer (1990), in a comparison of EMS and non-EMS contries in the floating rate period, also finds a strong correlation between changes in nominal and real exchange rates.
27. See also Alogoskoufis (1992), who attributes the increase in persistence to the accommodation by the monetary authorities of shocks. This evidence is also consistent with the results of Klein (1975).
28. Regression run was $\Delta\log P_t = B_0 + B_1\Delta\log P_{t-1} + \epsilon_t$. We ran the same regression for the GNP deflators, with similar results.
29. Eichengreen (1993) also presents these statistics for four of the countries.
30. Meltzer and Robinson (1989) present their results for levels, growth rates and permanent growth rates of the series. We present only growth rates, to make the results comparable to those in Table 9.1.
31. Both variables were rendered stationary by first differencing.
32. Specifically, four restrictions are placed on the matrix of the shocks: two are simple normalisations, which define the variances of the shocks to aggregate demand and aggregate supply; the third assumes that demand and supply shocks are orthogonal; the fourth is that demand shocks have only temporary effects on output, that the cumulative effect of demand shocks on the rate of change in output must be zero.
33. See Keating and Nye (1991).
34. The impulse response functions were calculated from VARs run for the separate regime periods. Because the number of observations was limited, the Bretton Woods regime could not be split into the two sub-periods shown in preceding tables.
35. The results for the G-11 in the inter-war period figures are similar to those reported for the USA by Cecchetti and Karras (1992), who estimate a three-variable VAR with monthly data. The late 1920s and early 1930s reveal a major negative demand shock consistent with Friedman and Schwartz's (1963) attribution of the onset of the Great Depression to monetary forces. After 1931, negative supply shocks predominate, consistent with Bernanke's (1983) and Bernanke and James (1991) explanation for the severity of the Great Depression that stresses the collapse of the financial system.
36. Though the shocks are smaller, the rankings by regime for the weighted average of individual country shocks are similar to the G-11 aggregate.
37. These results are very similar to those presented for the G-7 in Bordo (1993b) and by Eichengreen and Bayoumi (1994a).

Data Appendix

For the G-7 countries, see data appendix in Bordo (1993a), except for the following series on government expenditures and revenues.

Canada

1880–1988, B. R. Mitchell (1993) *International Historical Statistics: The Americas, 1750–1988* (New York: Stockton Press).

France

1880–1988, B. R. Mitchell (1992) *International Historical Statistics: Europe, 1750–1988* (New York Stockton Press).

Germany

1880–1988, Mitchell (1992).

Italy

1880–1988, Mitchell (1992).

Japan

1880–1975, B. R. Mitchell, (1991) *International Historical Statistics: Asia* (New York: Stockton Press); 1976–90, *IFS Yearbook,* 1987 and 1992.

UK

1880–1988, Mitchell (1992).

USA

1880–1988, Mitchell (1993).

Additional Countries

Argentina

(1) Population: 1880–1988, B. R. Mitchell (1993); (2) Money: 1884–1913, G. della Paollera (1988). *How the Argentine Economy Performed During the International Gold Standard: A Reexamination*, Doctoral dissertation, University of Chicago; 1914–84: D. F. Cavallo and Y. Mundlak, *Estadisticas de la evolucion economica de Argentina 1913–1984*;

1985–90, *International Financial Statistics Yearbooks 1992*. (3) Real GDP: 1884–1913, G. della Paollera (1988); 1914–88, B. R. Mitchell (1993). (4) GDP deflator: 1884–1913, G. della Paollera (1988); 1914–88, B. R. Mitchell (1993): (5) GPI: 1913–84: D. F. Cavallo and Y. Mundlak; 1985–90, *International Financial Statistics Yearbook 1992*. (6) Exchange rate: 1884–1912, G. della Paollera (1988); 1913–84, D. F. Cavallo and Y. Mundlak ; 1985–90: *World Currency Yearbook*. (7) Government expenditures and revenues: 1880–1988, B. R. Mitchell (1993).

Belgium

(1) Population: 1880–1988, B. R. Mitchell (1992). (2) Money: 1919–88, *International Financial Statistics Yearbook*, 1973, 1987 and 1992. (3) Real GNP: 1913–88, B. R. Mitchell (1992). (4) GNP deflator: 1913–88, B. R. Mitchell (1992). (5) CPI: 1880–1988, B. R. Mitchell (1992). (6) Exchange rate: 1880–1914, supplied by Marc Flandreau, Stanford University; 1915–40, International Monetary Fund; 1948–90, *International Financial Statistics Yearbooks*, 1973, 1987 and 1992 . (7) Government expenditures and revenues: 1880–1988, B. R. Mitchell (1993).

Brazil

(1) Population: 1880–1988, B. R. Mitchell (1993). (2) Money: 1880–1988, IBGE (1990), *Estatisticas Historicas do Brasil: Series Economicas, Demograficas e Sociais de 1550 a 1988* (Rio de Janeiro: IBGE). (3) Real GDP: 1880–1988, B. R. Mitchell (1993). (4) GDP deflator: 1880–1988, B. R. Mitchell (1993). (5) CPI: 1912–88, B. R. Mitchell (1993). (6) Exchange rate: 1889–1989, M. de Paiva Abreu (ed.) (1990) *A Ordem do Progresso: Cem Anos de Politica Economica Republicana, 1889–1989* (Rio de Janeiro: Editora Campus). (7) Government expenditures and revenues: 1880–1988, B. R. Mitchell (1993).

Chile

(1) Population: 1880–1988, B. R. Mitchell (1993). (2) Money: 1880–94, A. L. Rodriguez (1991) *Chile During the Gold Standard: A Successful Paper Money Experience* ; B. R. Mitchell (1993). (3) Real GDP: 1880–94, A. L. Rodriguez (1991); 1909–88, B. R. Mitchell (1993). (4) GDP deflator: B. R. Mitchell (1993). (5) CPI: 1913–88, B. R. Mitchell (1993). (6) Exchange rate: 1880–1988, *Sinopsis Geografico: Estadisticas de la Republica de Chile* (annual series). (7) Government expenditures and revenues: 1880–1988, B. R. Mitchell (1993).

Israel

(1) Population: 1948–90, *International Financial Statistics Yearbooks*, 1973, 1987 and 1992 . (2) Money: 1948–90, *IFS Yearbooks*, 1973, 1987 and 1992. (3) Real GDP: 1948–90, *IFS Yearbooks*, 1973, 1987 and 1992. (4) GDP deflator: 1984–90, *IFS Yearbooks*, 1973, 1987 and 1992. (5) CPI: 1948–90, *IFS Yearbooks*, 1973, 1987 and 1992. (6) Exchange rate: 1948–90, *IFS Yearbooks*, 1973, 1987 and 1992. (7) Government expenditures and revenues: 1948–90, *IFS Yearbooks*, 1973, 1987 and 1992.

Netherlands

(1) Population: 1880–1988, B. R. Mitchell (1992). (2) Money: 1880–1988, *International Financial Statistics Yearbooks*, 1973, 1987 and 1992. (3) Real GNP: 1900–88, B. R. Mitchell (1992). (4) GNP deflator: 1900–1988, B. R. Michell (1992). (5) CPI: 1880–1988, B. R. Mitchell (1992). (6) Exchange rate: 1880–1914: Marc Flandreau Stanford University; 1915–40: International Monetary Fund; 1948–90: *IFS Yearbooks*, 1973, 1987 and 1992. (7) Government expenditures and revenues: 1880–1988, B. R. Mitchell (1992).

Sweden

(1) Population: 1880–1988, B. R. Mitchell (1992). (2) Money: 1880–1988, *International Financial Statistics Yearbooks*, 1973, 1987 and 1992. (3) Real GNP: B. R. Mitchell (1992). (4) GNP deflator: B. R. Mitchell (1992). (5) CPI: 1880–1988, B. R. Mitchell (1992). (6) Exchange rate: 1880–1914: Marc Flandreau, Stanford University; 1915–47: International Monetary Fund; 1948–90; *IFS Yearbooks*, 1973, 1987 and 1992. (7) Government expenditures and revenues: 1880–1988, B. R. Mitchell (1992).

Switzerland

(1) Population: 1880–1988, B. R. Mitchell (1992). (2) Money: 1919–88, *IFS Yearbooks*, 1973, 1987 and 1992. (3) Real GNP: 1913–88, B. R. Mitchell (1992). (4) GNP deflator: 1913–88, B. R. Mitchell (1992). (5) CPI: 1890–1988, B. R. Mitchell (1992). (6) Exchange rate: 1880–1914, Marc Flandreau, Stanford University; 1915–47: International Monetary Fund; 1948–90: *IFS Yearbooks*, 1973, 1987 and 1992. (7) Government expenditure and revenue: 1880–1988, B. R. Mitchell (1992).

References

Alogoskoufis, G. S. (1992) 'Monetary Accommodation, Exchange Rate Regimes and Inflation Persistence', *Economic Journal*, 2 May, pp. 461–80.

Alogoskoufis, G. S. and R. Smith, (1991) 'The Phillips Curve, the Persistence of Inflation and the Lucas Critique: Evidence from Exchange-Rate Regime,' *American Economic Review*. vol. 81, no. 2, pp. 1254–73.

Barro, R. J. and D. B. Gordon (1983) 'Rules, Discretion and Reputation in a Model of Monetary Policy', *Journal of Monetary Economics*, vol. 12, pp. 101–21.

Barsky, R. B. (1987) 'The Fisher Hypothesis and the Forecastability and Persistence of Inflation', *Journal of Monetary Economics*, vol. 19, no. 1, January, pp. 3–24.

Baxter, M. and A. C. Stockman, (1989) 'Business Cycles and the Exchange-Rate Regime: Some International Evidence', *Journal of Monetary Economics*, vol. 23, May, pp. 377–400.

Bayoumi, T. and B. Eichengreen (1992) 'Is there a Conflict between EC Enlargement and European Monetary Unification?', NBER Working Paper No. 3950, January.

Bayoumi, T. and B. Eichengreen (1993) 'Shocking Aspects of European Monetary Unification'. in Francesco Torres and Francesco Giavazzi (eds), *Adjustment and Growth in European Monetary Union* (Cambridge: Cambridge University Press)

Bayoumi, T. and B. Eichengreen (1994a) 'Economic Performance under Alternative Exchange Rate Regimes: Some Historical Evidence'. in: Peter B. Kenen, Francesco Papadia and Fabrizio Saccomanni (eds.), (1994), pp. 257–97) *The International Monetary System* (Cambridge: Cambridge University Press).

Bayoumi, T. and B. Eichengreen (1994b) 'Monetary and Exchange Rate Arrangements for NAFTA'. *Journal of Development Economics*. Vol. 43, pp. 125–65

Bernanke, B. (1983) 'Non-Monetary Effects of the Financial Crisis in the Propagation of the Great Depression', *American Economic Review*, vol. 73, pp. 259–76.

Bernanke, B. and H. James (1991) 'The Gold Standard, Deflation and Financial Crisis in the Great Depression: An International Comparison', in R. Glenn Hubbard (ed.), *Financial Markets and Financial Crisis* (Chicago: University of Chicago Press), pp. 33–68.

Blanchard, O. and D. Quah (1989) 'The Dynamic Effects of Aggregate Demand and Aggregate Supply Disturbances', *American Economic Review*, vol. 79, September, pp. 655–73.

Bordo, M. D. (1981) 'The Classical Gold Standard: Some Lessons for Today', *Federal Reserve Bank of St. Louis Review*, vol. 63, May, pp. 2–17.

Bordo, M. D. (1984) 'The Gold Standard: The Traditional Approach', in M. D. Bordo and A. J. Schwartz (eds), *A Retrospective on the Classical Gold Standard, 1821–1931* (Chicago: University of Chicago Press).

Bordo, M. D. (1993a) 'The Bretton Woods International Monetary System: A Historical Overview', in M. D. Bordo and B. Eichengreen (eds), *A Retrospective on the Bretton Woods System: Lessons for International Monetary Reform* (Chicago: University of Chicago Press).

Bordo, M. D. (1993b) 'The Gold Standard, Bretton Woods and Other Monetary Regimes: A Historical Appraisal', *Federal Reserve Bank of St. Louis Review*, vol. 75, no. 2, March/April, pp. 123–91.

Bordo, M. D. and F. E. Kydland (1992) 'The Gold Standard As a Rule: An Essay in Exploration', *Explorations in Economic History* vol. 32, pp. 423–64.

Bordo, M. D. and A. J. Schwartz (1989) 'Transmission of Real and Monetary Disturbances Under Fixed and Floating Rates', in J. A. Dorn and W. A. Niskanen (eds), *Dollars, Deficits and Trade* (Boston Mass.: Kluwer Academic Publishers), pp. 237–58.

Bruno, M. (1993) *Crisis, Stabilization and Economic Reform* (Oxford University Press).

Cagan, P. (1984) 'On the "Report of the Gold Commission 1982 and Convertible Monetary System"', *Carnegie Rochester Conference Series on Public Policy*, no. 21, Autumn.

Cecchetti, S. G. and G. Karras, (1992) 'Sources of Output Fluctuations during the Interwar Period: Further Evidence on the Causes of the Great Depression', NBER Working Paper No. 4049, April.

Cukierman, A. (1992) *Central Bank Strategy, Credibility, and Independence: Theory and Evidence* (Cambridge Mass.: MIT Press).

Darby, M. R. and J. R. Lothian (1989) 'The International Transmission of Inflation Afloat', in M. D. Bordo (ed.), *Monetary, History, and International Finance: Essays in Honor of Anna J. Schwartz* (Chicago: University of Chicago Press), pp. 203–36.

Darby, M. R. and J. R. Lothian *et al.* (1983) *The International Transmission of Inflation* (Chicago: University of Chicago Press).

DeKock, G. and V. Grilli (1992) 'Fiscal Policies and the Choice of Exchange Rate Regime', CEPR Discussion Paper No. 631, March.

Dornbusch, R. (1976) 'Expectations and Exchange Rate Dynamics', *Journal of Political Economy* , vol. 84, pp. 1161–76.

Eichengreen, B. (1992a) 'Trends and Cycles in Foreign Lending', in Horst Siebert (ed.), *Capital Flows in the World Economy* (Kiel: World Institute of World Economics), pp. 3–28.

Eichengreen, B. (1992b) 'The Gold Standard Since Alex Ford'. in S. N. Broadberry and N F R Crafts (eds), *Britain in the International Economy 1870–1939* (Cambridge University Press), pp. 49–79.

Eichengreen, B. (1992c) 'History of the International Monetary System: Implications for Research in International Macroeconomics and Finance', in F. van der Ploeg (ed.), *Handbook of International Macroeconomics* (Oxford: Basil Blackwell).

Eichengreen, B. (1993) 'Three Perspectives on the Bretton Woods System', in M. D. Bordo and B. Eichengreen (eds), *A Retrospective on the Bretton Woods System* (Chicago. University of Chicago Press and NBER).

Edwards, S. and J. Santaella (1993) 'Devaluation Controversies in the Developing Countries: Lessons from the Bretton Woods Era', in M. D. Bordo and B. Eichengreen (eds), *A Retrospective on the Bretton Woods System: Lessons for International Monetary Reforms* (Chicago: University of Chicago Press), pp. 405–55.

Fishlow, A. (1987) 'Market Forces or Group Interests: Inconvertible Currency in Pre-1914 Latin America', University of California at Berkeley.

Flood, R. P. and P. Isard (1989) 'Monetary Policy Strategies', *IMF Staff Papers*, vol. 36, pp. 612–32.

Friedman, M. (1960) *A Program for Monetary Stability* (New York: Fordham University Press).

Friedman, M. and Schwarts, A. J. (1963) *A Monetary History of the United States, 1867–1960* (Princeton, NJ: Princeton University Press).

Giavazzi, F. and M. Pagano (1988) 'The Advantage of Tying One's Hands: EMS Discipline and Central Bank Credibility', *European Economic Review*, vol. 32, pp. 1055–82.

Giovannini, A. (1993) 'Bretton Woods and Its Precursors: Rules Versus Discretion in the History of International Monetary Regimes', in M. D. Bordo and B. Eichengreen (eds), *A Retrospective on the Bretton Woods System*. (Chicago: University of Chicago Press).

Grossman, H. J. and J. B. Van Huyck (1988) 'Sovereign Debt as a Contingent Claim: Excusable Default, Repudiation, and Reputation', *American Economic Review* vol.78 pp. 1088–97.

Grilli, V. and G. Kaminsky, (1991) 'Nominal Exchange Rate Regimes and the Real Exchange Rate: Evidence from the United States and Great Britain, 1885–1986', *Journal of Monetary Economics*, vol. 29, April, pp. 191–212.

Hall, R. E. (1982) 'Explorations in the Gold Standard and Related Policies for Stabilizing the Dollar', in Robert E. Hall (ed.), *Inflation: Causes and Effects* (Chicago: University of Chicago Press), pp. 111–22.

Hanke, S. H., and K. Schuler (1991) *Banco Central o Caja de Conversión*? (Buenos Aires: Fundacion Republica).

Hanke, S. H. L. Jonung and K. Schuler (1993) *Russian Currency and Finance. A Currency Board Approach to Finance* (London and New York: Routledge).

Heymann, D. (1987) 'The Austral Plan', *American Economic Review*. vol. 77, no. 2, pp. 284–7.

Heymann, D. (1991) 'From Sharp Disinflation to Hyperinflation, Twice: The Argentine Experience, 1985–1989', in M. Bruno, S. Fisher, E. Helpman and N. Liviatan (eds), *Lessons of Economic Stabilization and its Aftermath* (Cambridge: MIT Press).

Heymann, D. and A. Leijonhufvud (1995) 'High Inflations' (Oxford: Clarendon Press).

Jonung, L. (1979) 'Knut Wicksell's Norm of Price Stabilization and Swedish Monetary Policy in the 1930s', *Journal of Monetary Economics*, vol. 5, no. 4, October, pp. 459–96.

Keating, J. W. and J. V. Nye (1991) 'Permanent and Transitory Shocks in Real Output: Estimates from Nineteenth Century and Postwar Economies' St Louis: Washington University Working paper No. 160, July.

Klein, B. (1975) 'Our New Monetary Standard: Measurement and Effects of Price Uncertainty, 1880–1973', *Economic Inquiry*, vol. 13, pp. 461–84.

Kydland, F. E. and E. Prescott (1977) 'Rules Rather than Discretion: The Inconsistency of Optimal Plans', *Journal of Political Economy*, vol. 85, pp. 473–91.

Leijonhufvud, A. (1984a) 'Constitutional Constraints on the Monetary Powers of Government', in R. B. McKenzie (ed.), *Constitutional Economics*, (Lexington, Mass.: Lexington Books), pp. 95–107.

Leijonhufvud, A. (1984b) 'Inflation and Economic Performance', in Barry N. Seigel (ed.) *Money in Crisis: The Federal Reserve, the Economy and Monetary Reform* (Cambridge, Mass.: Ballinger) pp. 19–36.

Leijonhufvud, A. (1986) 'Rules with some Discretion', in C. D. Campbell and W. R. Dougan (eds), *Alternative Monetary Regimes*, (Baltimore Md: Johns Hopkins Press), pp. 38–43.

Leijonhufvud, A. (1987a) 'Rational Expectations and Monetary Institutions', in *Monetary Theory and Economic Institutions: Proceedings of a Conference held by the International Economic Association of Fiesole, Florence, Italy* (New York: St, Martin's Press), pp. 44–65.

Leijonhufvud, A. (1987b) 'Constitutional Constraints on the Monetary Powers of Government', in J. A. Dorn and A. J. Schwartz (eds), *The Search for Stable Money: Essays on Monetary Reform* (Chicago: University of Chicago Press), pp. 129–42.

Leijonhufvud, A. (1988) 'Foreword' to T. D. Willet. (ed.), *Political Business Cycles: The Political Economy of Money, Inflation and Unemployment, Pacific Research Institute for Public Policy* (Durham, NC: Duke University Press), pp. xv–xxii.

Leijonhufvud, A. (1990) 'Monetary Policy and the Business Cycle Under Loose Convertibility', *Greek Economic Review*, vol. 12, September. pp. 133–51.

Leijonhufvud, A. (1992) 'High Inflations and Contemporary Monetary Theory', *Economic Notes by Monte dei Paschi di Siena*, vol. 21, no. 2, pp. 211–24.

Levin, A. (1993) 'Exchange Rate Convertibility and Relative Price Adjustment in Argentina, 1991–1993', (mimeo) International Finance Division, Federal Reserve Board of Governors, Washington, DC, May.

Lucas, R. E., Jr (1993) 'On the Welfare Cost of Inflation' (mimeo), University of Chicago, July.

McKinnon, R. I. (1988) 'An International Gold Standard Without Gold', *Cato Journal*, vol. 8, Fall, pp. 351–73.

McKinnon R. I. (1993) 'International Money in Historical Perspective', *Journal of Economic Literature*, vol. xxxi, no. 1, March, pp. 1–44.

Meltzer, A. H. (1990) 'Some Empirical Findings on Differences between EMS and non-EMS Regimes: Implications for Currency Blocs', *Cato Journal*, 10 November, pp. 455–83.

Meltzer, A. H. and S. Robinson, (1989) 'Stability under the Gold Standard in Practice', in M. D. Bordo (ed.), *Monetary History and International Finance: Essays in Honor of Anna J. Schwartz*. (Chicago: University of Chicago Press), pp. 163–95.

Mussa, M. (1986) 'Nominal Exchange Rates and the Behavior of Real Exchange Rates: Evidence and Implications', *Carnegie Rochester Conference Series on Public Policy*, vol. 25. pp. 117–214.

Osband, K. and D. Villanueva (1992) 'Independent Currency Authorities: An Analytic Primer', IMF Working Paper, July.

Paldam, M. (1993) 'The Political Economy of Stopping High Inflation' (mimeo), Arhus, Denmark: Institute of Economics.

Redish, A. (1993) 'Anchors Away: The Transition from Commodity Money to Fiat Money in Western Economies', Innis Lecture at the Canadian Economics Association Annual Meeting, Ottawa, June.

Rockoff, H. (1984) 'Some Evidence on the Real Price of Gold. Its Cost of Production, and Commodity Prices', in M. D. Bordo and A. J. Schwartz (eds), *A Retrospective on the Classical Gold Standard, 1821–1931* (Chicago: University of Chicago Press).

Rostowski, J. and J. Shapiro "Secondary Currencies in the Russian Hyperinflation and Stabilization of 1921–1924', Discussion Paper No. 59, Centre for Economic Performance, London.

Santaella, J. A. (1993) 'Stabilization Programs and External Enforcement: Experience from the 1920s', IMF Working Paper, January.

Sargent, T. J. (1982) 'The Ends of Four Big Inflations', in R. E. Hall (ed.), *Inflation: Causes and Effects* (Chicago: University of Chicago Press) pp. 41–97.

Schwartz, A. J. (1984) 'Introduction', in M. D. Bordo and A. J. Schwartz (eds), *A Retrospective on the Classical Gold Standard, 1821–1931.* (Chicago: University of Chicago Press).

Schwartz, A. J. (1993) 'Currency Boards – Their Past, Present, and Possible Future Role', *Carnegie-Rochester Conference Series on Public Policy*, vol. 39, December, pp. 147–87.

Siklos, P. (1993) 'Tales of Parallel Currencies: The Early Soviet Experience,' (mimeo), Wilfred Laurier University, May.

Stockman, A. C. (1983) 'Real Exchange Rates under Alternative Nominal Exchange Rate System', *Journal of International Money and Finance*, vol. 2 August, pp. 147–66.

Stockman, A. C. (1988) 'Real Exchange-Rate Variability under Pegged and Floating Nominal Exchange-Rate System: An Equilibrium Theory', *Carnegie Rochester Conference Series on Public Policy*, vol. 29, Autumn, pp. 259–94.

Svensson, L. (1993) 'The Simplest Test of Inflation Target Credibility', NBER Working Paper No. 4604, December.

Vegh, C. (1992) 'Stopping High Inflation: An Analytical Overview', *IMF Staff Papers*, vol. 39, no. 3, September, pp. 621–95.

Wood, G. (1994) 'Central Bank Independence in New Zealand: Analytical, Empirical and Institutional Aspects', City University of London Business School (mimeo).

10 Financial Innovation and the Price Level: Wicksell and the New Monetary Economics[1]

Luis Hernandez-Aramburo

> Money of account, namely that in which debts and prices and general purchasing power are expressed, is the primary concept of a theory of money. (J. M. Keynes)

Axel Leijonhufvud has a deep sense of the meaning of history. About a man who can trace his genealogical roots to the Middle Ages, this observation is not suprising. He is also a rebel, an attitude for which I do not have a good explanation. These two features of his character help to explain why he usually approaches the labyrinths of economic theory with a sceptical eye and through some historical exploration of the subject. In this type of criticism, our profession regards him as a consummate master. In this chapter I try to bring into the foreground Axel's passionate belief in the uses of the past. It is only too evident from this work why I consider him to be 'my grandparent in the paths of errancy'.

FINANCIAL INNOVATION AND MONETARY THEORY

The international financial system has experienced a deep transformation in the past twenty-five years. The main causes of this transformation have been the disappearance of traditional borderlines among financial intermediaries, and technological advances which facilitated the world-wide transmission of capital flows as well as the creation of sophisticated financial instruments. These developments, in turn, have stimulated new ways of thinking about monetary policy. Among them, new monetary economics stands out for its polemical assertions.

The new monetary economics (NME) is not a school of thought in the proper sense of the word. Rather, it is a label given to a series of papers by

Fischer Black, Eugene Fama, Robert Greenfield, Robert Hall and Leland Yeagar that share a common view of the relationship between financial regulation and the stability of the price level. The basic tenet of the NME is that an economy in which institutional and technological developments render unnecessary the use of money as a physical medium of exchange is characterised by its price stability. Consequently, any regulation imposed on financial intermediaries for the sole purpose of controlling the price level is not only unnecessary, it is also harmful. It interferes with the market mechanism. A corollary of this proposition is that if the economy is not completely devoid of cash, the separation of the medium of exchange and the standard of value enhances the stability of the price level. In this case too, government regulation of financial intermediaries would be superfluous in principle and most probably damaging in practice.

One would presume that these controversial assertions have been the subject of a careful analysis. Surprisingly, this has not been the case. Consider, for example, the reviews by White (1984b) and McCallum (1985), by far the most popular and influential assessments of the NME. Neither author disputes that the price level would be anchored properly in a pure bookkeeping economy. Rather, they question whether it is feasible to separate the medium of exchange and the standard of value. Therefore, they never address the problem at the core of the NME, namely whether financial innovation increases or decreases the stability of the price level.

The analogy often drawn between Knut Wicksell's pure credit economy and Eugene Fama's bookkeeping economy provides a further example that the claims of the new monetary economics have not been properly assessed:

> The rapid development during the 1960s and 1970s of computer and telecommunications technologies has led some writers [for example, Fama, 1980] to contemplate economies – anticipated by Wicksell – in which virtually all purchases are effected not by the transfer of a tangible medium of exchange, but by means of signals to an accounting network ... If there were literally *no* medium of exchange, the wealth accounts being claims to some specified bundle of commodities, *the economy in question would be properly regarded and analyzed as a non-monetary economy, albeit one that avoids the inefficiencies of crude barter.* (McCallum and Goodfriend (1987); italics added)

There are indeed strong similarities between the economies envisioned by Wicksell and by Fama. Both of them try to understand the basic link between the development of credit institutions and the stability of the price system. However, McCallum and Goodfriend fail to emphasise that Fama and Wicksell arrive at strikingly opposing conclusions. According to Fama, the price level would be anchored firmly in a cashless economy, whereas for Wicksell the slightest disturbance coming from the real sector would lead to a cumulative inflationary process. Moreover, Wicksell explicitly rejects the proposition that a pure credit economy should be regarded and analysed as a frictionless barter economy.

This chapter attempts to sort out the differences between Wicksell and the NME, to shed light on the discussion of price stability in a cashless economy. This issue is important in any attempt to assess the NME's case for financial deregulation. The heart of the problem is to determine whether a cashless economy replicates the behaviour of a barter economy. To compare both economies one can follow two types of criteria. One of them, advocated by the NME, stresses the nature or backing of the standard of value. The other (Wicksell's) is based on the ability of each economy to co-ordinate saving and investment decisions.

The NME's approach is basically in tune with recent monetary theory. This observation could lead one to believe that the NME has the best part of the argument, and that it provides superior criteria to Wicksell's. This chapter does not support such conclusions. In fact, the arguments of Wicksell and the NME have never been confronted. Wicksell's contribution has been largely ignored. To explain why, it is useful to follow Leijonhufvud's (1991) methodological suggestion and to conceive of the development of monetary theory as a sequence of decisions, of choices leading up to the frontiers. Backtracking into the past, that is, following the evolution of particular debates in economics, allows one to understand why some approaches were abandoned and how they might shed light on the theoretical problems of the present.

For the problem at hand, it is convenient to reconsider the work of Don Patinkin. Patinkin represents a bifurcation, a fork in the road, in the discussion of price stability. His discussion of the determinacy of the price level set an agenda one of whose multiple branches evolved into the NME. At the same time, his interpretation of Wicksell dismissed the problem of intertemporal co-ordination in the determination of the price level. To regain a perspective of Wicksell's work and to explain why it represents an alternative to the NME, the rest of this chapter is organised into six sections. The first section discusses the problem of price level determinacy in Patinkin and the NME. The second points out some shortcomings or previous criticisms of the NME, and the third discusses Patinkin's interpretation of Wicksell's cumulative process. The fourth section criticises this interpretation and contrasts the approaches of Wicksell and the NME to be a problem of co-ordinating savings and investment. The fifth section looks into the issue of separating the standard of value and the medium of exchange, and the final section concludes with some remarks on the shortcomings of Wicksell's and the NME's analysis from the point of view of portfolio theory.

THE NEW MONETARY ECONOMICS AND THE DETERMINACY OF THE PRICE LEVEL

Price level determinacy first played a major role in the work of Patinkin. In his path-breaking work, *Money Interest and Prices* (1965 (2nd edn)), he follows Hicks (1935) in search of an integration of monetary and value theory. His main point was to demonstrate that the demand for money and the demand for other commodities had to be derived simultaneously. In other words, Patinkin criticised the use of two different theories, the quantity theory and general equilibrium theory, to determine the price level independently of relative prices.

Patinkin forcefully argued that the separation, or dichotomy, of the monetary and the real sectors had been justified incorrectly by appealing to the homogeneity postulate and the neutrality of money. He pointed out that if *all relative prices* increase proportionally, as stated by the homogeneity postulate, then *all commodities* become simultaneously more valuable. However, this is not possible: an increase in relative prices implies that at least one commodity, the standard, becomes less valuable. Patinkin concluded that this logical gap could only be bridged if the increase of *all* prices takes place in terms of a standard which is a fictitious commodity. The validity of the homogeneity postulate is thus limited to an *abstract* or *accounting* price system. Nevertheless logical consistency is gained at the expense of economic intuition, since accounting prices have no meaning for market participants, Patinkin (1965, p. 16)

Patinkin noticed that *money* prices are in effect *abstract* prices in general equilibrium models, because money does not belong to the commodity space. Therefore, as in the case of any other abstract unit, the price level in terms of money can be set at any arbitrary value. It cannot then be said that the quantity theory determines the price level. In order to pin down the price level, money must be integrated into value theory. Money must be regarded as one more commodity whose relative price in terms of the other goods has to be determined.

Patinkin (1965) resorted to the introduction of real money balances into the utility function to accomplish the integration of monetary and value theory. The stability of the price level is guaranteed by a wealth or real balance effect. Nevertheless, as John Gurley and E. C. Shaw pointed out: in a credit economy, Patinkin's approach has serious shortcomings. The change in the price level will only induce a wealth effect in Patinkin's model if the stock of money consists of outside money, that is, if money represents a net external or outside claim from the standpoint of the private sector. Otherwise, changes in the price level only redistribute wealth between debtors and creditors, Gurley and Shaw (1960, p. 73). Since the net worth of private debt must be positive for the wealth effect to be present, they labeled Patinkin's approach the 'net-money doctrine'.

The problem with the net-money doctrine is that the evolution of the financial system is characterised by rapid growth of inside money relative to outside money, Gurley and Shaw (1955). As the share of outside money in the money supply declines, the strenght of the wealth effect fades away. At the limit, outside money disappears. Consequently, there is no mechanism to pin down the price level:

> Net-money doctrine, as we have said, cancels out all private domestic debt against its counterparts in monetary and non-monetary form, so that when an economy contains no outside money or 'foreign' securities it becomes money-less and bond-less. *It is in effect a barter society, without a determinate price level.* (Gurley and Shaw (1960), p. 142; italics added)

Following Gurley and Shaw, Black (1970) criticised conventional monetary theory for its narrow focus on outside money; but, unlike them, he interpreted the parallelism between a barter and financial economy, and the concomitant indeterminacy of the price level, as a source of important insights:

> I maintain the views expressed by Vickery, Tobin, Gurley and Shaw, and Patinkin are the correct ones. In a world without controls on banking, the real sector will be independent of the financial sector, and the price level will be indeterminate. Traditional monetary theories will be inapplicable; in fact it will not be possible to define the quantity of money in meaningful terms. Finally, I claim this world will have several advantages, and few obvious disadvantages over our present economic and monetary system. (Black (1970), p. 10)

Black's main argument is that, if there is no outside money the money supply is not well defined. To analyse the mechanics of a world without outside money, Black conceives of an economy where money as a physical medium of exchange does not exist. In this economy, the major financial institutions are banks. All payments are made by cheque and there is a major corporation in charge of the clearing system. In fact, one could imagine that all payments are made through an electronic network. This electronic money functions as a medium of exchange; however, it would be impossible to reduce it to an unambiguously measurable concept. In this situation, 'there is nothing ... that can meaningfully be called a quantity of money' (Black (1970), p. 12). If money does not exist as a well-defined concept, there is no money the central bank can 'target' to influence the price level. Consequently, restrictions on the banks' activities such as the types of deposit banks can accept, the interest rate they can pay, or the minimum amount of reserve that they must keep, should be removed. This process of deregulation would certainly generate innovations, but not of the type that would lead to arbitrary jumps in prices, or to a drifting price level (Black (1970), p. 15).

Fama's two seminal papers (1980, 1983) extended Black's efforts to explain why 'separating price level control from financial intermediation is *feasible* and *desirable*' (Fama (1983), p. 21; italics added). In Fama's view, banks, the representative financial institutions, fulfil two major functions: providing transaction services, and portfolio management services. The thrust of his argument is that neither of these functions disturbs the stability of prices. For examples, the activity of portfolio management places banks at the centre of the saving and investment process. However, according to the Modigliani–Miller theorem, in a competitive equilibrium, firms' investment decisions are independent of financial decisions. Banks, or any other financial institutions, play a passive role because they cannot, by themselves, disturb the equilibrium between savings and investment. Consequently, they are not a threat to price stability. On the other hand, the transaction services performed by banks in a bookkeeping economy are purely an accounting exercise. Banks only record debit and credit operations in their clients' accounts. In both cases, there is nothing special about the demand and supply for banks' services that would upset price stability.

Although both Fama and Black argue that the financial system cannot generate any inflationary pressure by itself, they do differ on the issue of the determinacy of the price level. For Black, as well as for Gurley and Shaw, the price in a cashless economy is indeterminate. The financial and real sectors are completely independent. This reduces a cashless economy to a barter economy in which, as Patinkin notes, the money price level is undefined. By contrast, in Fama's view the price level is determined precisely because the cashless and the barter economy are so alike. He argues that if all money is inside money, banks are essentially mutual funds. All deposits are claims to a portfolio backed by some real asset. Therefore, the price level must be expressed ultimately in terms of one of the economy's real goods:[2]

> Since the economy in which we have embedded our competitive unregulated banking system is basically non-monetary, with some real good serving as a numeraire, price level determinacy reduces to a standard problem concerning the existence of a *stable* general equilibrium in a *non-monetary* system. (Fama (1980) p. 44; italics added)

In his arguments, Fama emphasises that the function of money relevant to the determination of the price level is the standard of value, not the medium of exchange. This observation corrects Black's assertion that the price level is indeterminate when the 'money supply does not exist'. The definition of the media of exchange, or its multiplicity, on which Black's criticism ultimately rests, is in principle a phenomenon independent of the definition of the standard.

A further step in the development of the NME arose out of Fama's remark that the *numeraire never enters the exchange mechanism.* To fully understand this claim, it is convenient to analyse the process of exchange in a bookkeeping economy:

> The general nature of an accounting system of exchange merits further illustration. Consider a market for securities like the New York Stock Exchange. Individuals keep accounts with brokers who trade on their behalf. Suppose a client signals his brokers to trade his holdings of security X for holdings of security Y. The broker need not find another securityholder who wants to both buy X and sell Y; that is, the desired transaction does not require a double coincidence of wants. The broker simply enters bids to sell X and buy Y and the market's quotation system solves the problem of finding a buyer for X and a separate seller of Y. At the end of each day, the brokers in the system make the appropriate accounting transfers to settle the day's chain of transactions. In short, as in any system of exchange, the combination of an efficient system for collecting bids and establishing security prices, along with an accounting system for completing transactions, does away with the need for a physical medium of exchange to settle transactions. (Fama (1983), p. 10)

Thus when one economic unit wishes to transfer a given amount of wealth to another, he signals his broker–banker with a check or some

> other modern way of accessing the bank's bookkeeping system ... All prices, including prices of securities, are stated in terms of a numeraire, which we have assumed is one of the economy's real goods, *but the numeraire never appears physically in the process of exchange described above.* The essence of the accounting system of exchange is that it operates through debits and credits, which do not require any physical medium. (Fama (1980), p. 42)

Fama stresses the independence of the medium of exchange and the standard of value to underline that the current regulatory framework directed at controlling the media of exchange is unnecessary to stabilise the price level. In fact, it is detrimental to the efficiency of the financial system. Greenfield and Yeager (1983) find in the union of the medium of exchange and the standard of value an extra source of economic instability. They identify the problem as follows:

> Market-clearing forces do not work very well to maintain or restore equilibrium between money's supply and demand because *money does not have a single price of its own that can adjust on a market of its own.* Instead the medium of exchange has a fixed price in the unit of account (each dollar of the money supply has a price exactly $1). With no specific price and market to impinge upon, imbalance between money's supply and demand must operate upon the dollar's purchasing power, that is, in the whole general price level. *This process requires adjustments in the markets and in the price of millions of individual goods and services,* leaving scope for quantities traded and produced to be affected. (Greenfield and Yeager (1983), p. 309; italics added)

Separating the medium of exchange from the standard of value corrects any imbalance in the supply and demand for money by changing the price of the medium of exchange in terms of the standard. Robert Hall, and Greenfield and Yeager, explore different mechanisms to separate the medium of exchange from the standard of value. The feasibility of their proposals is analysed in Cowen and Krozner (1994).

SOME CRITICISMS OF THE NEW MONETARY ECONOMICS

Most assessments of the New Monetary School rely in one way or another on the reviews by White (1984) and McCallum (1985). Nevertheless, these reviews overlooked some of the most controversial and interesting issues raised by the NME. The NME can be summarised succinctly in the following propositions:

1. A bookkeeping economy and a barter economy share the same price stability properties. There is no need to impose regulations to anchor the price system.
2. In a bookkeeping economy, the standard plays no role in the exchange of other goods.
3. In a cash economy, the separation of the medium of exchange from the standard of value would improve macroeconomic stability.

McCallum and White only consider whether the medium of exchange can be divorced from the standard of value, a problem related to point 2 in my list. According to them, the *feasibility* of this separation must be shown before its *desirability*, McCallum (1985) or *plausibility*, White (1984, p. 699) is discussed. In their opinion, one of the difficulties in analysing the separation of the medium of exchange from the unit of account is the lack of a theory to explain how the standard is chosen. In this regard,McCallum makes the following observation:

> there is nothing in this analysis or in most monetary analyses that determines the choices of the numeraire. Yet in talking about the price level, we are typically asking what determines prices in terms of the numeraire. (Fischer (1983), p. 8)

The importance of this criticism is that no member of the NME has clarified how agents choose a specific unit of account among many possible candidates, or how the authorities can induce, or force, the choice of one specific numeraire. If this selection process is not fully specified, the separation of the medium of exchange from the standard of value cannot be implemented. On the other hand, this criticism does not explain either why the standard of value must be the same as the medium of exchange. To this end, McCallum invokes the following 'substantive proposition':

> unless provided with inducements to do otherwise, *sellers will quote prices in terms of the medium of exchange* ... The validity of this proposition is *strongly* supported by the evolutionary and historical analysis presented by White (1984). (McCallum (1985), p. 17; italics added)

Belief in the truth of this proposition led both authors to reject the feasibility of the NME and to bypass any analysis of points 1 and 3 mentioned above. The problem with this 'substantive proposition' lies in the presumption that the medium of exchange is primary to all other functions of money (McCallum (1985), p. 16; White (1989), p. 4). Nevertheless, neither on analytical nor on historical grounds is it possible to conclude that the function of the standard of value can be subsumed into the function of the medium of exchange. Analytically, the need for a common standard is as pressing as that of a medium of exchange. Agents cannot engage in economic activities if the heterogeneity of standards prevents them from computing the value of transactions. Furthermore, the historical evidence is not as supportive of White's and McCallum's stand as they presume. For example, in his work on primitive money, Einzig (1949, p. 368) concludes: 'Allowing for all considerations, on balance it seems probable that more often than not the standard of value preceded the medium of exchange.'

The preceding arguments suggest that the problem of feasibility should not deter a comprehensive analysis of the NME. This issue is far from settled. In addition, the NME raises important analytical questions which go beyond the problem of feasibility. For example, should a bookkeeping economy in fact be regarded as a barter economy? Is it necessary – or is it a wise precaution – to regulate financial intermediaries to guarantee price stability? Finally, if a bookkeeping economy were feasible, would it be more susceptible to inflationary bursts than one with a physical medium of exchange? In the following, it will be argued that the work of Wicksell represents an alternative paradigm to shed light on these questions and is therefore a better starting point for a critical examination of the NME.

Wicksell's declared aim in *Interest and Prices* (1898) and *Lectures on Political Economy Vol. II* (1906) was to explain the deflationary trend of prices in the second half of the nineteenth century. He felt that this phenomenon could not be explained satisfactorily by the quantity theory. In particular, in view of the trend towards cashless systems, he considered the assumption of a constant velocity of circulation to be unacceptable. In Wicksell's opinion, money as a medium of exchange possesses a quantity dimension and a velocity dimension and therefore, the influence of both factors on the price level must be taken into account. Wicksell proceeded accordingly to examine two extreme cases: a cash economy and a credit economy. He justified his approach as follows:

> We intend therefore, as a basis for the following discussion, to imagine a state of affairs in which money does not actually circulate at all, neither in the form of coins (except perhaps as small change) nor in the form of notes, but where domestic payments are effected by means of

> the *Giro* system and bookkeeping transfers. A thorough analysis of this purely imaginary case seems to me to be worth while, for it provides a precise antithesis to the equally imaginary case of a pure cash system, in which credit plays no part whatever. The monetary systems actually employed in various countries can then be regarded as *combinations* of these two extreme types. If we can obtain a clear picture of the causes responsible for the value of money in *both* of these imaginary cases, we shall, I think, have found the right key to a solution of the complications which monetary phenomena exhibit in practice. (Wicksell (1898), pp. 70–1)

Wicksell's intention was to carry out a thought experiment, an abstraction of the forces he observed at work. Fama's work should be viewed in a similar light. From this perspective, the most important shortcoming of McCallum's and White's analyses is that they did not pursue this thought experiment. They stopped short at the problem of feasibility and did not go further to consider the lessons from these imaginary cases.

PRICE LEVEL DETERMINACY AND THE STANDARD OF VALUE

Wicksell's explanation of the secular trend of prices was based on the distinction between the market and the natural rate of interest. The natural and the market rate are two values of the interest rate. The former is the value at which savings are equal to investment, and the latter the value of the interest rate determined by the credit market. If the natural rate is above the market rate, investment is greater than savings. Under the assumption of full employment, the increase in aggregate demand is translated completely into higher prices. Wicksell studied the problem under two extreme assumptions. In the cash economy, the exchange of goods is carried out through a physical medium, which is identified as money. In the credit economy, all transactions are carried out by bookkeeping entries. In both cases, the discrepancy between the natural and the market rate drives the price level.

In a cash economy, Wicksell concluded, the cumulative process cannot last for ever. If prices rise, that is if the natural rate is above the market rate, the demand for gold increases. As a result, gold will drain out of the banks' reserves and banks will be compelled to raise the market rate. In the case of a credit economy, however, financial innovation weakens the link between the drain of reserves and the increase in the market interest rate. Higher prices do not increase the demand for gold, but rather the demand for alternative financial instruments such as bank deposits and giro accounts. The salient feature of these instruments is that they represent a withdrawal from one bank, as much as a deposit into another one. Consequently, in a centralised banking system, such as the one described by Black and Wicksell, there is no drain of reserves, and no force to bring the natural and the market rate together. Wicksell illustrated this situation with a famous analogy:

> the movement and equilibrium of actual *money prices* represent a fundamentally *different phenomenon*, above all in *a fully developed credit system* from those of *relative prices*. The latter might perhaps be compared with a mechanical system which satisfy the conditions of *stable* equilibrium, for instance a pendulum…
>
> The analogous picture for *money prices* should rather be some easily movable object, such as a cylinder, which rests on a horizontal plane in so-called neutral equilibrium. The plane is somewhat rough and a certain force is required to set the price-cylinder in motion and to keep it in motion. But so long as this force – the raising or lowering of the rate of interest – remains in operation, the cylinder continues to move in the same direction. (Wicksell (1898), pp. 100–1; italics added)

Fama argues that a bookkeeping economy is as stable as a barter economy because there is no distinction between money prices and relative prices. However, this distinction is the very point disputed by Wicksell in the previous quote. The similarity between a barter and a credit economy, according to Fama, is related to the nature of the standard. In the absence of cash, money prices must be expressed in terms of a standard which needs to be one of the economy's real goods, or at least to be backed by one of them. Furthermore, as in the case of a barter economy, this implies that the demand and supply functions 'in the market for the standard' are well defined. In other words, money prices are in fact relative prices, and therefore oscillate like Wicksell's pendulum.

Following the same logic, one must also determine the nature of *money* prices in Wicksell's *credit* economy, that is what is the backing of the standard? Apparently, there is no straightforward answer to this question. John Hicks, for example, wonders:

> Wicksell's model is best understood as a pure credit model; there is no money that is no credit. All money is someone's promise to pay; but to pay what? (Hicks (1977), p. 61)

Banking deposits cannot function as a standard of value either. The standard must be a well-defined item and David Laidler observes:

> Wicksell never extended his careful analysis of the motives which prompted the holding of currency in a pure cash economy into a parallel model of the demand for bank deposits in a pure credit economy. (Laidler (1991), p. 148)

The problem in Wicksell's bookkeeping economy is that, if a real commodity such as gold is not functioning as a standard, it is unclear what the nature of his price system is. This vagueness has led to some serious misunderstandings. For example, in Patinkin's influential interpretation, Wicksell's model refers to *abstract* prices (Patinkin, 1965, p. 581). The lack of a clearly defined role for money makes it impossible to define monetary prices:

> the [money] price level no longer appears as a variable of [the] system … and hence obviously cannot be determined by it. Indeed, what we have here is the indeterminacy of Wicksell's 'pure credit economy' in which all transactions are carried out by checks, while banks hold no reserves. The economic interpretation of this indeterminacy is straightforward: In order for the absolute price level to be determined by market-equilibrating forces, changes in it must impinge on *real* behavior in *some* market … Now, the joint assumption of a pure inside-money economy and the absence of distribution effects implies that there is no such impingement on the real demands of the private sector for commodities, bonds and money respectively. Similarly, the absence of reserve requirements (either legal or economic) implies that there is no impingement on the real demand and supply functions of the banking sector. Hence the economy does not generate resistance to any arbitrary change in the price level. Accordingly, there is nothing to prevent the frictionless flow of prices from one level to another. (Patinkin (1965), p. 303).

It is now possible to summarise the conventional wisdom on price determinacy in a credit economy. In Wicksell's economy, the price level is undetermined because the demand and supply functions of the monetary standard are not well defined. This is particularly true if one regards money as being a medium of exchange, since there is none in a bookkeeping economy except for banks' deposits, which, as Laidler notes, do not fulfil this requirement. If money is purely an accounting magnitude, following the homogeneity postulate, the price level is determined arbitrarily. However, if the standard is a real commodity, it becomes possible to determine monetary prices as relative prices. The general equilibrium theory of a so-called barter economy is applicable to a pure inside-money economy. The price level is anchored completely.

Patinkin's interpretation left Wicksell's cumulative process in a sorry state. It lost any explanatory power as a model of inflation. The metaphor of a cylinder was replaced by that of medieval ghost money. In other words, the dynamics of Wicksell's intertemporal equilibrium model was replaced by a static problem where the issue became whether the standard represents a real or an abstract commodity. Just as in the Middle Ages, there were fully-fledged coins and purely accounting – ghost – monies. To correct this interpretation, one must demonstrate that the cumulative process is not about the behaviour of abstract prices and, more importantly, that a credit economy cannot be reduced to a barter economy even though the price level is defined in terms of a standard with real backing.

INFLATION IN A CREDIT ECONOMY

Much of our understanding of the properties of general equilibrium models is based on the idealisation of a barter economy. Wicksell himself derived his condition for monetary equilibrium from comparing and contrasting a barter and a monetary economy. This condition states that the price level is stable when the rate of interest in the money market is equal to the rate of interest that would prevail if capital could be lent in kind, that is without the intervention of money. This rate would correspond to the yield on newly-created capital (Wicksell (1898), p. 102).

In his criticism of the cost production theory of money, Wicksell maintained that it is not possible to have inflation in a barter economy. According to him, an excess supply of one good implies an excess demand for other goods. Therefore, it is difficult to determine a priori if the price level will increase or decrease. To explain the phenomenon of inflation, one must spell out 'how and why the monetary or the pecuniary demand for goods exceeds or falls short of the supply of goods in given conditions'. (Wicksell, 1906, p. 160). A monetary economy is different from a barter economy because 'liquid real capital (i.e. goods) are never lent ... it is *money* which is lent, and the commodity capital is then *sold* in exchange for this money' (Wicksell, 1898, p. xxvi). Wicksell stressed the importance of this distinction in the following way:

> *It is not, as is often supposed, merely the form of the matter which is thus altered, but its very essence.* For real capital goods cannot longer be supposed to be actually borrowed and lent; they are now bought and sold. An increase in the demand for real capital goods is no longer a *borrowers'* demand which tends to raise the rate of interest, but a *buyers'* demand which tends to raise the price of commodities. (Wicksell (1898), p. 135; italics of the sentence added)

The existence of money as a unit of account means that in equilibrium the ratio of prices in terms of the standard must be consistent with the relative price which would be established if goods were bartered. In an intertemporal context this means that the money rate should be equal to the natural rate. One could express Wicksell's concerns more generally as the problem of consistency between the two different price systems. Under this interpretation, Wicksell's problem would simply disappear if one of these systems consisted of *abstract* prices.

Wicksell's insight was that the difficulty of co-ordinating the two price systems rested on the difference between the nature of money and other commodities. As he expressed it:

> 'supply' and 'demand', expressions so conveniently applied to almost everything under the sun, become obscure and, in reality, meaningless when applied to money. (Wicksell (1906), p. 20)

This attitude is exactly the opposite of Patinkin's and explains why Patinkin was so critical of the cumulative process:

> For what Wicksell is essentially saying is that the level of money prices is indeterminate as long as the quantity of money is not fixed, and that continuous changes in the quantity of money will cause continuous changes in the 'price' (value) of money relative to other commodities. But the same statement can be made of the relative price of potatoes – if the quantity of potatoes is continuously changing. Conversely, the equilibrium of money prices can be just as stable as that of relative prices – provided that in each case the initial quantities remain unchanged. (Patinkin (1965), p. 596).

Patinkin, however, fails to explain why the supply of money can be increased continuously, but not so the supply of potatoes. For Wicksell, money is different from other commodities. Although the exchange of goods generates a demand for (or supply of) money, this is not a demand for a quantity of money, but for some value. Any quantity of money could satisfy this demand if it circulates at the appropriate velocity. The only limit to the velocity of money, and therefore the only anchor for the price level, is the demand for money as an asset, as a store of value (Wicksell, 1906, p. 23). The problem is that, either as a medium of exchange or as a reserve asset, credit facilities function as a substitute for money. The development of banks increases the velocity of money and jeopardises the status of money as a nominal anchor.

Wicksell understood correctly that all money was ultimately credit. However, methodologically, he found it to be more meaningful to view the use of credit instruments as influencing the velocity of the monetary base, rather than the size of the money supply. An advantage of this approach was to unveil the powerful influence of the banking system on the velocity of money. In a pure credit economy, for example, the banks' control is complete, since money (gold) never exits the banking system. Every agent's withdrawal corresponds to someone else's deposit:

> Banks are not, like private persons, restricted in their lending to their own funds or even to the means placed at their disposal by savings. By the concentration in their hands of private cash holdings, which are constantly replenished by in-payments as fast as they are depleted by out-payments, they posses a fund of loans which is always elastic and, on certain assumptions, inexhaustible. With a pure credit system the banks can always satisfy any demand whatever for loans and at rates of interest however low, at least as far as the internal market is concerned (Wicksell (1906), p. 194)

The NME, by contrast, disputes the view that banks can sustain any discrepancy at all between the natural and the market rate, or consequently, any price level. This stand closely resembles the so-called 'real bills doctrine'. According to this doctrine, a 'desirable behavior of the money supply will be assured if banks, including the central bank, restrict their loans to the discounting of "real bills", i.e. to the financing of non-speculative investments' (McCallum 1986). If they find this requirement to be satisfied, banks should accommodate any request for loans without fear of inflationary consequences. This suggestion is not different from Black's:

> The banks will make loans to individuals, businesses, and governments. They will probably establish a schedule of interest charges for each borrower, and will then allow him to write checks on his account that *increase the amount of his loan whenever he needs the money* ... Since

> *banks will not be restricted in making loans to businesses,* they will be able to supply the bulk of the loans that businesses need, both short term and long term. (Black (1970), pp. 10–12; italics added)

White (1984b) identifies two problems with the real bills doctrine. One, which he labels the *nominalist* fallacy, does not apply properly to the NME, but has been the objection most frequently pointed out in the literature, for example Mints (1947, pp. 30–1), Patinkin (1965, p. 309) and Humphrey (1982). The problem is the following: if the money supply accommodates the 'needs of trade', but these are a function of the price level, then the money supply also becomes dependent on the price level. The system becomes unstable because a price increase induces an expansion of the money supply; this expansion pushes prices up, which in turn steps up the money supply. In an economy where the standard is a real commodity, this problem does not arise. The money supply is limited by the physical quantity of the standard.

The second fallacy is known as the *inelastic-supply-of-bills*. It arises from the belief that the needs of trade are exclusively a function of real output, and ignores the effects of expectations of investors and the influence of the financial system through the rediscount rate. The belief that the productive structure of the economy leads to constant, stable credit requirements is the distinctive belief shared by the proponents of the real bills doctrine and the NME. According to Selgin (1989), the problem with both approaches is that they assume a steady state in which the financial system appears only as a facilitator of barter exchanges of goods already produced. This is a denial of the very essence of Wicksell's tradition: the problem of intertemporal co-ordination, Leijonhufvud (1981).

The validity of the Modigliani–Miller theorem is the argument offered by the NME to abstract from savings and investment considerations in the determining of the price level. The Modigliani–Miller theorem states that the present value of a project is independent of the source to finance it because the discount rate is unique. In Wicksell's credit economy this need not be the case. Bank loans can make a project worthwhile if the banks set the market rate below the natural rate:

> Now it must be a matter of indifference to the individual business man whether he derives his profits from higher gross receipts, from lower costs, in the narrow sense of the word, or from cheaper credit. It follows obviously that unless the fall in the rate of discount *is neutralized by simultaneous changes elsewhere*, it must, when it has persisted long enough to exert a depressing influence on long-term rates of interest,

> provide a stimulus to trade and production, and alter the relation between supply and demand of goods and productive services in such a way as necessarily to bring about a rise in all prices. (Wicksell (1898), p. 89)

It is noteworthy that this paragraph appears at the point in which Wicksell criticises those who deny 'the possibility of the banking policy being responsible in any way for such fluctuations in prices as occur in practice' (Wicksell, 1898, p. 85). More important, however, is to emphasise that since there is no medium of exchange in a credit economy, the importance of money as a numeraire becomes clearer in explaining the discrepancy between the natural and the market rate of interest:

> It has already been mentioned that this possibility [the discrepancy] arises out of the fact that the transfer of capital and the remuneration of factors of production do not take place in kind, but are effected in an entirely indirect manner as a result of the intervention of money. (Wicksell (1898), p. 135)

In Wicksell, the source of price instability is the mismatch between the yield of the project and the cost of financing it. The mismatch is intrinsic in a world where investment and savings are carried out by different agents. It is compounded in an economy where banks, *as a whole*, can freely extend the amount of credit. From this perspective, one cannot claim that the similarity of a barter and a financial economy rests on the nature of the standard. Such assertions can only be made if the ability of both economies to solve the problem of intertemporal co-ordination is the same. This issue, and not whether the standard is an abstract or a real commodity, is the dividing line between Wicksell and the NME. The NME assumes rather than proves the validity of the Modigliani–Miller theorem. This attitude neglects the savings–investment problem and this omission weakens the case for deregulation made by the NME.

THE STANDARD OF VALUE AND MACROECONOMIC ADJUSTMENT

The core of the NME is the proposition that the financial system should be completely deregulated. The justification for this policy prescription is based on the premise that a bookkeeping economy resembles a barter economy in which the price system is well anchored. If some transactions need to be carried out with the help of a physical medium of exchange, this medium should be different from the standard of value. In this case too, it would be unnecessary to regulate the institutions operating the payment mechanism. The analysis of Wicksell's work in the last section casts doubts on the proposition that a pure credit economy mimics such a barter economy. This section assesses the desirability of separating the standard of value from the payment mechanism and the role of the numeraire in Wicksell's work.

Patinkin and Fama reduced the problem of inflation to a static problem, one of definition or determinacy of the price level. Wicksell's analysis of inflation was inherently dynamic and this allowed him to recognise that the selection of the standard is one more element of monetary policy. This was one of the reasons why he emphasised the legal character of the numeraire. Wicksell's multiple testimonies about the conventional or arbitrary character of money should not be understood as stating that the standard is an 'abstract', valueless entity as in Patinkin's interpretation; rather, they should be read as saying that it is a subject of 'choice', of policy-making:

> Absolute prices on the other hand – money prices – are a matter in the last analysis of pure convention, depending on the *choice of a standard of price* which it lies in our own power to make. (Wicksell (1898), p. 4)

Wicksell frequently pointed out that it was senseless to take the absolute level of the interest rate as a policy objective. The level of the market rate is not important in itself, only its relation to the natural rate. Since the natural rate is not directly observable, the relative level of the natural and the market rate can only be inferred from the movement of prices. A price level increase, for example, implies that the market rate is below the natural rate; therefore, the monetary authorities have to move the market rate upwards to halt the increase. Similarly, in the case of deflation, the authorities must bring down the market rate until the price level stabilises.

Once the policy rule is defined, the next issue is how to measure changes in the price level. Wicksell devoted some attention to this problem and was perfectly aware that any index was always a conventional and imperfect measure, Wicksell (1898, ch. 2; 1906, ch. 4). A crucial problem was to decide which commodities should be included in the index basket.

If the natural rate is not unique, different baskets correspond to the selection of different *average* natural rates. For example, if, according to one basket, the price index moves upwards, it is an indication that the natural rate is above the market rate. Consequently, the market rate should rise. Alternatively, if another basket is selected and the price index moves downwards, this means that the natural rate is below the market rate. The market rate should fall. The selection of a particular index becomes perforce a policy choice (Chiodi (1991), ch. 3). A good example of this type of dilemma is found in the debate on the productivity price norm between Wicksell and Davidson (see Uhr, 1960).

Wicksell's concern about the price index reveals that he was sensitive to some of the issues related to the unit of account. However, he centred most of his attention on the components of the basket and not on the selection of the standard used to bundle them into an index. On the other hand, he did ponder the NME proposal of separating the medium of exchange and the standard of value:

> It is sometimes said to be feasible to *base* a monetary system upon gold and yet to dispense entirely, or almost entirely with the employment of gold both in circulation and in the banks' reserves. This would be done by extending the use of cheques, by the issue of notes of which the cover is of a purely banking nature, and so on. This view, which is held by some of the most prominent writers on monetary questions, must be regarded as utopian. In such a system the value of money would be *directly* exposed to the effects of every fortuitous incident on the side of the production of the precious metals and every caprice on the side of its consumption. It would undergo the same violent fluctuations as do the values of most other commodities ... But it would be quite possible to maintain a stable value of money without the use of reserves of a precious metal. Only it would be necessary for the *metal to cease to serve as a standard of value.* (Wicksell (1898), pp. 34–5)

Wicksell recognised accurately that demonetising the standard, as epitomised by Hall's composite commodity, could lead to higher price variability. In Wicksell's model, this occurs because banks can only adjust their reserves by modifying the rate of consumption of gold, and this can only be achieved by changing the price level. For example, if banks have excess gold reserves, they must reduce the interest rate to increase the price level. The increase in the price level (a fall in the relative price of gold) stimulates gold consumption and reduces bank reserves. However, as gold consumption proceeds, reserves dwindle beyond their optimal level. Banks now are forced to deflate (increase the relative price of gold) in order to stimulate the extraction of gold.

Wicksell thought that a fiat standard was an escape from this roller-coaster, a clear recognition of the standard of value as a policy decision, but did not go deeper into details. Wicksell's remarks on the nature of the standard were succinct, because he thought that more general and powerful forces were at work. His preoccupation with the development of credit institutions and its impact on the velocity of money may have overridden, or led him to simplify, the problem of defining the standard of value. For example, he noted that as the velocity of circulation increases:

> the money which would cancel these claims revert to its starting point. If in such cases the use of ready money becomes quite superfluous, *whilst metallic money remains a measure of value* for the payment in question, it may be regarded as an *infinitesimally small* amount of money circulated with infinite velocity in accordance with the formula $0 \times \infty$ (nil multiplied by infinity) which according to the circumstances may signify any quantity whatever. (Wicksell (1906), p. 70; italics added)

This observation shows clearly that Wicksell's credit economy is better understood as a limit argument. It is not that he imagined that the quantity of gold is actually zero, only that its importance becomes very small. His system is in fact anchored to gold, but could just as well be any other real commodity. Therefore, the price level is not as stable as the NME contends *independently* of the nature of the standard. The development of the financial system delays the convergence of the natural and the market rate. The only exception Wicksell acknowledged was the case of a fiat standard, but he did not fully work out its implementation.

CONCLUSIONS

One of the lessons one draws from this work, is that care must be taken in comparing a barter and a monetary economy. If one regards money as a medium of exchange, then a bookkeeping economy presents the problem that it is impossible to define money prices. In this regard, the NME has emphasised correctly that the determination of the price level is related to the standard of value and not to the medium of exchange. However, the NME does not accept fully that the standard of value is a monetary problem. The NME still hinges on the view that money is primarily a medium of exchange. Consequently, the NME identifies a bookkeeping economy not as a monetary, but as a barter, economy.

Wicksell's work also has as a starting point the function of money as a medium of exchange. However, his efforts to understand the determinants of the velocity of circulation led him to stress the importance of credit, and therefore the units in which it is reckoned. In his analysis, the distinction between a barter and a money economy is no longer the existence of a medium of exchange. The difference now lies in the conditions that must be fulfilled to maintain the equilibrium of savings and investment. From this point of view, the case for deregulation made by the NME depends critically on the validity of the Modigliani–Miller theorem. This is a point that has not been stressed either by the NME or by its critics.

This chapter has also established that some of the criticisms made against Wicksell are not valid. In particular, Patinkin's interpretation is incorrect in arguing that the cumulative process is built on a fictitious or abstract standard. On the other hand, it is true that Wicksell could have done more to clarify the role of the standard. For example, had he paid more attention to the standard of value, he could have established the basis of a portfolio theory. Wicksell had a view of the economy as a medieval fair where one gold sovereign could circulate fast enough to perform all the transactions in the market. A moment of reflection on the price system, however, reveals that possessing a piece of gold is not the same as holding a bill of exchange. Prices are quoted in terms of gold, rather than bills. Wicksell's decision to regard bills of exchange and gold sovereigns as perfect substitutes appear as a shortcoming today, particularly in light of J. M. Keynes' and John Hicks' contributions to portfolio theory. Indeed, the link between liquidity preferences and the standard of value was recognised explicitly by Hicks:

> The crucial characteristic of money (as it appears in this portfolio theory) is that its resultant yield is certain … What is meant by *certainty*? The certainty of the yield on money (whether zero or not) is no more than a certainty of money itself. Why should that signify? I[t] can only signify because money is accepted as a 'measure of value'. (Hicks (1967), pp. 8–28)

Money is liquid because its yield is certain. Certitude arises from being the standard in which debts and prices are reckoned. In this regard, Wicksell's analysis suffers from the same shortcomings as the NME. The standard in a credit economy is portrayed as an immovable point of reference. There is a complete disregard for the liquidity quality of the standard. On the other hand, one advantage of Wicksell's approach is that by focusing on the concept of velocity it brings to the foreground the importance of institutions and the problems of co-ordination. From this point of view, it is clear why Axel Leijonhufvud has insisted that monetary theory has yet to encompass the lessons of the past, but also to what extent his own theoretical work has kept alive the Swedish tradition coming down from Wicksell.

Notes

1. I would like to thank A. Leijonhufvud, K. Velupillai, D. Vaz and W. Gebauer for their comments to a previous draft.
2. In principle, multiple price levels can co-exist, since each good or basket of goods could function as a standard. Fama does not explain satisfactorily why the economy will settle on using one standard. His justification is based on the premise that using one standard saves information costs, since using one standard in the case of n goods leads to $(n - 1)$ prices, rather than $(n)*(n - 1)/2$. Unfortunately he does not specify the nature of these costs. This is the subject of the first chapter of my UCLA doctoral dissertation. *The Standard of Value and the Complexity of Arbitrage.*

References

Black, F. (1970) 'Banking and Interest Rates in a World Without Money', *Journal of Bank Research*, Autumn.

Chiodi, G. (1991) *Wicksell's Monetary Theory* (London: Macmillan).

Cottrell, A. (1989) 'Price Expectations and Equilibrium when the Interest Rate is Pegged', *Scottish Journal of Political Economy*, vol. 36, no. 2.

Cottrell, A. and M. S. Lawlor (1991) '"Natural Rate" Mutations: Keynes, Leijonhufvud and the Wicksell Connection', *History of Political Economy*, vol. 23, no. 4.

Cowen, T. and R. Krozner (1994) *Explorations in New Monetary Economics* (Cambridge, Mass.: Basil Blackwell).

Eatwell, J., M. Milgate and P. Newman (1987) *The New Palgrave: A Dictionary of Economics*, (London: Macmillan).

Einzig, P. (1949) *Primitive Money in its Ethnological, Historical, and Economic Aspects* (London: Eyre & Spottiswoode).

Fama, E. (1980) 'Banking in the Theory of Finance', *Journal of Monetary Economics*, vol. 12.

Fama, E. (1983) 'Financial Intermediation and the Price Level Control', *Journal of Monetary Economics*, vol. 6.

Fischer, S. (1983) 'A Framework for Monetary and Banking Analysis', *Economic Journal.*

Fischer, S. (1986) 'Friedman versus Hayek on Private Money: A Review Essay', *Journal of Monetary Economics*, vol. 17.

Friedman, M. and A. J. Schwartz (1986) 'Has Government Any Role in Money?', *Journal of Monetary Economics*, vol. 17.

Greenfield, R. and L. Yeager (1983) 'A Laissez Faire Approach to Monetary Stability', *Journal of Money, Credit and Banking*, vol. 15.

Gurley, J. and E. Shaw (1955) 'Financial Aspects of Economic Development', *American Economic Review*, vol. XLV, no. 4.

Gurley, J. and E. Shaw (1960) *Money in a Theory of Finance* (Washington, DC: The Brookings Institution.)

Hall, R. (1982) 'Explorations in the Gold Standard and Related Policies for Stabilizing the Dollar', in R. Hall (ed.), *Inflation* (Chicago: University of Chicago Press).

Heinsohn, G. and O. Steiger (1983) 'Private property, Debts and Interest or: The Origin of Money and the Rise and Fall of Monetary Economics', *Studi Economici*, no. 21.

Hicks, J. (1935) 'A Suggestion for Simplifying the Theory of Money', *Economica*. Reprinted in *Critical Essays in Monetary Theory* (Oxford: Clarendon Press).

Hicks, J. (1967) 'The Two Triads' in *Critical Essays in Monetary Theory* (Oxford: Clarendon Press).

Hicks, J. (1977) 'Monetary Experience and the Theory of Money', in *Economic Perspectives: Further Essays on Money and Growth* (Oxford: Clarendon Press).

Hicks, J. (1989) *A Market Theory of Money* (Oxford: Clarendon Press).

Hoover, K. (1988) *The New Classical Macroeconomics* (Oxford and New York: Basil Blackwell).

Howitt, P. (1992) 'Interest Rate Control and Nonconvergence to Rational Expectations', *Journal of Political Economy*, vol. 100, no. 41.

Humphrey, T. (1982) 'The Real Bills Doctrine', *Federal Reserve Bank of Richmond Economic Review*, September/October.

Humphrey, T. (1986) 'Cumulative Process Models From Thornton to Wicksell', *Federal Reserve Bank of Richmond Economic Review*, May/June.

Johnson, H. (1968) 'Problems of Efficiency in Monetary Management', *Journal of Political Economy*, vol. 76.

Laidler, D. (1972) 'On Wicksell's Theory of Price Level Dynamics', *Manchester School*, vol. 50.

Laidler, D. (1984) 'Misconceptions about the Real-Bills Doctrine: A Comment on Sargent and Wallace', *Journal of Political Economy*, vol. 92, no. 11.

Laidler, D. (1991) *The Golden Age of the Quantity Theory* (Princeton, NJ: Princeton University Press).

Leijonhufvud, A. (1973) 'Effective Demand Failures', *Swedish Journal of Economics*, March. Reprinted in Leijonhufvud (1981).

Leijonhufvud, A. (1981) *Information and Coordination* (Oxford and New York: Oxford University Press).

Leijonhufvud, A. (1987) 'Natural Rate and Market Rate', in Eatwell *et al.* (1987).

Leijonhufvud, A. (1991) 'The Uses of the Past' (mimeo) UCLA

Leijonhufvud, A. and D. Heyman (1995) *High Inflations*, Arne Ryde Memorial Lectures (Oxford: Clarendon Press).

McCallum, B. (1981) 'Price Level Determinacy with an Interest Rate Policy Rule and Rational Expectations', *Journal of Monetary Economics*, vol. 8.

McCallum, B. (1985) 'Bank Deregulation, Accounting Systems of Exchange and the Unit of Account: A Critical Review', *Carnegie–Rochester Conference Series on Public Policy*, vol. 23.

McCallum, B. (1986) 'Some Issues Concerning Interest Rate Pegging, Price Level Determinacy, and the Real Bills Doctrine', *Journal of Monetary Economics*, vol. 17.

McCallum, B. and M. Goodfriend (1987) 'Demand for Money: Theoretical Studies', in Eatwell *et al.* (1987).

Mints, L. (1947) *A History of Banking in Great Britain and the United States.* (Chicago: University of Chicago Press).

Myrhman, J. (1991) 'The Monetary Economics of the Stockholm School', in L. Jonung (ed.), *The Stockholm School of Economics Revisited* (Cambridge: Cambridge University Press).

Patinkin, D. (1952) 'Wicksell's Cumulative Process', *Economic Journal*, vol. LXII.

Patinkin, D. (1961) 'Financial Intermediaries and the Logical Structure of Monetary Theory', *American Economic Review*, vol. LI.

Patinkin, D. (1965) *Money Interest and Prices; An Integration of Monetary and Value Theory* (2nd ed.) (New York: Harper & Row).

Sargent, T. and N. Wallace (1982) 'The Real Bills Doctrine versus the Quantity Theory: A Reconsideration', *Journal of Political Economy*, vol. 90, no. 61.

Sargent, T. and N. Wallace (1975) '"Rational Expectations", the Optimal Monetary Instrument and the Optimal Money Supply Rule' *Journal of Political Economy*, vol. 83, no. 2.

Selgin, G. (1989) 'The Analytical Framework of the Real Bills Doctrine', *Journal of Institutional and Theoretical Economics*, no. 145.

Sraffa, P. (1932) 'Dr. Hayek on Money and Capital', *Economic Journal*, March.

Thornton, H. (1991/1802) *An Enquiry into the Paper Credit of Great Britain*, reprint ed. by F. A. Hayek (New York: A. M. Kelly).

Tobin, J. (1963) 'Commerical Banks as Creators of "Money"', in D. Carson (ed.), *Banking and Monetary Studies* (Homewood, Ill.: Irwin).

Uhr, C. G. (1960) *Economic Doctrines of Knut Wicksell* (Berkeley and Los Angeles: University of California Press).

Viner, J. (1937) *Studies in the Theory of International Trade* (New York: Harper).

White, L. H. (1989) *Competition and Currency: Essays in Free Banking and Money* (New York and London: New York University Press).

White, L. H. (1984a) 'Competitive Payments Systems and the Unit of Account', *American Economic Review*, vol. LXXIV, no. 4, reprinted in White (1989).

White, L. H. (1984b) *Free Banking in Great Britain: Theory, Experience and Debate* (Cambridge University Press).

White. L. H. (1986) 'A Subjectivist Perspective on the Definition and Identification of Money', in I. Kirzner (ed.), *Subjectivism, Intelligibility and Economic Understanding*, reprinted in White (1989) (New York and London: New York University Press).

Wicksell, K. (1898/1936) *Interest and Prices* (New York: Macmillan).

Wicksell, K. (1906/1935) *Lectures on Political Economy Vol. II* (New York: Macmillan).

Yeager, L. (1983) 'Stable Money and Free Market Currencies', *Cato Journal* vol. 3.

Part IV

11 Evidence on Stabilising Speculation: Leijonhufvud on Keynes*

Richard Sweeney

INTRODUCTION

One focus of Leijonhufvud's work is the interaction of expectation and speculation. In his understanding of Keynes, for example, speculators transmit real shocks to aggregate demand to movements in real variables by not allowing 'the' (real) interest rate to adjust to its new long-run equilibrium value.[1] A fall in the investment schedule, accompanied by a fall in the business sector's excess supply of bonds, puts upward pressure on bond prices (downward pressure on bond yields and interest rates). Instead of the interest rate falling to keep the level of aggregate demand equal to the full-employment level of aggregate supply, deficient aggregate demand arises because speculators keep the interest rate close to the initial level that is now too high. Speculators cause this problem, not out of perversity, but because they expect the current downward pressure on bond yields soon to reverse; they act on these beliefs in an attempt to profit. Similarly, speculators play a key role in determining the nominal (and real[2]) exchange rate.

In Leijonhufvud's view, the same sort of speculator behaviour can in one environment be appropriate from the speculators' wealth-maximising view and beneficial from society's view, but inappropriate and harmful in another. For example, in an environment where central banks commit credibly to maintain fixed parities, speculators stabilise exchange rates at these parities, making the banks' task easier. In particular, government fiscal policies will probably have larger multipliers with, than without, speculation that supports the parity. If central banks

*Prepared for *Inflation, Institutions and Information: A Conference in Honour of Axel Leijonhufvud*, sponsored by The Central Bank of Uruguay and The Center for Computable Economics, UCLA 6–7 September 1993, Montevideo, Uruguay. Thanks are due to participants in the Conference for insightful and helpful comments.

convince speculators misleadingly that government policies will always adjust to support the parities, speculators stabilise rates at levels that are unsustainable and in this way hinder markets' adjustment to new equilibrium values.[3]

There is a huge theoretical and empirical literature on the behaviour of speculators and the nature of speculative markets; in particular, finance puts much emphasis on both, and many open-economy macro models make strong assumptions about speculation. Virtually none of the contributions speaks explicitly to Leijonhufvud's concerns. To be sure, the vast majority of finance models assume that speculators process economically available information efficiently and thus market prices reflect this information appropriately;[4] in this view, any actual risk-adjusted profits for the marginal speculator are evanescent and have a zero expected value.[5]

Many open-economy macro models rely on assumptions about the results of speculation. For example, in Mundell–Fleming models, a key issue is the 'mobility' of capital; it is sometimes assumed that there is sufficient speculation to set home-country interest rates equal to those in world markets. Similarly, models of (real and nominal) exchange-rate overshooting often assume that speculation ensures that the Fisher open condition holds, or in deterministic versions of overshooting, appreciation of the home-country exchange rate equals the gap between world and home interest rates; in these models, a current disturbance leads to sharp changes in the exchange rate and the home-country interest rate, with gradual trend movements in both back to steady-state levels (Dornbusch, 1975).

Leijonhufvud is a Keynesian with a difference.[6] Much of his work shows little support for ongoing fine-tuning. He does, however, see the case that extraordinary shocks may occasionally hit the economy with such severity that markets' co-ordination mechanisms do not function as well as in ordinary circumstances; in such cases, activist policies have the potential[7] to push the economy back to a range where markets once again co-ordinate activity in a workably efficient manner – not perfectly, but well enough that one might be sceptical of intervention to improve performance. In the dominant views in finance in the mid-1990s, financial markets process information efficiently: it is a short, though not necessary, step to argue that intervention is ineffectual or counterproductive in most circumstances; further, even in extraordinary circumstances, in a world where markets process information efficiently, one might well be sceptical of claims that intervention would be useful.

This chapter surveys briefly some of the finance and open-economy macro literature on speculative markets; in particular, it looks at empirical results that might provide evidence for and against some of Leijonhufvud's concerns.

EVIDENCE ON MARKET EFFICIENCY

Speculators' behaviour and the behaviour of speculative prices are the heart of modern finance. Current finance orthodoxy is that markets are efficient, or current prices embody the market's best information on the value of assets. In this view, though markets are revealed *ex post* to make many mistakes, the mistakes are essentially random, they are not persistent and are not cumulative; and *ex ante* the markets' mistakes are not forecastable. In this view, the government can improve speculative-market behaviour only to the extent that it has information about its own or other governments' future policies that is unavailable to the market; if the government released the information, markets would react at least as well as if the government kept the information and acted itself to aid market adjustment. Certainly, some observers provide evidence that markets may in some ways and on some occasions behave in an inefficient manner, but this is distinctly a minority view.[8]

The efficient markets hypothesis (EMH) is the maintained assumption in much of modern finance and has proved to be an immensely fruitful way of looking at the world. For example, many event studies assume the EMH holds and, conditional on this assumption, measure the effects on asset holders' wealth of a particular type of surprise; examples include the effect on equity values of earnings surprises, issuance of new debt or equity, and dividend changes. Further, the EMH can be thought of as simply the zero profit condition in economic theory – under competition, the marginal actor makes zero profits; markets 'should' offer no easy profits for the marginal actor.[9]

The evidence for the EMH is in some ways weak. Empirical results in the 1950s and early 1960s that changes (or rates of return) in speculative prices were random led to the EMH, where the randomness was taken as support for efficiency. is Ironically, most observers are now sceptical of whether returns are in fact random empirically; instead, many observers believe that expected rates of return may well be time-varying because of time-varying risk premiums. It is now widely accepted that any test of market efficiency is a joint test of the EMH and the risk model used; any rejection may be of the EMH or the risk model; many observers have strong priors that any rejection is of the risk model.

Beginning in the late 1970s, much evidence rejects simple models of risk. For example, after adjusting for market-beta risk, higher than expected returns are found for small firms, firms with high ratios of book-to-market equity, and firms with high earnings–price ratios; in these cases, predetermined variables appear to offer scope for super-normal profits. Further, small firms tend to have higher than expected returns in January, these abnormal returns tend mainly to come in a few days around the turn of the year, and there are other calendar-related effects (Keim, 1989). Rather than being taken as evidence for rejecting the EMH, these results are called 'anomalies', with the assumption that a richer risk model would show the predetermined variables and calendar effects as proxies for risk factors that explain the apparent super-normal profits.

The remainder of this section discusses evidence on short-run patterns in financial asset prices that might be taken as evidence of inefficiency, and also discusses longer term patterns.

Short run patterns

Early work on the profitability of mechanical trading strategies focused on so-called filter rules, as used by Alexander (1961, 1964). In the *locus classicus*, Fama and Blume (1966) compare the rates of return from taking speculative positions based on signals from filters, to rates of return from simply buying and holding an asset. They use filter rules of the form: buy if the asset's price rises *Y* per cent or more from its previous local low, sell if the price falls *Y* per cent below its previous local high. They find that gross profits are positive and largest for small filters – half of 1 per cent gives the largest profits – but that transaction costs give negative net profits for all filters. Praetz (1976, 1979) and Sweeney (1988) argue that Fama and Blume's measure of profits is biased against finding a filter effect. Further, Richard Sweeney argues that Fama and Blume average together firms where the filter seems not to work with firms where the filter seems to work; he shows that the stocks that appear profitable for small filters in Fama and Blume's paper for the late 1950s and early 1960s also appear profitable over the 1970s and early 1980s. Further, for twelve years starting in 1970, Sweeney (1990) finds significant profits from half of 1 per cent filters where the fifty stocks traded each year are the fifty that offered the largest filter-rule profits for the preceding year. Sweeney's (1988, 1990) test assumes that the risk premium on days when the filter has the speculator 'out' of a given stock is the same as the average risk premium over the period; this does not require a constant risk premium, but rather that the speculator's 'buy' and 'sell' decisions are uncorrelated with movements in the risk premium. Of course, Sweeney may underestimate transaction costs, the trades may not be executable in practice, and he may have mis-estimated risk so that the measured profits are not economic profits once risk is accounted for correctly. In particular, the measured profits may arise from having long positions in a stock when its risk, and hence expected return, is higher than average, and being out of the stock when its risk, and hence expected return, is below average.

Thus, the evidence may be interpreted by one observer as showing inefficiency, and by another as showing time-varying risk premiums. This ambiguity is capable of resolution, even in principle, only if observers can agree on a model of time-varying risk premiums, leaving aside the issue of whether there exist sufficient data of adequate quality to allow for a discriminating test.[10]

Atkins and Dyl (1990), and Bremer and Sweeney (1991, 1993) find evidence that stocks which suffer a large one-day fall in price tend to rebound substantially over following trading days.[11] For example, Bremer and Sweeney (1993) find that stocks that have a one-day price fall 10 per cent or more on average rebound a bit less than 4 per cent over the next five days. Brown *et al.* (1988,1993) use smaller price falls, find small but significant price reversals, and argue their results are consistent with the EMH because of varying risk premiums; in contrast, Bremer and Sweeney (1993) argue that the returns they find are too large to be explained by risk premium variations.[12,13]

In foreign exchange markets, simple mechanical trading rules – the Alexander filter, as well as single and double moving average approaches – also seem to be profitable. Dooley and Shafer (1976, 1983) provide evidence of measured profits from speculation, though they do not test the significance of the profits; similarly, Taylor (1986) reports economically important profits from bond and currency speculation, based on time series models' signals, though he provides no test of significance. Sweeney (1986) and Surajaras and Sweeney (1992) find speculative profits and also report significant test results.[14] These test results show economically important and statistically significant profits that appear to have reasonable temporal stability (see Mori and Murray (1990) for a sample that includes the currencies in Surajaras and Sweeney plus other currencies, and for a later time period). Just as with measured profits from trading rule strategies in US equities markets, there is ambiguity of implications for the reliability of markets for processing information; in addition to questions about data adequacy and whether trades can be executed at the prices used, observers can view the results as showing either inefficiency or time-varying risk premiums.

These results for equities and currencies focus on short-term price movements and speculations that take only relatively short time periods. For example, Sweeney (1990) finds that the average time an investor is in or out of a stock is approximately three days. In Surajaras and Sweeney (1992), the average time in or out of a currency varies substantially across currencies and also periods; this is because they use different size triggers for different currencies, change the triggers over time, and encounter more or less exchange-rate volatility over different time periods. Still, for many currencies, the average period is approximately a month, and for some is perhaps two weeks.

Leijonhufvud is concerned with speculation in two types of case. In the first, speculators may hinder adjustment over a substantial period, causing important problems for the economy. An example cited above is the case where a fall in the investment demand schedule requires a fall in interest rates to keep the economy on the full-employment growth path, but speculators keep the interest rate close to the old, too-high level, putting downward pressure on output. However, speculators must keep the interest rate too high for a year or two to cause substantial macroeconomic problems. Short-term inefficiencies have little to say directly about this case. The view that markets work efficiently over all time periods, and hence there is little case for intervention to offset counterproductive speculation, relies mainly on evidence from studies of short-term price behaviour. Hence, evidence on short-term inefficiencies at least partly undermines the view that counter-productive longer-term speculation is a non-issue.

Many monetarists argue that the source of disturbances is not the private sector – fluctuations in investment and consumption demand – but rather the public sector, in particular, fluctuations in the money stock because of the central bank either causing or allowing these fluctuations. Further, monetarists focus on the growth in the money stock as the measure of the tightness of monetary policy rather than the level of the interest rate, the measure many Keynesians prefer. Such monetarist views make most sense if financial markets are efficient, rather than inefficient in a way that the government can ameliorate.

The monetarist view that government macro policy is more likely to *cause* rather than ameliorate problems is related to an explanation some observers offer for the measured speculative profits detected in foreign exchange markets, that speculators are simply profiting from misguided, loss-making central bank intervention. Consistent with this view, Taylor (1982) and Schwartz (1990) argue that central banks lose money from foreign exchange market intervention. These losses are, however, sensitive to the domestic and foreign interest rates used, and do not adjust for the premiums earned by bearing the risk inherent in open foreign exchange positions; by analogy, investors who put their funds in T-bills rather than the stock market earn less on average, but this shortfall may be interpreted as arising from a less risky position rather than as a loss. Sweeney (1994) finds that, adjusting for risk and using daily intervention data, the US Fed did not make losses and perhaps made profits over the period 1985–91. Using the same approach, Sjöö and Sweeney (1994) find similar results for Swedish Riksbank intervention during the period 1986–90.[15] On the one hand, those arguing that speculators' profits arise from central bank losses do not claim that the profits are concentrated during currency crises; central bank losses cannot be concentrated in crisis periods if these losses are to explain measured speculative profits from mechanical trading rules found during non-crisis periods. On the other hand, the September 1992 and July/August 1993 European Community (EC) currency crises arose in substantial part because speculators did not view EC parities and official commitment to parities as credible; this is precisely the role Leijonhufvud envisions for speculators in the case where government policies are inconsistent with stable parities.

Longer-term patterns

Over longer-term horizons (for example, using the latest three years to explain the coming three years), Fama and French (1988), and Poterba and Summers (1988) provide evidence for cycles in real stock returns for the US stock markets,[16] and Huizinga (1987) for cycles in real exchange rates. Among other ways, these apparent cycles can be attacked as being spurious or as explained by time-varying risk premiums. If risk premiums show cyclical patterns, caused perhaps by cycles that arise from exogenous disturbances to the real economy, in efficient markets the expected rate of return on risky assets will also show cycles; actual rates of return will then tend to show the same sorts of cycle. For those sceptical of the EMH, these results strengthen doubts; for those with strong priors that markets are efficient, these results show that risk premiums vary importantly over time.

De Bondt and Thaler (1985, 1987) argue that investors overreact to news, both good and bad. In their view, stocks that do particularly badly over a period of, say, two or three years, do worse than is justified and will probably outperform the market over the next two or three years as investors come to see that the shares are undervalued; stocks that did particularly well probably did so in part through overreaction to good news, and will on average underperform the market in the following years. De Bondt and Thaler present evidence to support the overreaction hypothesis.

There are many counter-arguments and some counter-evidence. For example, Chan (1988) presents evidence to support the argument that stocks which perform poorly also become riskier and thus should be expected on average to outperform the market; stocks that have done very well become less risky and should be expected on average to underperform the market. Further, if returns are calculated in different ways from De Bondt and Thaler, their results are weakened.

As De Bondt and Thaler note, their results are driven by Januarys; if only February-to-December's returns are examined, the data are not consistent with overreaction. It is possible that De Bondt and Thaler's results arise from January effects that are well-recognised in finance. Many observers argue that January effects arise from tax-loss selling or from mismeasurement; if so, the De Bondt and Thaler results may simply be an artefact of the January effect. The January effect, however, works mainly on small firms – those in the lowest one or two deciles of firm size; firms chosen with the same approach as De Bondt and Thaler use tend to be in the middle range of deciles and are not small firms (see Zarowin, 1989, 1990).

Shiller (1981a, 1981b; 1989), Campbell and Shiller (1987, 1988) and West (1988a, 1988b), among others, argue that stock market indices (in real terms) show too much variability relative to movements in underlying economic fundamentals.[17,18] Their work assumes a constant real risk-adjusted discount rate for valuing equities. Cochrane (1991, 1992) argues that these results may simply be evidence on how variable the real discount rate must be to generate the observed variability in market indices.[19]

Bubbles

Since before the Mississippi Bubble, the South Sea Bubble and Dutch Tulipmania, many economists have believed that asset markets are subject to bubbles or fads, periods when speculation drives asset prices far away from what economic fundamentals justify, with such episodes ending with a crash.[20] These views found a new lease of life with the rational expectations revolution. In rational expectations models, solutions to the equilibrium time path typically have multiple roots, with some roots giving explosive time paths. Often, the modeller ignores explosive roots. Other modellers focus on the explosive paths as examples of 'rational' bubbles; in this view, the economy may move along a path with explosive asset prices, with the economy eventually crashing, and asset prices returning to levels that fundamentals justify.

Rational bubbles models have (at the very least) two problems. First, the bubbles may imply logical contradictions. Brock (1975) shows conditions in a representative agent optimising model under which explosive paths can be ruled out. In deterministic models, paths are ruled out if they lead eventually to a position where the equilibrium no longer holds; decision-makers will not go down a time path that must ultimately be abandoned.[21] Many (but hardly all) rational bubbles models are macro models, where it is difficult to see if micro contradictions are implied that rule out explosive paths. For example, macro models often have explosive roots that imply the price level goes to infinity and real balances to zero; if there comes a point where real balances are so small that the economy reverts to barter, the smooth behaviour implied by the explosive solution is violated; the explosive solution cannot hold.

Second, tests for bubbles in the literature are not persuasive to many observers who have priors that bubbles do not occur. Flood and Hodrick (1990) argue 'that no econometric test has yet demonstrated that bubbles are present in the data. In each case, misspecification of the model or alternative market fundamentals seems the likely explanation of the findings'.

The literature on bubbles, then, reinforces the priors of those who lean toward the view that destabilizing speculation is an important phenomenon, but has little effect on the priors of those who do not.

OPEN-ECONOMY MACRO MODELS

The case for government macro policy is that there are disturbances that markets do not handle well, even if markets can, in principle, handle them. An example is a collapse in investment demand. In principle, prices – including wage and interest rates – can adjust to keep the economy on an equilibrium time path; in practice, these prices might not adjust appropriately. In many Keynesian models, the transmission mechanism from a real disturbance, such as a change in investment demand, is that prices and wage rates adjust sluggishly;[22] in such cases, government action may compensate for inadequate private reactions.

Assumptions about speculation, and conclusions about its implications, are more transparent in open- rather than closed-economy macro models. Partly this is so because there are identifiable speculative episodes in foreign-exchange markets, partly because open-economy models must take some account of international parity conditions. The major speculative crises are in equities and foreign-exchange markets, for example, the stock market 'crashes' of October 1987 and October 1989, and the September 1992 and July/August 1993 crises in the EC's exchange rate mechanism (ERM). Macro models typically do not include equities markets; thus, foreign exchange markets are the financial markets where speculation is discussed. Further, open-economy models must handle the covered interest-rate parity condition, that the forward premium on one currency relative to another must just offset the interest-rate differential; and the uncovered interest-rate parity condition (or the 'Fisher' open condition), that under some circumstances speculation ensures that the expected rate of spot exchange-rate appreciation equals the interest-rate differential.[23] Empirically, covered interest parity holds for Euro interest rates; models of industrialised countries must include this condition. One approach to the uncovered parity condition is to assume that the country analysed is small; that the foreign interest rate is exogenous; that actors expect no change in the exchange rate; and that the Fisher open condition holds. In this case, the home country interest rate equals the world rate. This is a common assumption in Mundell–Fleming models. Overshooting models of interest-rate and exchange-rate determination are an alternative to Mundell–Fleming models.[24] Many overshooting models are deterministic and assume that uncovered interest-rate parity holds; in a typical overshooting conceptual experiment, after a surprise increase in the home-country money stock causes both exchange-rate depreciation and a fall in the home-country interest rate, the interest-rate differential equals exchange-rate appreciation, with the home-country interest rate rising over time to equal the foreign rate as the rate of currency appreciation goes to zero (Dornbusch, 1975).

Open-economy macro models have an abysmal record in explaining the behaviour of nominal and real exchange rates on a month-to-month or quarter-to-quarter basis in empirical work based on Mundell–Fleming models or models derived from the monetary approach to the balance of payments (MABOP). Though modellers have often been successful in fitting Mundell-Fleming and MABOP models in sample, the models break down out of sample. For example, Meese and Rogoff (1983, 1988) investigate the out-of-sample properties of a number of exchange rate prediction models based on Mundell–Fleming and MABOP models. They compare the forecasting results of these models to the assumption of random walk behaviour, and find that random walk forecasts are generally better than model forecasts, even though model forecasts use actual, contemporaneous values of explanatory variables that would not in practice be available to the forecaster until later.

Over longer-term horizons, there is some evidence that some MABOP models give better forecasts than do random walks (Mark, 1992). One difficulty in using long-horizon models is that the cycles they are supposed to detect are long enough relative to the data that the data contain only a few cycles, making it difficult to detect patterns, or to measure them reliably.[25,26]

Interpretation of empirical inadequacy of open-economy macro models depends in many cases on the observer's priors about speculation. On the one hand, the models' failures may arise from destabilising speculation that obscures underlying relationships that are much as the models posit. The empirical shortcomings might thus be taken as evidence that misguided speculation dominates international financial markets; in this case, one might want to take action to reduce the effects of speculation, reasonably using the profession's understanding of relationships as embodied in the models that fare so poorly empirically. On the other hand, observers who think speculation is stabilising tend to view the poor empirical results as arising from the failure of economists to understand what drives open economies. The empirical shortcomings of the models must raise doubts about using the theoretical models as a basis for conducting policy.

SOME CONCLUSIONS

Leijonhufvud raises the possibility that speculation may play a major role in destabilising the macro economy. In the examples he discusses, the destabilising speculation does not occur daily, but rather may occur at crucial times when price variables must change substantially to re-equilibrate the economy; speculators stabilise price variables at levels that are now too high for equilibrium. Further, the destabilising speculation may persist for a substantial time.

In the efficient markets paradigm of modern finance, destabilising speculation does not arise as an important, operational issue. Under the efficient markets hypothesis (EMH), financial markets make many mistakes from day to day, but these mistakes are random; cannot be forecast; are corrected as soon as detected (and are soon detected); and generally cannot be improved by government intervention. None of the large number of empirical papers in finance has looked directly at the possibilities that Leijonhufvud raises. Rather, these possibilities are ruled out in the minds of many observers because of the seemingly strong evidence supporting the EMH. Put another way, there seems to be an irreconcilable conflict between destabilising speculation explaining some major macroeconomic movements and the evidence that financial markets work as if dominated by stabilising speculation.

The EMH has been a fertile approach to understanding the behaviour of financial markets; it is not to be discarded lightly. This is especially so because paradigms of inefficiency do not generate strong predictions, as the EMH does and hence provide little guidance for the researchers or the policy-maker.

Some important financial markets show signs of inefficiency on a day-to-day basis, and possibly on a longer-term basis. This evidence can do no more than shift the observer's priors about the EMH; evidence that can be interpreted as a sign of inefficiency can also be interpreted as arising from time-varying risk premiums (or from mismeasurement, among other possibilities). Even taking the evidence at face value, this does not support the case for frequent and ongoing government intervention to improve the behaviour of financial markets. Instead, I take the results as arguing that financial markets are fallible enough in ordinary times that one might question seriously whether they always behave efficiently in extraordinary times.

Notes

1. This is a simplification of Leijonhufvud's analysis. In fact, the current market-clearing value of the interest rate is not the value it will take on in steady-state growth, but is instead the current market-clearing value *on the path to* the steady state. The point is, major disturbances to aggregate demand require sharp changes in the current interest rate, changes that are large relative to variations in the interest rate along the equilibrium path to steady-state growth.
2. Decomposing the changes in the real rate shows that, over the period of several months, these real-rate changes arise in large part from changes in the nominal exchange rate rather than from changes in the ratio of national price levels.
3. Thus, destabilising speculation may move prices away from unchanged equilibrium values, or may keep prices unchanged when equilibrium values have changed.
4. In the absence of asymmetric information, prices are assumed to reflect information that is available to any actor. With asymmetric information, prices may not fully reflect information available to some but not all actors; typically, these models have no role for government intervention to overcome asymmetries. A substantial amount of work discusses how markets evolve mechanisms for overcoming asymmetries, for example, the literature on agency costs.
5. Infra-marginal speculators make rents; their profit-motivated activities are what make markets efficient. If evaluating assets is a constant cost activity, then both marginal and infra-marginal speculators make only (risk-adjusted) normal returns.
6. During the 1992 US presidential campaign, candidate Bill Clinton described himself as a Democrat with a difference.
7. Accompanying the possibility of market failure is the possibility of government failure, where the government adopts counter-productive policies, as many monetarists warn regarding Federal counter-cyclical policy.

8. See, for example, De Bondt and Thaler's (1989) review article on theory and evidence of stock market overreaction.
9. In the short run, as actors enter and leave financial markets there can easily be (positive or negative) rents earned. If the rents arise from skills of individuals who are mobile among a number of financial firms that are more or less fixed in the short run, the rents will mainly accrue to the individuals, and the market values of financial firms will show little effect of such rents.
10. Sweeney (1991a, 1991b), for US equities and foreign exchange, uses a market model to test for whether there are shifts in beta risk on days the speculator is in the market; he finds that beta changes are often not statistically important and are in any case far too small to explain measured profits. Of course, one may object that the market model is not an adequate vehicle for measuring risk and detecting risk changes, an objection illustrating the text's point. Roll (1977) argues that for the capital asset pricing model (CAPM) the data for testing its validity do not exist currently and are highly unlikely ever to exist; the CAPM remains, however, the most used model in finance for estimating the effects of risk.
11. Brown and Harlow (1988) and Howe (1986) examine larger falls and longer holding periods.
12. See Lehman (1990) for consistent tests weekly on data.
13. A rich behavioural literature is consistent with the view that these empirical results arise from over-reaction; see Kahneman and Tversky (1973, 1979); Kahneman *et al.* (1990); and Tversky and Kahneman (1981, 1986).
14. For earlier work that finds evidence of speculative profits in foreign-exchange markets but that provides no test of significance, see Logue *et al.* (1978); Dooley and Shafer (1976); and Cornell and Dietrich (1978).
15. Riksbank foreign exchange intervention during the September 1992 EC currency crisis apparently generated substantial loses, but this period is not included in the data Sjöö and Sweeney (1994) use.
16. Lo and MacKinley (1988) find evidence consistent with these results from variance ratio tests.
17. Shiller (1981a, 1981b) and LeRoy and Porter (1981) are 'first generation' variance bounds papers where the trend stationarity of real dividends is a major issue regarding the validity of the tests; see Flavin (1983) and Kleidon (1986a, 1986b). Campbell and Shiller (1987, 1988) and West (1988a, 1988b) are 'second generation' tests where dividends need not be trend stationary.
18. Shiller (1979) and Singleton (1980) examine variance bounds tests for interest rates. The finding of excess volatility for interest rates is less impressive than that for stocks, and more easily explained in terms of statistical bias (Flavin, 1983); presumably this is why debate focuses on excess volatility in stock markets.
19. Grossman and Shiller (1981) argue that the necessary variation in the real discount rate required to explain the empirical results is implausibly large. Sweeney (1992) argues that the assumption of a real discount rate is more reasonable when applied to the firm as a whole rather than to its equity, along the lines of the Miller-Modigliani theory about the effects of varying leverage on the value of the firm. It is possible that the observed fluctuations

in market indices are consistent with each firm having a constant discount rate but also having time-varying leverage.

20. But see Garber's (1990) fundamentals explanations of these famous bubbles.
21. Stiglitz (1990) argues that this calls for infinitely-lived actors or complete markets. Shleifer and Summers (1990) argue that if there is uncertainty about when the bubble will burst in a world of risk-averse short-lived actors, bubbles need not be eliminated completely.
22. But we know from Leijonhufvud's work that this price-wage sluggishness is not Keynes' transmission mechanism; see Leijonhufvud (1968, 1983, 1992).
23. In many international capital asset pricing models, this condition does not hold because a (systematic) risk premium causes a gap; this risk premium may be time-varying in response to variations in economic fundamentals.
24. Or one may view overshooting models as dynamic variants of Mundell–Fleming models.
25. This is similar to the problem that Fama and French (1988) face in looking for long-term patterns in (real) US stock market rates of return, and Huizinga (1987) in real exchange rates.
26. Meese (1986), using a MABOP-type model, finds evidence of bubbles in exchange rates; Flood and Hodrick (1990) argue that his tests are too restrictive.

References

Alexander, S. S. (1961) 'Price Movements in Speculative Markets: Trends or Random Walks', *Industrial Management Review*, vol. 2, May, pp. 7–26.

Alexander, S. S. (1964) 'Price Movements in Speculative Markets: Trends or Random Walks, No. 2', *Industrial Management Review*, vol. 5, Spring, pp. 25–46.

Atkins, A. and E. Dyl (1990) 'Price Reversals, Bid-Ask Spreads, and Market Efficiency', *Journal of Financial and Quantitative Analysis*, vol. 25, December, pp. 535–47.

Bremer, M. and R. Sweeney (1991) 'The Reversal of Large Stock-Price Decreases', *Journal of Finance*, vol. 46, June, pp. 747–54.

Bremer, M. and R. Sweeney (1993) *Short-run Rebounds after Large Stock-Price Decreases: The Virtue of Resisting Panic Selling* (Washington, DC: Georgetown University School of Business).

Brock, W. (1975) 'A Simple Perfect Foresight Monetary Model', *Journal of Monetary Economics*, vol. 1, pp. 133–50.

Brown, K. and W. V. Harlow (1988) 'Market Overreaction: Magnitude and Intensity', *Journal of Portfolio Management*, Winter, pp. 6–13.

Brown, K., W. V. Harlow and S. Tinic (1988) 'Risk Aversion, Uncertain Information, and Market Efficiency', *Journal of Financial Economics*, vol. 22, December, pp. 355–85.

Brown, K., W. V. Harlow and S. Tinic (1993) 'The Risk and Required Return of Common Stock Following Major Price Innovations', *Journal of Financial and Quantitative Analysis*, vol. 28, March, pp. 101–16.

Campbell, J. and R. Shiller (1987) 'Cointegration and Tests of Present Value Models', *Journal of Political Economy*, vol. 95, October, pp. 1062–88.

Campbell, J. and R. Shiller (1988) 'Stock Prices, Earnings, and Expected Dividends', *Journal of Finance*, vol. 43, July, pp. 661–76.

Chan, K. C. (1988) 'On the Contrarian Investment Strategy', *Journal of Business*, vol. 61, April, pp. 147–63.

Cochrane, J. (1991) 'Volatility Tests and Efficient Markets', *Journal of Monetary Economics*, vol. 27, pp. 463–85.

Cochrane, J. (1992) 'Explaining the Variance of Price-Dividend Ratios', *The Review of Financial Studies*, vol. 5, pp. 243–80.

Cornell, B. and J. K. Dietrich (1978) 'The Efficiency of the Market for Foreign Exchange Under Floating Exchange Rates', *The Review of Economics and Statistics*, vol. 60, February, 111–20.

De Bondt, W. and R. Thaler (1985) 'Does the Stock Market Overreact?', *Journal of Finance*, vol. 40, July, pp. 793–805.

De Bondt, W. and R. Thaler (1987) 'Further Evidence on Investor Overreaction and Stock Market Seasonality', *Journal of Finance*, vol. 42, July, pp. 557–581.

De Bondt, W. and R. Thaler (1989) "A Mean Reverting Walk Down Wall Street", *Journal of Economic Perspectives*, vol. 3 Winter, pp. 189–202.

Dooley, M. and J. Shafer (1976) 'Analysis of Short Run Exchange Rate Behavior: March 1973 to November 1975', International Finance Discussion Papers No. 76 (Washington, DC: Federal Reserve System) February.

Dooley, M. and J. Shafer (1983) 'Analysis of Short Run Exchange Rate Behavior: March 1973 to November 1981', in *Exchange Rate and Trade Instability*, David Bigman and Teizo Taya (eds), (Cambridge, Mass.: Ballinger), pp. 43–69.

Dornbusch, R. (1975) 'Expectations and Exchange Rate Dynamics', *Journal of Political Economy*, vol. 84, December, pp. 1161–76.

Fama, E. and M. E. Blume (1966) 'Filter Rules and Stock-Market Trading', *Journal of Business*, vol. 39, January, pp. 226–41.

Fama, E. and K. French (1988) 'Permanent and Temporary Components of Stock Prices', *Journal of Political Economy*, vol. 96, April, pp. 246–73.

Flavin, M. (1983) 'Excess Volatility in the Financial Markets: A Reassessment of the Empirical Evidence', *Journal of Political Economy*, vol. 91, February, pp. 89–111.

Flood, R. P. and R. J. Hodrick (1990) 'On Testing for Speculative Bubbles', *Journal of Economic Perspectives*, vol. 4, Spring, pp. 85–101.

Garber, P. M. (1990) 'Famous First Bubbles', *Journal of Economic Perspectives*, vol. 4, Spring, pp. 35–54.

Grossman, S. and R. Shiller (1981) 'The Determinants of the Variability of Stock Prices', *American Economic Review, Papers and Proceedings*, vol. 71, May, pp. 222–27.

Howe, J. (1986) 'Evidence on Stock Market Overreaction', *Financial Analysts Journal*, July–August, pp. 74–7.

Huizinga, J. (1987) 'An Empirical Investigation of the Long-Run Behavior of Real Exchange Rates', Carnegie-Rochester Conference Series on Public Policy, Autumn, pp. 149–214.

Kahneman, D. and A. Tversky (1973) 'On the Psychology of Prediction', *Psychological Review*, July, pp. 237–51.

Kahneman, D. and A. Tversky (1979) 'Prospect Theory: An Analysis of Decision Under Risk', *Econometrica*, March, pp. 263–90.

Kahneman, D., P. Slovic and A. Tversky (1990) 'The Causes of Preference Reversal', *American Economic Review*, vol. 80, March, pp. 204–17.

Keim, D. (1989) 'Trading Patterns, Bid-Ask Spreads, and Estimated Security Returns, The Case of Common Stocks at Calendar Turning Points', *Journal of Financial Economics*, vol. 23, November, pp. 75–98.

Kleidon, A. (1986a) 'Variance Bounds Tests and Stock Price Valuation Models', *Journal of Political Economy*, vol. 94, October, pp. 953–1001.

Kleidon, A. (1986b) 'Bias in Small Sample Tests of Stock Price Rationality', *Journal of Business*, vol. 59, April, pp. 237–61.

Lehmann, B. (1990) 'Fads, Martingales, and Market Efficiency', *Quarterly Journal of Economics*, February, pp. 1–28.

Leijonhufvud, A. (1968) *On Keynesian Economics and the Economics of Keynes* (Oxford: University Press).

Leijonhufvud, A. (1983) 'What was the Matter with IS–LM?, in J. -P. Fitoussi (ed.) *Modern Macroeconomic Theory*, (New York: Basil Blackwell).

Leijonhufvud, A. (1992) 'Keynesian Economics: Past Confusions, Future Prospects', in A. Vercelli and N. Dimitri (eds), *Macroeconomics: A Survey of Research Strategies* (Oxford University Press).

LeRoy, S. and R. Porter (1981) 'The Present Value Relation: Tests Based on Implied Variance Bounds', *Econometrica*, vol. 49, January, pp. 97–113.

Lo, A. and C. MacKinley (1988) 'Stock Market Prices Do Not Follow Random Walks: Evidence from a Simple Specification Test', *The Review of Financial Studies*, vol. 1, March, pp. 41–66.

Logue, D. E., R. J. Sweeney and T. D. Willett (1978) 'Speculative Behavior of Foreign Exchange Rates during the Current Float', *Journal of Business Research*, vol. 6, May, pp. 159–74.

Mark, N. (1992) 'Exchange Rates and Fundamentals: Evidence on Long-Horizon Predictability and Overshooting', Ohio State University Working Paper, Columbus, Ohio.

Meese, R. A. (1986) 'Testing for Bubbles in Exchange Markets: The Case of Sparkling Rates', *Journal of Political Economy*, vol. 94, April, pp. 345–73.

Meese, R. A. and K. Rogoff (1983) 'Empirical Exchange Rate Models of the Seventies; Do They Fit Out of Sample? *Journal of International Economics*, vol. 14, February, pp. 3–24.

Meese, R. A. and K. Rogoff (1988) 'Was It Real? The Exchange Rate–Interest Rate Differential Relation over the Modern Floating-Rate Period', *Journal of Finance*, September, pp. 933–48.

Mori, C. A. and M. P. Murray (1990) *Still Beating the Foreign Exchange Market* (Lewiston, Maine: Bates College).

Poterba, J. and L. Summers (1988) 'Mean Reversion in Stock Prices: Evidence and Implications', *Journal of Financial Economics*, vol. 22, October, pp. 27–59.

Praetz, P. D. (1976) 'Rates of Return on Filter Tests', *Journal of Finance*, vol. 31, March, pp. 71–5.

Praetz, P. D. (1979) 'A General Test of a Filter Effect', *Journal of Financial and Quantitative Analysis*, vol. 14, June, pp. 385–94.

Roll, R. (1977) 'A Critique of the Asset Pricing Theory's Tests. Part I: On Past and Potential Tests of the Theory', *Journal of Financial Economics*, vol. 4, pp. 129–76.

Schwartz, A. (1990) Presentation to the Shadow Federal Open Market Committee.

Shiller, R. (1979) 'The Volatility of Long Term Interest Rates and Expectations Models of the Term Structure', *Journal of Political Economy*, vol. 87, December, pp. 1190–219.

Shiller, R. (1981a) 'Do Stock Prices Move Too Much to be Justified by Subsequent Changes in Dividends?', *American Economic Review*, vol. 71, June, pp. 421–36.

Shiller, R. (1981b) 'The Use of Volatility Measures in Assessing Market Efficiency', *Journal of Finance*, vol. 36, May, pp. 291–311.

Shiller, R. (1989) *Market Volatility* (Cambridge, Mass.: MIT Press).

Shleifer, A. and L. H. Summers (1990) 'The Noise Trader Approach to Finance', *Journal of Economic Perspectives*, vol. 4, Spring, pp. 19–33.

Singleton, K. (1980) 'Expectations Models of the Term Structure and Implied Variance Bounds', *Journal of Political Economy*, vol. 88, December, pp. 1159–76.

Sjöö, B. and R. J. Sweeney (1994) *Do Central Banks Beat the Foreign Exchange Market? Evidence from Sweden* (Washington, DC: Georgetown University School of Business).

Stiglitz, J. E. (1990) 'Symposium on Bubbles', *Journal of Economic Perspectives*, vol. 4, Spring, pp. 13–18.

Surajaras, P. and R. J. Sweeney (1992) *Profit-Making Speculation in Foreign Exchange Markets* (Boulder, Col.: Westview Press).

Sweeney, R. J. (1986) 'Beating the Foreign Exchange Market', *Journal of Finance*, vol. 41, March, pp. 163–82.

Sweeney, R. J. (1988) 'Filter Rule Tests: Methods and Results', *Journal of Financial and Quantitative Analysis*, pp. 285–300.

Sweeney, R. J. (1990) 'Evidence on Short-Term Trading Strategies', *Journal of Portfolio Management*, Fall, pp. 20–6.

Sweeney, R. (1991a) *Time-Varying Risk Premia in Equities Markets* (Washington, DC: Georgetown University School of Business).

Sweeney, R. (1991b) *Time-Varying Risk Premia in Spot Foreign Exchange Markets* (Washington, DC: Georgetown University School of Business).

Sweeney, R. (1992) *Excess Volatility, Dividend Smoothing, and Leverage* (Washington, DC: Georgetown University School of Business).

Sweeney, R. (1994) *Does The Fed Beat the Foreign-Exchange Market?* (Washington, DC: Georgetown University School of Business).

Taylor, D. (1982) 'Official Intervention in the Foreign Exchange Market, or Bet Against the Central Bank', *Journal of Political Economy*, vol. 90, pp. 356–68.

Taylor, S. (1986) *Modelling Financial Time Series* (Chichester, Sussex: John Wiley).

Tversky, A. and D. Kahneman (1981) 'The Framing of Decisions and the Psychology of Choice', *Science*, 30 January, pp. 453–8.

Tversky, A. and D. Kahneman (1986) 'Rational Choice and the Framing of Decisions', *Journal of Business*, October, pp. S251–78.

West, K. (1988a) 'Dividend Innovations and Stock Price Volatility', *Econometrica*, vol. 56, January, pp. 37–61.

West, K. (1988b) 'Bubbles, Fads and Stock Price Volatility Tests: A Partial Evaluation', *Journal of Finance*, vol. 43, July, pp. 639–56.

Zarowin, P. (1989) 'Short-run Market Overreaction: Size and Seasonality Effects', *Journal of Portfolio Management*, Spring, pp. 26–9.

Zarowin, P. (1990) 'Size, Seasonality and Stock Market Overreaction', *Journal of Financial and Quantitative Analysis*, vol. 25, March, pp. 113–25.

12 From the Hayek Hypothesis to Animal Spirits: The Phase Transition Based on Competitive Experimental Market

Shu-Heng Chen, Jenn-Shyong Kuo and Chu-Chia Lin

> We might start, then by asking *how believable simple people cope with incredibly complex situations*. If we knew a bit about that, we could then go on to study the conditions under which market interaction will and will not configure the complex system into one that is incredibly smart. (Leijonhufvud (1993), p. 2)

MOTIVATION AND INTRODUCTION

In mathematical economics, the nature of the convergence of market behaviour has been studied in two different types of model. The first type pays no explicit attention to the specification of *the distribution of information* and the *communication structure*. The Arrow–Debreu model is an example. Physically speaking, in this type of model, there is no real market which can function as a *channel*[1] to transmit information. Because of this property, the convergence analysis under these models may reflect little about the information transmission among traders and about their learning processes.[2] Therefore, the *Hayek hypothesis*[3] might not be studied effectively under this types of model.

In contrast, the second type of model, which we shall call the *parallel distributed processing* (PDP) model,[4] gives explicit specification on both the *distribution of information* and *the communication structure*. Models of this type are exemplified by Kirman (1983), Haller (1990) and Ioannides (1990).[5]

The essential feature of PDP is that *no single trader possesses all relevant information in the beginning, and that is distributed over all traders.*

As exhibited by Figure 12.1, before trading, the information is distributed among all traders, which can be characterised by:

$$\Omega_i^b(0) = \{X_i^b, \succeq_i^b, \omega_i^b\} \quad i=1, 2, 3, 4 \tag{1}$$

or

$$\Omega_j^s(0) = \{Y_j^s\} \quad j = 1, 2, 3, 4 \tag{2}$$

where $\Omega_i^b(0)$ is the information set[6] of the ith buyer at trading period 0, that is before trading. X_i^b, the commodity space for the ith buyer, $\succeq_i^b$ the ith buyer's preference relation on X_i^b, ω_i^b the initial endowment of the ith buyer, $\Omega_j^s(0)$ the information set for the jth seller at trading period 0, and Y_j^s his/her production set. Equations (1) and (2) simply say that, before trading takes place, each trader only has information about her/himself and knows nothing about the others.

Pooling all buyers' and sellers' information, we gain access to the largest information set, that is,

$$\Omega(0) = \cup_{i,j}\Omega_{i,j}^{b,s}(0) \tag{3}$$

Figure 12.1 Market as a parallel distributed processor to solve allocation problems

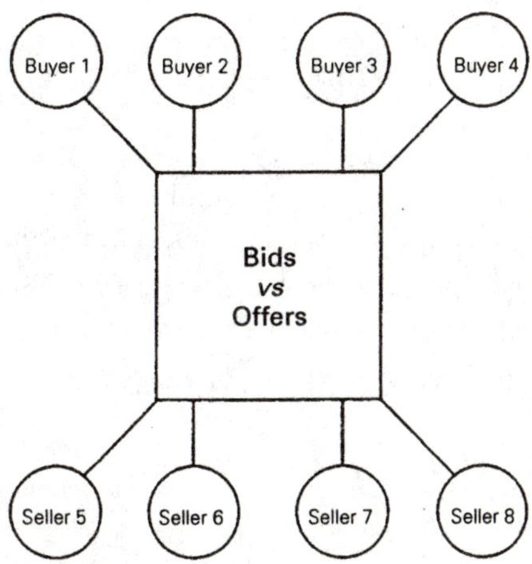

The information about $\Omega(0)$ starts to disseminate among traders through a sequence of public trading processes, that is, bidding or offering a price, taking the price, making a deal and so on. Because of learning, it is reasonable to assume:

$$\Omega_i^b(t) \subseteq \Omega_i^b(t+1) \quad \forall i \tag{4}$$

and[7]

$$\Omega_j^s(t) \subseteq \Omega_j^s(t+1) \quad \forall j \tag{5}$$

Efficient allocation of resources does not require traders to learn everything about $\Omega(0)$. Instead, it only requires traders to learn *sufficient statistics*[8] about $\Omega(0)$, that is, Ω^*. Clearly,

$$\Omega^* \subseteq \Omega(0) \tag{6}$$

Now, the Hayek hypothesis can be restated as follows.

Hayek Hypothesis in Terms of the Walrasian Equilibrium

There exists a Ω^* such that for t sufficiently large

$$\Omega_{i,j}^{b,s}(t) \to \Omega^* \quad \forall i, j \tag{7}$$

and

$$\{p(t), q(t)\} \to \{p^*, q^*\} \tag{8}$$

where $\{p^*, q^*\}$ are the *competitive equilibrium* (Walrasian equilibrium) price and quantity. The Walrasian equilibrium can be converted to a notion of *Pareto efficiency* by the first fundamental theorem of welfare economics.[9] Thus, another way to state the Hayek hypothesis is as follows.

Hayek Hypothesis in Terms of Pareto Efficiency

There exists a Ω^* such that for t sufficiently large

$$\Omega_{i,j}^{b,s}(t) \to \Omega* \quad \forall i, j \tag{7}$$

and

$$E(t) \rightarrow E^{*} \tag{9}$$

where $E(t)$ denotes the realised market efficiency at time t and E^{*} represents market efficiency in terms of Pareto optimal allocation.

By Equations (7), (8) and (9), the Hayek hypothesis can be studied as a convergence analysis in mathematical economics.

To our excitement, the recent development of experimental economics in double-auction markets[10] has lent support to the Hayek hypothesis. However, surprisingly enough, studies of double-auction markets have been mainly confined to the situation where there is no uncertainty, and hence, whether the validity of the Hayek hypothesis is robust enough for the case of uncertainty basically is unknown. Therefore, this thesis will contribute to the study of the Hayek hypothesis under uncertainty. In particular, we study the effect of *demand uncertainty* on the Hayek hypothesis.

By demand uncertainty, we mean that $\Omega_i^b(0)$ can be characterised as follows:

$$\Omega_i^b(0) = \left\{X_i^b, \geq_i^b (S), \omega_i^b\right\} \tag{10}$$

where S is a random variable with a specified probability mass function. For example,

$$S = \begin{cases} X, & \text{with probability } 0.5 \\ Y, & \text{with probability } 0.5 \end{cases}$$

That is, consumers' preference is *state-dependent*. When uncertainty is introduced, the competitive equilibrium price and quantity have to be adjusted. In economics, this can be done by the method of certainty equivalence of Simon (1956).[11] The certainty-equivalence adjusted equilibrium price and quantity will then be denoted as p^{*} and q^{*}. Similarly, we use E^{*} to denote the certainty-equivalence adjusted E^{*}.[12] After this modification, the Hayek hypothesis can be restated as follows:

Hayek Hypothesis under Uncertainty

There exists a Ω^{*} such that for t sufficiently large.

$$\Omega_{i,j}^{b,s}(t) \rightarrow \Omega^{*} \quad \forall i, j \tag{11}$$

and

$$\{p(t), q(t)\} \rightarrow \{p^*, q^*\} \tag{12}$$

or

$$E(t) \rightarrow E^* \tag{13}$$

The purpose of this thesis is to design competitive experiments based on oral double auctions and to test the validity of the three convergence properties described in Equations (11)–(13). But, what do we mean by a *test*? To answer this question, a few points have to be clarified:

- First of all, there is no observable data corresponding to $\Omega_{i,j}^{b,s}(t)$. Hence, a direct test for Equation (11) is not feasible.

This challenge leads us to test only Equations (12) and (13) in the analysis below.

- Secondly, it is very difficult to test the convergence property defined in the mathematical analysis, since all our experiments are implemented in finite periods, and convergence, in terms of mathematical analysis, cannot be defined in a finite period.

For this challenge, we have to admit that we cannot give a formal test for Equations (12)–(13) in the mathematical sense. Instead, we follow what the literature used to do, that is, simply displaying a time series plot of $p(t)$ or $E(t)$ as the evidence of convergence, at least in the visual sense. In the third section, corresponding to Equation (13), the uncertainty-adjusted efficiency will be defined respectively, and, as we shall see from the plots, almost all the sequences of efficiency based on experiments 1 to 4 have no tendency to converge.[13] This lead to the next challenge:

- The last challenge concerns the dynamic aspect of the Hayek hypothesis. For example, one might be interested in the following issue: does the degree of uncertainty have any impact on the effectiveness of using prices as a device for learning or for the communication of information.

For this challenge, if the plot has a tendency to converge, we can estimate how fast it does so. That is, we can estimate the speed of convergence by running the Daniels–Plott regression.[14] From that we can infer how effective prices can be devices for learning, and how this effectiveness will be affected under the impact of different degrees of uncertainty. If, however, the plot has no tendency to converge, which is exactly the situation troubling us, the speed of convergence cannot be defined in the same way Daniels and Plott (1988). To cope with this problem, some other measure has to be developed. The means by which we propose[15] to solve this problem is to use the *inf function* and the *sup function* borrowed from the mathematical analysis.[16]

These two functions not only have a nice mathematical property, that is, they always converge, but also have important economic implications, for example, their geometry can reflect the effectiveness of learning.[17] For example, for those sequences which have no tendency to converge, the inf function of efficiency can be used to decide whether the progress is observable in a weak sense, while the sup function of efficiency can be used to decide whether the progress is strongly observable. Economic theories based on a strong sense of rationality, such as rational expectations, might require learning to be both weakly and strongly observable, whereas those based on a weak sense of rationality, such as animal spirits, might require weak observability only.

We do not think that the study of the Hayek hypothesis under uncertainty is just a trivial extension of the study under certainty. To put it briefly, the work is motivated by a series of issues which can be classified as the phenomenon of phase transition appearing in the mathematical dynamic systems.

1. The first issue, as proposed by Leijonhufvud (1973), concerns the extent to which the economy may properly be regarded as a self-regulating system. This concern can be further exemplified by his famous *cosmological working hypothesis* or *corridor theory*:

> Within some range from the path (referred to as 'the corridor' for brevity), the system's homeostatic mechanisms work well and deviation-counteracting tendencies increase in strength. Outside that range these tendencies become weaker as the system becomes increasingly subject to 'effective demand failures'. If the system is displaced sufficiently 'far out', the forces tending to bring it back may, on balance, be so weak and sluggish that – for all *practical* purposes – the Keynesian 'unemployment equilibrium' model is as

> sensible as representation of its state as economic statics will allow. (Leijonhufvud (1981), pp. 109–10)

2. The second issue, as proposed by Davidson (1982), concerns the extent to which expectations are formed as the way rational expectation hypothesis (REH) predicts. This concern can be further exemplified by his usage of Shackle's *principle of cruciality*:

   > Admittedly it would appear that the REH is a useful analytical tool for studying *noncrucial* decision-making involving small (i.e., almost costless) differences in outcomes, for then choice can be easily replicated. The choice of whether to purchase a Winesap or a Delicious apple may be approximated by the REH analogy, but it is very doubtful that choices between expensive and far-reaching commitments (e.g., at the microlevel the purchase of durables which cannot be resold without significant costs, or at the macrolevel choices between public policies) can represented by the REH analogy. (Davidson (1982), p. 196)

3. The third issue is still in the current development of macroeconomics, that is, the existence of multiple equilibria. When a market has multiple equilibria, some of them may not be Pareto efficient. Moreover, there is no guarantee that the market can always converge to the Pareto-efficient equilibrium, no matter where it starts. In other words, the Pareto-efficient equilibrium may have a very narrow domain of attraction or may not be stable at all. While theoretical models from animal spirits, mimetic contagion, self-fulfilling processes, path-dependence and the lock-in effect[18] have enhanced our understanding about the cause of the phase transition,[19] the experimental economics provides us with an alternative approach to it.

As we shall see, our experimental results do indicate strong evidence for the appearance of the phase transition when uncertainty is introduced into the market. In Experiment 1, where there is no uncertainty, the mean efficiency of the equilibrium can be as high as 97 per cent, that is, there is only 3 per cent of efficiency loss. However, when uncertainty is introduced (Experiments 2–4), the higher the degree of uncertainty,[20] the greater the efficiency loss. For example, efficiency loss in Experiment 4 is up to 8 per cent, while the efficiency loss in Experiment 2 is more than 12 per cent.

Moreover, the convergence analysis based on the pattern of the sup of efficiency also questions the stability of the Pareto-efficient equilibrium and whether learning can be regarded as rational learning in the sense of neoclassical economics under uncertainty. In Experiment 1, the sup of efficiency is almost a flat line around 99 per cent, which means the equilibrium under certainty is not only Pareto optimal but also stable. Also, the learning effect is both strongly and weakly observable. However, when uncertainty is introduced, these properties are shaken. Especially in Experiment 2, which has the highest degree of uncertainty, the sup of efficiency drops three times and falls from 94 per cent to 89 per cent, which means that the Pareto-efficiency equilibrium is not stable and it is difficult for the corresponding learning to be rational in the sense of neoclassical economics because its progress is not strongly observable.

Altogether, this evidences that *the degree of uncertainty might be responsible for the phase transition from the Hayek hypothesis to Keynes' animal spirits*. This is the most important finding in this chapter. It has profound implications for the current controversy and near-rationality (bounded rationality) between neoclassical economics and new Keynesian economics and deserves further studies.

Whether the market behaves as the Hayek hypothesis or animal spirits predict depends on the degree of uncertainty. When the degree of uncertainty is low, the market behaves just as the Hayek hypothesis predicts, and neoclassical economics characterised by the rational expectations hypothesis might be a useful tool with which to model the economy. However, when the degree of uncertainty is high, Keynes' animal spirits might tell us more about the 'real world' than can neoclassical economics. Therefore, the traditional way to model the economy under uncertainty by the method of certainty equivalence is not appropriate when the market functions as parallel distributed processors. In the final analysis, Keynes' contribution to the concept of the *true uncertainty* and its significance for resource allocations may be worthy of revisiting.

The rest of this chapter will be organised as follows. The second section gives a description of the experimental design and procedures; the third section presents our analysis of the results, and this is followed by the concluding remarks in the fourth section.

EXPERIMENTAL DESIGN AND PROCEDURE

Experimental Design

A total of four experimental markets were studied (Experiments 1, 2, 3 and 4), and each experiment has 15–18 trading periods. Experiment 1 was implemented under no uncertainty, where the market was characterised by a fixed supply curve for a fixed demand curve (Figure 12.2).[21] Experiments 2–4 were implemented under uncertainty where the market was characterised by a fixed supply curve and stochastic demand curve. This stochastic demand curve can be described by a finite-state probability mass function. The number of states is two, three and four respectively for Experiments 2, 3 and 4.

From Table 12.1, the mass function chosen for Experiment 1 was 1 for state *Y*; for Experiment 2 as 0.5 for state *X*, and 0.5 for state *Y*; for Experiment 3 was 0.11 for state *X*, 0.78 for state *Y*, and 0.11 for state *Z*; and for Experiment 4, we had one more state, namely, *W*, and the mass function chosen for those four demand curves appearing in Figure 12.2 is 0.06 for states *X*, *Z* and *W*, and 0.82 for state *Y*. These states correspond to the three demand curves appearing in Figure 12.2. The state was determined by a draw from a bingo cage at the end of each trading period.

All these induced demand and supply curves are derived based on the redemption value of buyers and cost of sellers, respectively. The induced

Figure 12.2 The competitive market structure of supply and demand functions

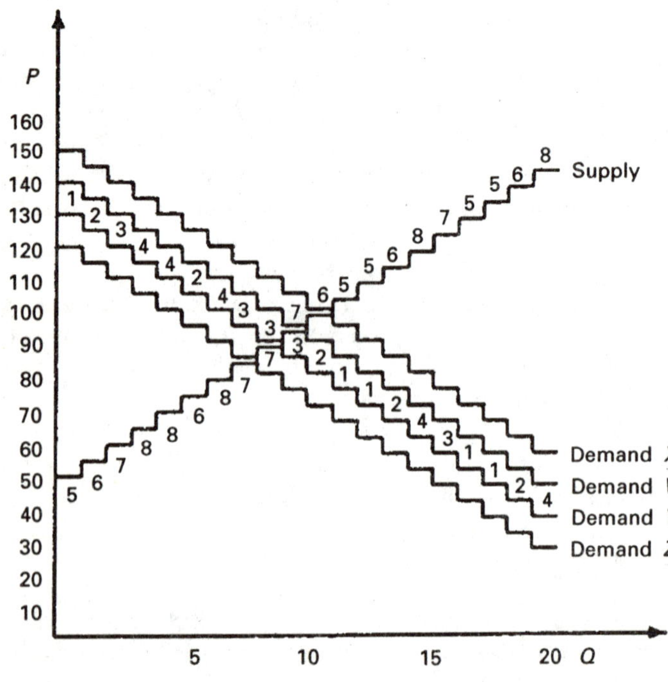

demands are derived based on the redemption value of each unit among different buyers in different states. The induced supply curve is derived based on the cost of each unit among different sellers. Each buyer can buy at most five units and each seller can sell no more than five units.[22] The incentive for selling or buying comes from the welfare triangle, that is, the sum of consumers' and producers' surplus. If the traders are profit maximisers, those demand curves and supply curves should enable us to decide a theoretical equilibrium price, P^e, and a theoretical equilibrium quantity, V^e, for Experiments 1–4. The theoretical equilibrium price is defined over a price interval corresponding to a unique theoretical equilibrium quantity. These P^es and V^es are listed in the last two columns of Table 12.1.

The market is organised as an oral double auction market and operated in a manner similar to the so-called 'New York rule'. Briefly, during the process of transaction, bids and offers are tendered verbally to an experimenter sequentially, and the experimenter will record these bids and offers with the traders' name[23] on a blackboard. In addition to bid/offer, traders can also accept the existing bids/offers or cancel or change their own bids/offers before someone else accepts them. Once a bid or offer is accepted, that is, a contract is formed, the experimenter will also mark this successful transaction on the blackboard. Thus all bids, offers and contracts are observed by all participants.[24]

Experimental Procedure

The subjects in our four experiments were drawn from a large number of applicants who were either freshmen or sophomores of National Chengchi University and had studied one or two semesters of Econ. 1 but with no expe-

Table 12.1 Experimental design

Exp.	*State of demand*	$V_t(P_k^e)$	P^e	V^e
1	P(Y) = 1	0	[90 92]	9
2	P(X, Y) = (0.5, 0.5)	25	[95 97]	10
3	P(X, Y, Z) = (0.11, 0.78, 0.11)	13.45	[90 92]	9
4	P(X, Y, Z, W) = (.06, .82, .06, .06)	8.64	[90 92]	9

Notes: P^e is the theoretical competitive equilibrium price.
V^e is the theoretical competitive equilibrium quantity.
$V_k(P^e)$ is the variance of P^e in Experiment *k*.
P_k^e is the expectation of P^e in Experiment *k*, *k* = 1, 2, 3, 4.

rience in experimental markets. To prepare for the experiments, a pilot experiment (Experiment 0) was conducted, but the results are not reported here.

Each experiment took about three and a half hours including the time spent on the instructions about trading and pre-trading practice,[25] and the money that subjects earned during each experiment was paid to them after the day of the experiment.

The markets conducted were all multiple unit oral double auctions. Each subject participated in just one experiment, and acted as either a buyer or a seller of a homogeneous, unspecified commodity. In each experiment there were four buyers and four sellers. The markets proceeded in a sequence of trading periods, each lasting five minutes. A bell was rung once one minute before the end of each trading period, and twice when there were only thirty seconds remaining.

During each trading period, subjects could bid/offer, but they had to raise their hands and get permission to do so. Once they got the permission, they presented their proposal orally in the following order: first, their number,[26] then the price, and how much they want to sell (buy) at that price. For some additional details, see Chen *et al.* (1993).

EXPERIMENTAL RESULTS

In this section, we shall first summarise some basic statistics with a simple description and then present the analysis of our experimental data.

Overview

Efficiency: E(t)

Figure 12.3 and Table 12.2 display the time series of the efficiency among all experiments. The sequence is in the order in which the period occurred. The efficiency of Experiment 1 for each period is above 90 per cent, with the exception of the second period, and is, on average higher than the efficiency of Experiments 2–4 (Table 12.3). Further examination of the results also indicates that the mean efficiency of Experiment 1 is statistically higher than that of Experiments 2–4 (see Table 12.4: Test 1–Test 3). Moreover, as can be seen in Figure 12.4, a negative relationship seems to exist between the mean efficiency and the degree of uncertainty, that is, the higher the degree of uncertainty, the lower the efficiency.

If we divide the fifteen periods equally into three stages, we can see that the mean efficiency for Experiments 1 and 4 keep on improving, and their variance keeps on decreasing, or, to put it another way, the inverse of their

Figure 12.3 The time series of efficiency

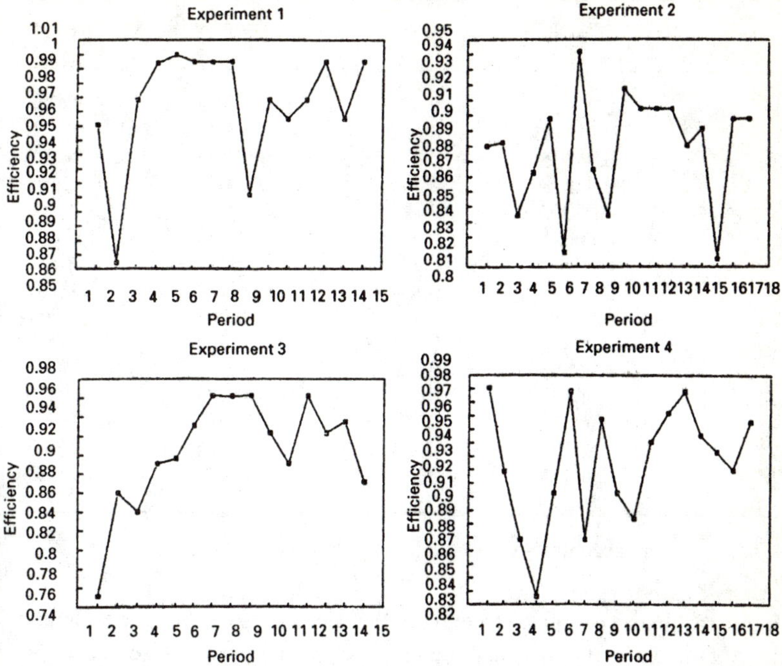

coefficient of variation keep on increasing. Experiments 2 and 3, the experiments associated with higher degree of uncertainty do not, however, share these two properties.

Contract prices: p(t)

Figure 12.5 is the time series plot of the contract prices of all the experiments. As can been seen, the actual equilibrium price is 2–4 francs less than the theoretical competitive equilibrium. As opposed to the variance of the contract price of Experiments 2–4, Experiment 1 has a stronger tendency to decline (see Table 12.5). If we divide the 15 periods equally into three groups, the variance of contract prices decline in all experiments (see Table 12.5). This property is consistent with previous research.

Efficiency

In this section we shall compare the efficiency under certainty (Experiment 1) with that under uncertainty (Experiments 2–4). We already

Table 12.2 The efficiency of all experiments

Period	*Exp.1*	*Exp. 2*	*Exp. 3*	*Exp. 4*
1	0.95080	0.87998	0.75094	0.98038
2	0.85449	0.88231	0.85995	0.91850
3	0.96825	0.83406	0.83976	0.86832
4	0.99408	0.86242	0.89071	0.82593
5	1.00000	0.89764	0.89615	0.90268
6	0.96825	0.80984	0.93095	0.97808
7	0.99470	0.94157	0.96255	0.86832
8	0.99470	0.86469	0.96254	0.95787
9	0.90211	0.83388	0.96255	0.90267
10	0.96825	0.91708	0.92327	0.88377
11	0.95503	0.90428	0.89071	0.94102
12	0.96825	0.90428	0.96255	0.96225
13	0.99470	0.90428	0.92326	0.97807
14	0.95503	0.88048	0.93544	0.94605
15	0.99471	0.89126	0.87149	0.93347

Note: E_k = the uncertainty-adjusted efficiency of experiment *k*.

know that under certainty the oral double auction institution is considerably more efficient than other institutions[27] such as the posted price auction, or the sealed second price auction, and it can take almost full advantage of trading by charging appropriate prices and making enough transactions. It would be interesting to know whether the oral double auction institution under uncertainty can perform just as well as when under certainty.

The definition of efficiency under certainty was given by Plott and Smith (1978):

> These markets are perfectly efficient if and only if the maximum amount of money is extracted by the participants from the experimenter. (p. 139)

By this statement, efficiency (E) can be defined as:

$$E(t) = \frac{\pi t}{\pi *}$$

where

π_t: The actual total profit obtained from the trade at period *t*.

Table 12.3 The mean, variance and inverse of coefficient of variation of efficiency

Exp.	*Exp. 1 mean (variance) 1/c.v.*	*Exp. 2 mean (variance) 1/c.v.*	*Exp. 3 mean (variance) 1/c.v.*	*Exp. 4 mean (variance) 1/c.v.*
1–15	0.965986 0.001537 (24.3506)	0.880536 0.001138 (26.1021)	0.904188 0.003142 (16.1308)	0.923158 0.002057 (20.3544)
1–5	0.953524 0.002767 (18.1270)	0.871282 0.000471 (40.1466)	0.847502 0.002754 (16.1495)	0.899162 0.002659 (17.4373)
6–10	0.970892 0.001287 (27.0634)	0.873412n 0.002446 (17.6600)	0.948372 0.000307 (54.1264)	0.918142 0.001814 (21.5571)
11–15	0.973544 0.000321 (54.3379)	0.896916 0.000092 (93.5100)	0.91669 0.001043 (28.3844)	0.952172 0.000256 (59.5108)

Note: c.v. coefficient of variation.

Table 12.4 Test for the equality of mean of efficiency

Test	*Exp.*	*Mean*	*Std Dev.*	*T-stat*	*Variance*	*DF*	*Prob* > |*T*|
1	1	0.96422	0.03982180	6.1192	unequal	27.5	0.0001
	2	0.88054	0.03492540	6.1192	equal	28	0.0000
2	1	0.90419	0.03982180	3.3040	unequal	24.8	0.0029
	3	0.88054	0.05802334	0.3040	equal	28	0.0026
3	1	0.96422	0.03982180	2.5832	unequal	27.3	0.0155
	4	0.92316	0.04695634	2.5832	equal	28	0.0153
4	2	0.88054	0.03492540	−1.3526	unequal	23	0.1894
	3	0.90419	0.05802344	−1.3526	equal	28	0.1870
5	2	0.88054	0.03492540	−2.8208	unequal	23	0.1894
	4	0.92316	0.04695634	−2.8208	equal	25.9	0.0091
6	3	0.90419	0.05802344	−0.9843	unequal	26.8	0.3338
	4	0.92316	0.04695634	−0.9843	equal	28.0	0.3334

Note: Std Dev = Standard deviation.
DF = Satterthwaite's approximation for degrees of freedom.

Figure 12.4 The distribution of mean efficiency according to variance

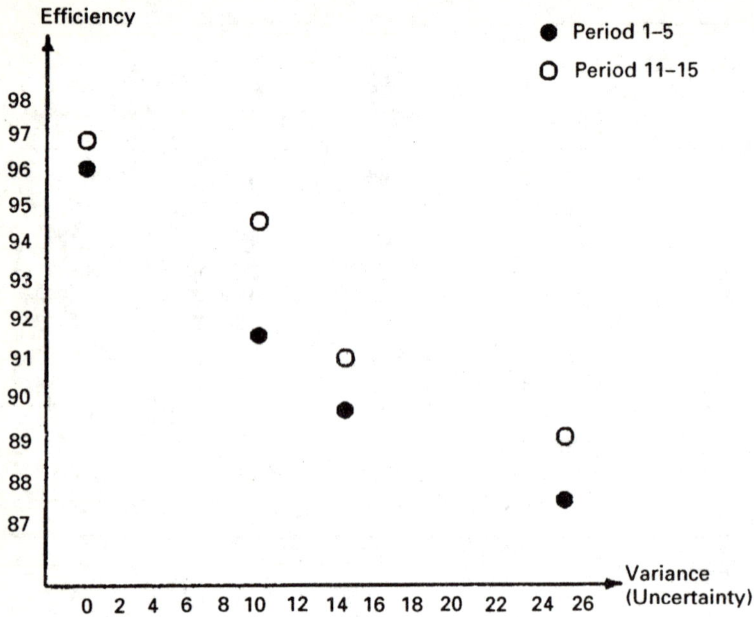

π^*: The theoretical surplus under certainty.
$E(t)$ is the ratio of the actual surplus extracted to the theoretical surplus. Thus, $E(t)$ is between 0 and 1.

However, the introduction of uncertainty complicates the definition of efficiency. An easy way to solve this problem is to assume that no risk aversion is present and that the objective of subjects is to maximise expected profit.[28] Under these assumptions, we can define an uncertainty-adjusted efficiency by applying the principle of certainty equivalence. Mathematically speaking, the definition is as follows:

$$Q(p) = \sum_{J=1}^{4} \text{card}\ (\{x : E(R(I(J), x)) \geq P\})$$

where
card: Cardinality.
$R(I(J), x, j)$. The redemption value of xth unit to $I(J)$th buyer at state j.

$$E(R(I(J),x)) = \Sigma_{j\in s}\, r_j R(I(J), x, j)$$

Figure 12.5 The time series of contract price

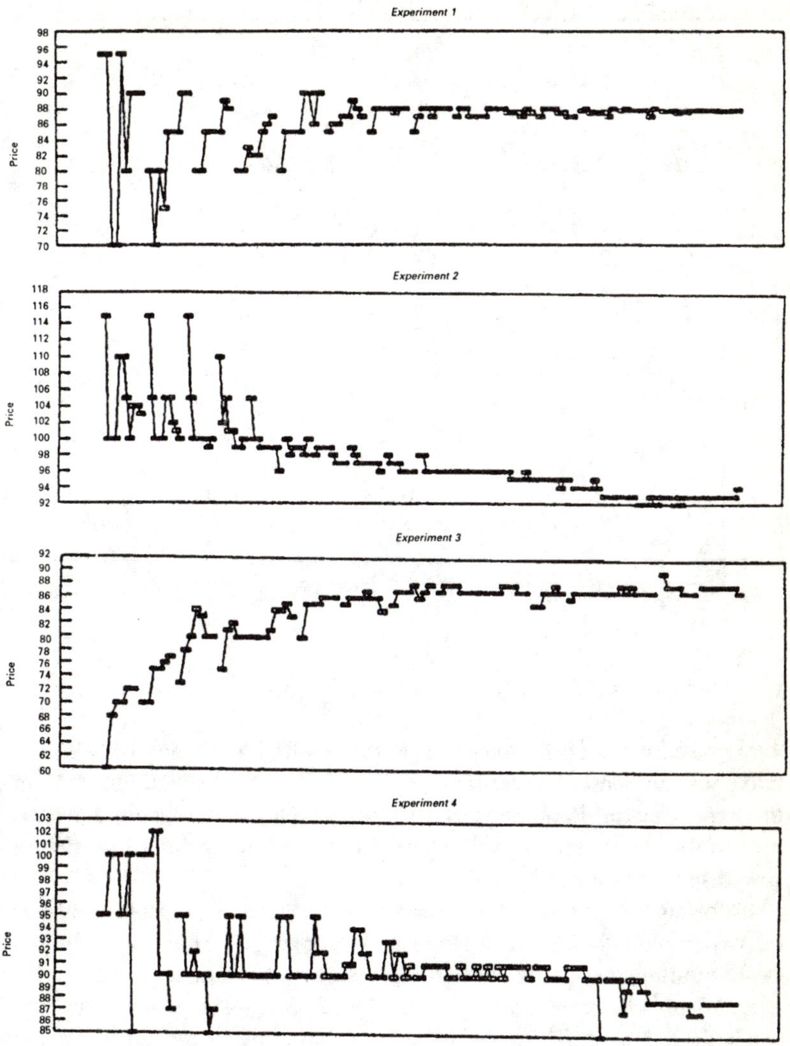

r_j: The probability with which the jth state occurs.
s = set of states.
x : 1, ..., 5

Based on the $Q(P)$ and supply curve, we then compute the uncertainty-adjusted efficiency as the ratio of the actual surplus extracted to

Table 12.5 The mean, variance and the inverse of coefficient of variation of contract price

Period	*Exp. 1* *Mean* *variance* *1/c.v.*	*Exp. 2* *Mean* *variance* *1/c.v.*	*Exp. 3* *Mean* *variance* *1/c.v.*	*Exp. 4* *Mean* *variance* *1/c.v.*
	86.45528	98.22058	83.98148	91.45238
1–15	14.71352	18.34839	30.74039	9.878293
	(22.5389)	(22.9300)	(15.1471)	(29.0974)
	84.48837	102.5217	77.125	92.93478
1–5	35.27312	20.29300	33.04687	20.40879
	(14.2257)	(22.7584)	(13.4162)	(20.5717)
	87.3625	97.39534	86.46153	90.93478
6–10	0.799843	1.494862	2.145956	1.539224
	(97.6838)	(79.6595)	(59.2181)	(73.2958)
	87.6625	94.76595	87.29729	89.26315
11–15	0.086343	1.200543	0.749452	1.419550
	(298.3323)	(86.4895)	(100.8390)	(74.9198)

Note: c.v. coefficient variation.

the theoretical surplus as in the certainty case. This ratio is also between 0 and 1.

Markets under uncertainty are perfectly efficient if, and only if, the maximum amount of *expected* money is extracted by the participants from the experimenter. If trading resulted in less than he maximum expected profit (and commissions) for the group as a whole, gains from trading would still be possible.[29]

Unfortunately, a simple time series plot of efficiency shows that none of the experiments has a strong tendency to converge – Figure 12.3. In this case, running the Daniels–Plott regression is meaningless. Therefore, alternative techniques have to be developed so that some useful analysis can still be made. Here we introduce the ideas of sup and inf, the definitions of sup and inf functions being as follows:

Let $\{x_n\}$ be a sequence of extended real numbers; we define the *superior of* x_n as:

$$sup\ (x_n) = sup_{j \geq n} x_n$$

and the *inferior of* x_n as:

$$inf\,(x_n) = inf_{j \geq n} x_n$$

The inf function and weakly observable learning

According to the definition above, the inf function *inf(y)* of *f(x)* informs us of the smallest *f(x)* when *x* is greater than or equal to *y*.[30] Therefore the *inf(t′)* of *E(t)* indicates the lowest efficiency attained by our experimental market after trading period *t′*. By definition, this function is non-decreasing; and hence, geometrically speaking, there are only two possibilities. In the first case, the inf function is flat, and, in the second case, the inf function is an upward step function.

First of all, let us consider an extreme case: that is, the *inf(t′)* function is flat. This means that the worst case happened in the last trading period and although learning improves market efficiency, it cannot prevent the market from breaking the worst record. Roughly speaking, the flat inf function simply means that the market keeps on losing what it has learned in the previous trading periods.[31] Therefore, if the market behaves as if it has a flat inf function of efficiency, the learning effect is too weak to be observed. It is in this sense that we suggest the following definition:

> The learning effect reflected in an experimental market is said to be *weakly observable* if the inf function of market efficiency is *not flat*, equivalently, it must be an *upward step function.*

The sup function and strongly observable learning

Similarly, the sup function *sup(z)* of *f(x)* informs us of the largest *f(x)* when *x* is greater than or equal to *z*. Therefore, the *sup(t′)* of *E(t)* indicates the highest efficiency attained by our experimental market since trading period *t′*. By definition, this function is non-increasing; and hence, geometrically speaking, there are only two possibilities. The first case is that we have a flat sup function and the second is a downward step function. The former implies that the highest market efficiency is attained in the last period, while the latter implies that the highest market efficiency was once achieved, but in the long run was lost. Therefore, the former suggests no evidence of retrogression; while the latter does. It is from this perspective that we have the following definition:

> The learning effect reflected in an experimental market is said to be *strongly observable* if the sup function of market efficiency is flat.

The sup and inf of E(t) and rational learning

To see the relationship between rational learning and the geometry of the sup and inf function of $E(t)$, consider the following problem. The agent in the experimental market attempts to learn about the competitive equilibrium price, P^e in a Bayesian fashion:

$$\hat{P}_t^e = E(P^e|\Omega_t)$$

Where $E(P^e|\Omega_t)$ is the conditional expectation of P^e, given Ω_t.
Furthermore, let $\hat{P}_t^e - P^e$ ($= E(P^e|\Omega_t) - P^e$) $= e_t$.
We state and prove the following theorem
Theorem 1 (Record Values)
Let $x_1, x_2, \ldots x_n$ be independent, identically distributed random variables. Let A_n be the event that a *record* occurs at time n, i.e., $A_n = \{X_n|X_n = \max_{k \le n} x_k\}$,
Then $A_1, A_2, \ldots$ are independent and $P(A_n) = \frac{1}{n}$
Proof:
Since $\{x_n\}$ are independent and uniformly distributed random variables and let $x = (x_1, \ldots, x_n)$
τ_i be the cyclic permutation
$\ni \tau_i(x_1, \ldots, x_n) \rightarrow (x_i, x_{i+1}, \ldots, x_{i-1})$
By assumption, $\tau_i x$ and x have the same distribution. One of the possible permutations is the following.
$\tau^*(x_1, \ldots, x_n) = (x_{(1)}, x_{(2)}, \ldots, x_{(n)})$
for which $x_{(1)} < x_{(2)} < \ldots < x_{(n)}$.
Since for every sequence x, there are $n!$ possible permutations, each with same density.
Then $P(x_n = \max_{k \le n} x_k) = \frac{(n-1)!}{n!} = \frac{1}{n}$ Q.E.D
Corollary 1:
Let $x_1, x_2, \ldots x_n$ be independent, identically distributed random variables. Let $y_i = |x_i|$ and B_n be the event that a record occurs at time n, i.e., $B_n = \{y_n|y_n = \max_{k<n} y_k\}$. Then $B_1, B_2, \ldots$ are independent and $P(B_n) = \frac{1}{n}$.
Corollary 2:
Let $x_1, x_2, \ldots x_n$ be independent, identically distributed random variables. Let $y_i = |x_i|$ and C_n be the event that a record occurs at time n, i.e., $C_n = \{y_n|y_n = \min_{k<n} y_k\}$. Then $C_1, C_2, \ldots$ are independent and $P(C_n) = \frac{1}{n}$.
Now we shall show that a modification of Theorem 1 can be applied to $\{|e_t|\}$ by checking the i.i.d. conditions.

(a) Independence
Since the conditional expectation operator $E(P^e|\Omega_t)$ is the Gram–Schmidt orthogonal process applied to a functional vector

space (Hilbert space), $\{e_t\}$ are serially uncorrelated. Moreover, if $\{e_t\}$ are multivariate normal, then $\{e_t\}$ are also independent.

(b) Identical distribution

However, by the next theorem, $\{e_t\}$ are not identically distributed.

Theorem 2:

Suppose $\Omega_{t-1} \subseteq \Omega_t$,

then $E[P^e - E[P^e|\Omega_t]]^2 \leq E[P^e - E[P^e|\Omega_{t-1}]]^2$

By Theorem 2,

$Var(e_t) \leq Var(e_{t-1})$

Even though $\{e_t\}$ are i.n.i.d. (independent but not identically distributed), it does not mean that Theorem 1 is of no use, as can be shown by Theorem 3. In fact, we state the following theorem without proof.

Theorem 3:

Let $x_1, x_2, \ldots, x_n$ be multivariate independent random variables, whose first and second moments are described as follows:

(a) $E(x_t) = 0$

(b) $Var(x_t) < Var(x_{t-1})$

Let B_n be defined as in Corollary 1 and C_n be defined as in Corollary 2, then $P(B_n) \leq \frac{1}{n}$ and $P(C_n) \geq \frac{1}{n}$.

What Theorem 3 says is that if the learning behaviour of the representative agent is described by Bayesian learning, the chance of making the biggest mistake is decreasing at most in the order of $\frac{1}{n}$ and the chance of making the smallest mistake is increasing at least in the order of $\frac{1}{n}$.

Since bigger mistakes correspond to greater loss of efficiency,[32] the theorem of record values can be applied to our efficiency measure E(t) too. This makes perfect sense, because if the probability of making the biggest mistake is decreasing at least in the order of $1/n$, then the probability of making the lowest profit is also at most in the order of $1/n$. Based on this observation, it is clear that the sup and inf function of $E(t)$ appearing in Type A of Figure 12.6 is more probable than that appearing in Type B of Figure 12.6.

Moreover, Type A observed in Figure 12.6 satisfies the definition of strong learning and weak learning, that is, it has a flat sup function and an upward inf function. *Therefore, rational learning is expected to meet the standard of both strong and weak learning.*

The sup function of $E(t)$ in Experiment 1 is much closer to Type A, while that in Experiment 2 seems to be closer to Type B (see Figure 12.7).

In Experiment 1, the sup of efficiency is almost a flat line around 99 per cent which means that the learning effect is both strongly and weakly observable. However, when uncertainty is introduced, these properties

Figure 12.6 Rational learning and the sup and inf function of efficiency, $E(t)$

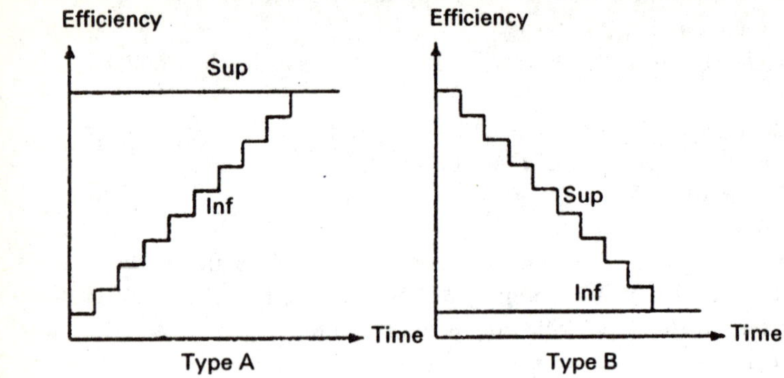

Figure 12.7 The sup and inf function of efficiency

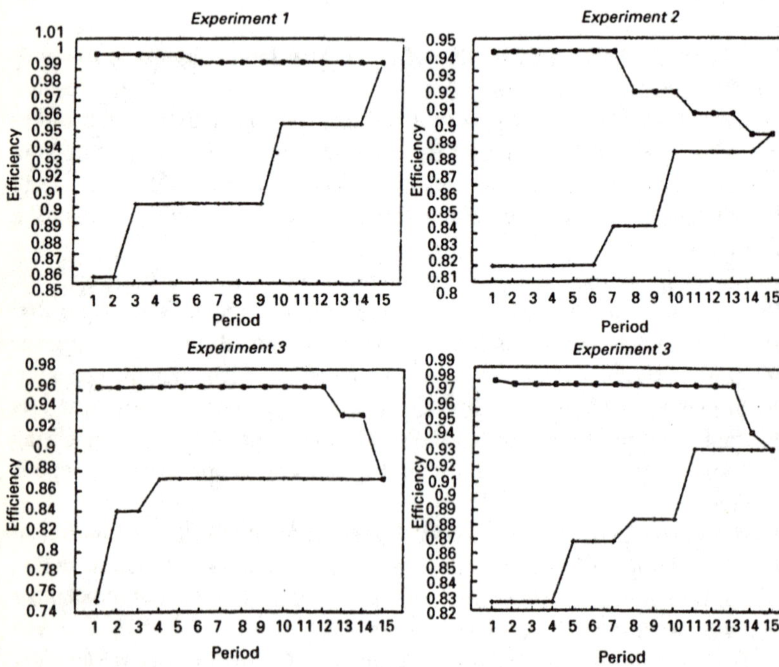

are shaken. In Experiment 2, which has the highest degree of uncertainty, the sup of efficiency drops three times and falls from 94 per cent to 89 per cent, which means that the learning effect is only weakly observable.

Table 12.6 The shape of the sup and inf of $E(t)$

	Flat sup	*Upward sloping inf*
Experiment 1	Yes	Yes
Experiment 2	No	Yes
Experiment 3	No	Yes
Experiment 4	No	Yes

The results have been summarised in Table 12.6. The patterns seem to suggest two things:

(a) First, learning behaviour under certainty (Experiment 1) is obviously different from that under uncertainty (Experiments 2–4). More specifically, the rationality of the former seems to be stronger than that of the latter.

(b) Secondly, rationality seems to be the weakest in Experiment 2, which has the highest degree of uncertainty. However, we are not sure whether the evidence is strong enough to suggest a negative relationship between rationality and uncertainty.

Conclusion

Based on the analysis above, we shall conclude this section by:

- reviewing the questions addressed in this chapter;
- providing the answers based on our experimental evidence; and
- suggesting the possible direction for further studies.
- *Q1: Does uncertainty have any effect on the validity of the Hayek hypothesis?*

Generally speaking, our experimental results do not suggest that the validity of the Hayek hypothesis is well-established in the case of uncertainty. The direct evidence against the validity of the Hayek hypothesis under certainty is *mean efficiency loss*. As can be seen from Tables 12.3 and 12.4, the mean efficiency loss of the market under uncertainty is generally greater than that of the market under certainty.

One may wonder, however, whether it takes longer to learn under uncertainty. If such is the case, then as time goes by the divergence of efficiency loss between the market under uncertainty and that under certainty will vanish. Thus, in the long run, the Hayek hypothesis is valid even with markets under uncertainty. Unfortunately, 'in the long run' is a not a working concept in the experimental markets, so we certainly cannot argue about what will happen in the long run. But, by comparing the last five-period mean efficiency loss, our attitude toward the long-run argument is sceptical, if not pessimistic.

- *Q2: If the Hayek hypothesis is not valid for the market under uncertainty, what kind of explanation might we be able to offer? Equivalently, can we enhance our understanding of economics by explaining why the Hayek hypothesis fails to hold in markets under uncertainty?*

Questions such as these might be difficult to find in mathematical economics and can be difficult for econometricians to answer, but by means of experimental economics, it is easier for us to search for an explanation, since we are much more confident of the environment from which the results were generated. The answer we feel bold enough to propose is the following: *The rationality in markets under certainty might be different from that under uncertainty.*

The evidence supporting this explanation is the analysis based on the method which had never been used in experimental economics, that is, *the sup and inf function* borrowed from mathematical analysis. Based on the theory of records value in probability theory, we have constructed a connection between rational learning (defined in terms of conditional expectations) and the behaviour of the sup and inf function of E(t). Surprisingly enough, only markets under certainty meet all requirements of weak and strong learning, while markets under uncertainty can only satisfy weak learning, but not strong learning.

- *Q3: If the Hayek hypothesis is not valid for markets under uncertainty, and the answer for Q2 lies in the relationship between rationality and uncertainty, is this relation regular? For example, does higher degree of uncertainty result in weaker rationality, and hence, weaker learning?*

We do not feel that our evidence is strong enough to give an unequivocal answer to this question, though we do have some evidence showing that the higher the degree of uncertainty, the greater the mean efficiency loss. More careful experiments need to be designed to get a closer look at this issue.

- *Q4: If the degree of rationality does have a functional relationship with the degree of uncertainty as described in Q3, how can we offer an explanation with microfoundations?*

Up to now, our experiments have not said anything about this. However, recent developments in computable economics may suggest an answer for us. According to the theory of computability and complexity and its connection to information theory, that is, the algorithmic information theory, bounded rationality can, in effect, be bounded by the degree of uncertainty. In terms of computable economics, this means that the computational constraints can be determined by the degree of uncertainty.[33] Therefore, there is a bound or constraint beyond which rational expectations may give way to animal spirits. Thus, experimental economics armed with computable economics might shed new light on the foundations of economics.

CONCLUDING REMARKS

As Leijonhufvud's corridor hypothesis correctly points out, macroeconomic theory which can allow exclusively one side of life to appear, no matter which side: Hell or Heaven, efficient or inefficient, full employment or mass unemployment, is not a good one. A good macroeconomic theory must be 'balanced' in the sense that both sides of the world can be exhibited. Moreover, it should enable us to predict under what circumstances we can have order, that is, markets can regulate themselves well; and under what circumstances we are left with nothing but chaos, that is, coordination failures.

> This system will *work quite well under most conditions but will work badly under some (extreme) conditions.* Qualitatively, this is as it should be, I think. All other self-regulating systems that we know in nature or from engineering have *bounded homeostatic capabilities.* Macroeconomists ought to wed themselves to models that work 'for richer for poorer, in sickness or in health'. What we have inherited, however, are models of just two types of systems – those that work without fail, and those that always fail to work. (Leijonhufvud (1993), p. 8)

In terms of modern mathematical language, Leijonhufvud's corridor hypothesis is precisely a statement about phase transition. In fact, Keynes (1936) proposed a theory to account for the phase transition from the perfect market to a handicapped one. His theory is based on quite a different perspective: to treat *uncertainty*, called the *true uncertainty*, by Davidson[34] and the post-Keynesians.

Unfortunately, no matter how profound Keynes' 'true uncertainty' is, the statistical foundation for his 'true uncertainty' has not been well established as yet. Hence it is by no means an easy job to test the significance of this 'true uncertainty' by econometrics. However, the approach based on experimental economics gives us an excellent chance to implement such a test, and this chapter, to the best of our knowledge, is the first paper to use the experimental method to test whether the degree of uncertainty, as predicted by Keynes, would be responsible for the phase transition from the Hayek hypothesis or rational expectations to animal spirits. The answer based on our results seems to be positive. Our experimental results offer evidence that bounded rationality can, in effect, be bounded by the degree of uncertainty. In modern terminology,[35] this means that the *computational constraints* can be determined by the degree of uncertainty. Therefore, there is a bound or constraint beyond which rational expectations may give way to animal spirits.

The conclusions made above are based on two different but related analyses: the mean efficiency; and the shape of the sup and inf functions of the efficiency of each market associated with different degrees of uncertainty. We found that, in terms of these two criteria, the market with the best performance is always Experiment 1, that is, the market without uncertainty, whereas the worst case is always Experiment 2, that is, the market with the highest degree of uncertainty. For example, when there is no uncertainty, the mean efficiency loss is less than 3.5 per cent. However, when the degree of uncertainty increases to 25, the average efficiency loss is high: up to 12 per cent.

The use of the shape of the sup and the inf of the efficiency is a new technique and, to the best of our knowledge, has not been explored by any other experimental economists. These techniques are extremely useful for convergence analysis, especially for those sequences which show no tendency to converge. In fact, they give an alternative way to define or describe what rational expectations or rational learning is. The flat sup of efficiency in Experiment 1 seems to reflect the market behaviour characterised by rational expectations or rational learning in neoclassical economics. On the other hand, the decreasing sup of efficiency in Experiment 2 seems to reflect the market behaviour characterised by bounded rationality or animal spirits in the new Keynesian economics. Thus, the phase transition from the Hayek hypothesis to animal spirits can exist even in such simple oral double auction markets.

Notes

1. I mean the *channel* used in information theory, such as Cover and Thomas (1991). O'Neill (1987) appeared to be the first application of *Shannon's*

coding theorem in the area of economic modelling. He showed that Shannon's coding theorem allows one precisely to define the sense in which *information is transmitted by an economic market (channel).* With the help of *rate distortion theory*, he was able to show that the Grossman–Stiglitz equilibrium (1980) requires signalling rates in excess of the channel capacity.

2. The essential force supporting convergence of this type of models is a variant of Newton's algorithm, such as Scarf's algorithm (1967, 1973) and Merrill's algorithm (1972). It turns out that the convergence property is mathematised as a recursive mapping under which the Brouwer fixed point or the Kakutani fixed point theorem can be realised. Therefore, the snapshot or instantaneous description of the computation of these algorithms seems to reflect nothing about the snapshot of information transmission among traders and about their learning processes.
3. The Hayek hypothesis asserts that a market can work correctly even if the participants have very limited knowledge about their environment or about other participants, see Smith (1982). Whether the Hayek hypothesis is the upshot of Hayek (1945) is still a moot point. See Thomsen (1992).
4. One might use different names, such as neural-network models, models of cellular automata, and graphical models.
5. Kirman (1992) did a brief survey of these models.
6. Formally speaking, it is required to be a σ – *algebra.*
7. Formally speaking, we are assuming that the sequence of information sets constitute a *filtration.*
8. For the interpretation of Hayek's view in terms of prices as sufficient statistics, see Koopmans (1957) and Williamson (1975).
9. The first fundamental theorem of welfare economics states that an allocation which is supported by a non-zero price (in particular a Walrasian equilibrium) is Pareto optimal.
10. That is, markets where both buyers and sellers can negotiate prices with each other. This type of market in experimental economics is distinguished from 'posted-offer' markets where sellers determine their prices and buyers are allowed only to decide whether or not to buy from them. Ketcham, *et al.* (1984) have shown that prices tend to be higher and efficiency lower under posted-offer pricing relative to double auction. Here, we only consider double auction markets. The surprising efficiency of these markets has been well established by Williams (1980), Plott (1982), and Friedman (1984).
11. Certainty equivalence means that in the stochastic problem we only need to know the first moment of the probability distribution of stochastic disturbance.
12. When uncertainty exists, the efficiency measure usually assumes no risk aversion.
13. See Table 12.2 on page 298 and Figure 12.3 on page 301.
14. See Daniels and Plott (1988)
15. Literature leave us no answer to this challenge. This is because most of their results in terms of contract prices have the property of convergence.
16. The definition of these two functions will be given in the third section of this chapter (page 306–7).
17. See the third section of this chapter (pp. 307–11).
18. See Arthur *et al.* (1987), Howitt and McAfee (1992), Topol (1991) and Woodford (1991).

19. That is, the transition from the existence of the global attractor to the existence of multiple equilibria. Bifurcation is a typical example.
20. The degree of uncertainty is measured by variance rather than by entropy.
21. Here, we employed the same induced supply and demand configuration as Daniels and Plott (1988).
22. For more details of individual redemption value and cost, please see Chen *et al.* (1993).
23. In our experiment, this is represented as a number.
24. See Plott (1982) for more details of the market process.
25. Two-period simulated trading was conducted to make sure that everybody understood the trading instructions.
26. Buyers were numbered 1, 2, 3 and 4. Sellers were numbered 5, 6, 7 and 8.
27. There are several papers discussing this issue, such as Plott and Smith (1978), Williams and Smith (1984), Smith (1982), and Plott (1982, 1989).
28. This assumption can be justified by the *law of large numbers*, in which case the appearance of uncertainty is trivial because of the property of *almost-sure convergence*.
29. The reason is that extra-marginal units have been traded or intra-marginal units and marginal units have not been traded. See Plott and Smith (1978).
30. In our case, x is allowed only to be taken on integer. So the previous definition applies straightforwardly.
31. Of course, in the real world, we cannot do so. We cannot blame decision-makers for having learned nothing if they make another record-breaking mistake, because the environment in which they have made the decisions has also changed. But in the experiment, the institutions and the environment of the market did not change from the beginning to the end.
32. We do not know exactly how $E(t)$ can be related to e_t. Here, we only assume $E(t)$ is a non-decreasing function of $|e_t|$.
33. For research taking this direction, see Norman and Shimer (1994).
34. See Davidson (1991) and Fitzgibbons (1988).
35. See Spear (1989).

References

Arthur, W. B., Y. M. Eromliev and Y. M. Kaniovski (1987) 'Path-Dependent Processes and the Emergence of Macro-Structure', *European Journal of Operational Research*, vol. 30, pp. 294–303.

Cover, T. M. and J. A. Thomas (1991) *Elements of Information Theory*, (New York: John Wiley)

Chen, S., J. Kuo and C. Lin (1993) 'From Hayek Hypothesis to Animal Spirits', Working paper No. 9403, National Chengchi University.

Daniels, B. and C. R. Plott (1988) 'Inflation and Expectation in Experimental Markets', in R. Tietz, W. Albers and R. Selten (eds), *Bounded Rational Behavior in Experimental Games and Markets*, Lecture Notes in Economics and Mathematical Systems, No. 314 (Berlin/Heidelberg: Springer-Verlag).

Davidson, P. (1982) 'Rational Expectations: A Fallacious Foundation for Studying Crucial Decision-Making Processes', *Journal of Post Keynesian Economics*, vol. V, no. 2, pp. 182–98.

Davidson, P. (1991) 'Is Probability Theory Relevant for Uncertainty? A Post Keynesian Perspective', *Journal of Economic Perspectives*, vol. 5, no. 1, pp. 129–143.

Fitzgibbons, A. (1988) *Keynes's Vision – A New Political Economy* (Oxford University Press).

Friedman, D. (1984) 'On the Efficiency of Experimental Double Auction Markets', *American Economic Review*, vol. 74, pp. 60–71.

Grossman, S. J. and J. E. Siglitz (1980) 'On the Impossibility of Informationally Efficient Markets', *American Economic Review*, vol. 70, pp. 393–408.

Haller, H. (1990) 'Large Random Graphs in Pseudo-Metric Spaces', *Mathematical Social Science*, vol. 20, pp. 165–93.

Hayek, F. A. (1945) 'The Use of Knowledge in Society', *American Economic Review*, vol. 54, pp. 519–30.

Howitt, P. and R. P. McAfee (1992) 'Animal Spirits', *American Economic Review*, vol. 82, pp. 493–507.

Ioannides, Y. M. (1990) 'Trading Uncertainty and Market Form', *International Economic Review*, vol. 31, pp. 619–38.

Ketcham, J., V. L. Smith and A. W. Williams (1984) 'A Comparison of Posted-Offer and Double-Auction Pricing Institutions', *Review of Economic Studies*, pp. 595–614.

Keynes, J. M. (1936) *The General Theory of Employment Interest and Money*, (London: Macmillan).

Kirman, A. P. (1983) 'Communication in Markets – A Suggested Approach', *Economics Letters*, vol. 12, pp. 101–8.

Kirman, A. P. (1992) 'Whom or What Does the Representative Individual Represent?', *The Journal of Economic Perspectives*, vol. 6, pp. 117–36.

Koopmans, T. C. (1957) *Three Essays on the Sate of Economic Science* (New York: McGraw-Hill).

Leijonhufvud, A. (1973) 'Effective Demand Failures', *Swedish Journal of Economics*, vol. 75, pp. 27–58, reprinted in Leijonhufvud (1981).

Leijonhufvud, A. (1981) *Information and Coordination – Essays in Macroeconomic Theory* (New York: Oxford University Press).

Leijonhufvud, A. (1993) 'Towards A Not-Too-Rational Macroeconomics', *Southern Economic Journal*, vol. 60, no. 1, pp. 1–13.

Merrill, O. H. (1972) 'Applications and Extension of an Algorithm that Computes Fixed Points of Certain Upper Semi-Continuous Point to Set Mappings', PhD thesis, Department of Industrial Engineering, University of Michigan.

Norman, A. L. and D. W. Shimer (1994) 'Risk, Uncertainty and Complexity', *Journal of Economic Dynamics and Control*, vol. 18, pp. 231–49.

O'Niell, W. D. (1987) 'An Application of Shannon's Coding Theorem to Information Transmission in Economic Markets', *Information Science*, vol. 41, pp. 171–85.

Plott, C. R. (1982) 'Industrial Organization Theory and Experimental Economics', *Journal of Economic Literature*, vol. 20, pp. 1485–527.

Plott, C. R. (1989) 'An Updated Review of Industrial Organization Application of Experimental Methods', in R. Schmalensee and R. D. Willig (eds), *Handbook of Industrial Organization*, vol. II (Amsterdam: North-Holland), pp. 1111–76.

Plott, C. R. and V. L. Smith (1978) 'An Experimental Examination of Two Exchange Institutions', *Review of Economic Studies*, vol. 45, pp. 133–53.

Scarf, H. E. (1967) 'The Approximation of Fixed Points of a Continuous Mapping', *Journal of Applied Mathematics*, vol. 15, pp. 1328–43.

Scarf, H. E. (with T. Hansen) (1973) *The Computation of Economic Equilibria* (New Haven Conn.: Yale University Press).

Simon, H. A. (1956) 'Dynamic Programming Under Uncertainty with a Quadratic Criterion Function', *Econometrica*, vol. 24, pp. 74–81.

Smith, V. L. (1982) 'Microeconomic Systems as An Experimental Science', *American Economic Review*, vol. 72, pp. 923–55.

Smith, V. L. and A. W. Williams (1982) 'The Effects of Rent Asymmetries in Experimental Auction Markets', *Journal of Economic Behavior and Organization*, vol. 3, pp. 99–116.

Spear, S. (1989) 'Learning Rational Expectations under Computability Constraints', *Econometrica*, vol. 57, no. 4, pp. 889–910.

Thomsen, E. F. (1992) *Price and Knowledge – a Market-Process Perspective* (London: Routledge).

Topol, R. (1991) 'Bubbles and Volatility of Stock Prices: Effect of Mimetic Contagion', *The Economic Journal*, vol. 101, pp. 786–800.

Williams, A. W. (1980) 'Computerized Double-Auction Markets: Some Initial Experimental Results', *Journal of Business*, vol. 53, pp. 235–58.

Williams, A. W. and V. L. Smith (1984) 'Cyclical Double-Auction Market With and Without Speculation', *Journal of Business*, vol. 57, pp. 319–51.

Williamson, O. E. (1975) *Markets and Hierarchies: Analysis and Antitrust Implications* (New York: Free Press).

Woodford, M. (1991) 'Self-Fulfilling Expectations and Fluctuations in Aggregate Demand', in N. G. Mankiw and D. Romer (eds), *New Keynesian Macroeconomics* (Cambridge, Mass.: MIT Press).

Part V

13 Synchronising Markets, Mutual Information and the Price Level: Co-ordination in a Non-General Equilibrium World

Arthur DeVany

INTRODUCTION

In his paper 'Information costs and the division of labor', Axel Leijonhufvud (1989) describes a system of production in which firms show increasing returns and every firm uses inputs that are produced by other firms in the system. Such a system of interrelated firms will show increasing returns if they overcome the co-ordination failures to which the system is subject. Co-ordination failures are failures of timing and synchronisation: when a critical input is not supplied on time this affects the production and delivery of other products throughout the system. They are brought on when specialised assets become the subject of opportunistic hold-ups, and this problem is also a failure of timing and market illiquidity.

A hold-up of just one input or output can cascade through the system, leaving many firms without an essential input, who are then unable to supply their outputs to other firms, who consequently fail to deliver to still others and so on. Co-ordination failures produce highly complex dynamics that include recessions, bubbles, liquidity constraints, sluggish prices, and excess demand or supply. They break down the mutual information in prices so that the prices of different commodities or assets contain different information, some stale, some fresh, some reflecting liquidity constraints, and others reflecting unanticipated states of the world.

To achieve the co-ordination required for a system of interrelated production to work well is a delicate balancing act. Contracts and hold-up problems must be solved. Markets must be organised so that asset trading is not too thin (another form of the hold-up problem). External shocks drive the evolution of the system through time. Endogenous uncertainty is intrinsic to the system, because the outcome depends on the interlinked actions of all the components. The system is complex and highly non-linear, and it is likely to have chaotic dynamics.

The system set out in Leijonhufvud (1989) contains all the ingredients for the co-ordination and information failures that are at the core of his research programme. All the important questions he has asked can be posed there in fruitful ways. What sorts of institutional arrangement are required to co-ordinate decentralised actions in this interrelated system? How are prices made, and what sorts of expectation will form? How will such a system detect inflation, and what institutions would it devise to cope with it? How does a price level emerge in this system? What effects does inflation have on real variables? In this chapter, we want to try to develop a model of the world which Axel describes in his 1989 paper, and show how it can be made to shed light on these questions.

We begin by showing why the institutions of a general equilibrium world are not up to the tasks of Axel's world. Then we pose an institutional set-up that co-ordinates trading in this world and show that it requires that markets be hierarchically organised. Pricing and the formation of expectations are shown to be emergent properties of the system; rational expectations emerge as a collective property of individual reservation price strategies – a form of Axel's corridor hypothesis. In the rational expectations equilibrium, no agent knows the distribution of outcomes or models the price-making process: they follow their own reservation price strategies and rational expectations is an emergent property of the interactions among agents and the rules of the market. Then we tackle the problem of synchronising markets that are operating on different frequencies, and show how this leads to temporal market hierarchies and finite price velocities. Shocks propagate through the system at a pace that is tied to the temporal frequencies of the underlying markets. Sustained and variable shocks destroy the mutual information in prices and have real effects.

THE CO-ORDINATION PROBLEM

The co-ordination problem we consider is among a set N of firms. Firm i's output $Y_{i,t}$ at time t is given by its production function:

$$Y_{i,t} = f_i(Y_{j \neq i,\, t-1}, K_{t-1}), \forall i \in N \tag{1}$$

i's output depends on the past outputs of all the other firms, denoted as the vector $Y_{j \neq i,\, t-1}$, and on the capital stocks of all firms, i as well as all the others, in the prior period; we denote this as the capital vector K_{t-1}.

These production relationships hold for all firms. Firms supply final outputs that are consumed as well as outputs that are inputs to other firms. We assume that these firms have been organised to capture economies of scale in the production of their own outputs. Even if individual production functions do not show non-convexities, the system in the aggregate will, in general, be non-convex (Milgrom and Roberts, 1990). Positive and negative complementarities make the interactions complex and raise the need for co-ordinating actions.

Strong complementarities might lead to mergers among activities, but this can lead to a loss of economies of scale. Scale economies require firms to rely on other suppliers with whom their production rates must be co-ordinated. All these trade-offs will be influenced by the cost of markets, contracts and internal controls. The sizes and shapes of firms and how they relate to one another will evolve over time towards higher levels of effectiveness and co-ordination. The organisation and co-ordinating mechanisms co-evolve. Axel argues that the organisation of the system evolves toward a coalition-like structure that is sustained by agreements and social institutions. The system evolves recursively through iterations of Equation 1. The output vector is a mapping from itself to itself; it is highly non-linear and subject to complex feedback.

GENERAL EQUILIBRIUM INSTITUTIONS WILL NOT WORK

The natural way to try to solve the co-ordination problems which Axel's complex structure presents is to put them in a general equilibrium framework. In the institutional and informational setting of a general equilibrium economy, with its time-dated commodities, state contingent assets and competitive markets, all the present and future prices in all possible states could be solved for simultaneously. There would be no co-ordination problem. In his 'Toward a Not-Too-Rational Macroeconomics', (Leijonhufvud, 1993) Axel argues why a general equilibrium framework is not suited to this problem. Among the criticisms he makes are that the computational burden on the agents is overwhelming, the dynamical system is too complex for the optimisation problem to be solved, and the expectations requirements are unreasonable.

Endogenous Uncertainty

We agree with this assessment, but we want to argue that there is a deeper contradiction in the general equilibrium framework created by the endogenous uncertainty that is intrinsic to Axel's system. The non-linear and complex system generates its own sources of uncertainty, and the effects of external shocks are not predictable, because they depend both on the state and the history of the system when they occur. The contractual and hold-up problems associated with asset specificity and interdependencies among firms at all levels is another source of internal uncertainty. State-contingent assets are not up to these problems.

When the operation of markets is the source of their own uncertainty, they cannot hedge one another simultaneously, as Chichilniski (1992) has shown. This self-referential loop has no solution when the futures and spot prices must be computed simultaneously. Some form of symmetry breaking is required. Markets must be arranged in a hierarchy to break the self-referential loop. Markets with a hierarchical structure supply information from lower to higher levels and they do it sequentially, which gives a temporal structure to actions. That is why futures markets clear after their underlying spot markets clear, and similarly for index trading.

Timeless Markets

In the general equilibrium model, trading is essentially timeless; all trades now and in the future take place at a single point in time. After that, the economy moves through a fixed state space according to the random shocks it receives, triggering the state-contingent contracts as it goes. The dates in the model are essentially the settlement dates of the contracts. The settlement dates are all the same for all commodities, and every market must operate on the same frequency. The shocks are nicely timed to occur once per period for all commodities.

But, in Axel's complex production system, the shocks are not so nicely timed and they come from exogenous and endogenous sources. If external shocks arrive according to a stochastic process, for example, more of them will impinge on commodities that have longer production periods. Futures contracts must have settlement dates that are appropriate to the length of the production period, because this is the period over which the hedge must be operable. This means that the apparent synchronisation of markets that the general equilibrium process achieves is an artefact of the common dating of all commodities. These common dates will not usually make good settlement dates for every spot and futures market.

The number of shocks generated exogenously and endogenously will not respect the general equilibrium requirement that there be just one per period; some periods will have many, others few and these will impinge differentially on commodities according to their production periods. The general equilibrium common time-dating of commodities may require some to trade at too high a frequency. Higher than optimal frequency will cause these markets to be too thin at the common general equilibrium contract settlement dates. Hence, their prices will be less liquid at these frequencies and will not reflect competitive conditions. For others, the general equilibrium frequency will be too low and traders will bear unnecessary inventory costs and risks.

Evolving State Space

Lastly, the timeless trading of the general equilibrium model may work for a world where all possible shocks are known and have definite probabilities, but that is not the interesting world of Axel's complex system. That world is driven by a stochastic dynamic in which the possibilities unfold as the world moves through its states. The state space is not stationary, as in general equilibrium, but evolves recursively, possibly even chaotically, always generating new possibilities as it goes.

MARKET TIMING, PRICING AND LIQUIDITY

Matching temporal sequences and hierarchies is the prime institutional problem to be solved in co-ordinating Axel's complex system of interrelated production. In general, markets will trade according to their own 'rhythms' and their periods must be synchronised in order to supply the right inputs at the right times and for prices to inform trading in the next hierarchy. Consider the simple case of bread. The wheat market clears annually. The inventory is drawn down monthly. It can be hedged with three-month futures contracts. Bakers rent their shops on five-year contracts. Rental contracts are adjusted yearly, contingent on the price level. Bakers own their equipment in perpetuity, but borrow on three-year notes to buy it. The commercial note market clears hourly. Equipment orders take six months to produce. The used equipment market is thin; its transactions are episodic and few. Bakery employees are hired weekly. Bread is baked and consumed daily.

How are these rhythms to be synchronised so that these markets will work effectively? They cannot all clear at once, for some supply information to others. This is the essence of Axel's point in (Leijonhufvud (1981), 258), where he says: 'In some lines of business, moreover, the practice will be to adjust prices in fairly small steps at fairly frequent intervals; in others, to use a larger step-size with longer intervals of posted "fixprice".'

Information and Sequencing

Consider the simplest possible form of Leijonhufvud's interrelated production system. Let there be just one output, which we label commodity *C*. Just one firm produces the commodity. The producer buys one unit of an input *I* for each unit of *C* produced. Production of *C* takes two weeks, and when the unit is finished, it is sold in the commodity market. The commodity market meets bi-weekly. The input *I* takes one week to produce and its market meets weekly. The commodity market meets on Friday and the input market meets on Monday. The producer of *C* sells on Friday and goes to the input market on Monday to replenish inventory for the next production run. S/he knows last Friday's price when s/he goes to Monday's market. Then, two weeks later, when another unit of *C* is finished and the producer goes to Friday's output market, s/he knows the input price from two Mondays before, when s/he last purchased, and the output price for two Fridays before, when the market for his/her product *C* last met. When s/he goes to the input market on alternate Mondays, s/he knows the output price from the previous Friday and the input price from two Mondays before.

The producer's information in the input and output markets is of different vintage. S/he knows more about input than output prices when it is time to sell his/her output because his/her input price information is just two days old, and his/her output price information is fourteen days old. When it is time to buy an input, s/he knows an input price from twelve days before and an output price from fourteen days before. His/her information is always more current in the input market than in the output market. Of course, this is just a consequence of the timing of the markets, and this situation would be reversed if the output market met on Fridays and the input market on Mondays.[1]

There are many other ways the markets could be co-ordinated. They could operate one after another, morning and afternoon (this is no different from the prior example), or they could operate simultaneously. No market we know of permits bids that are conditional on prices in another market, so, even if the markets are open the same day, a seller or buyer must go from one to the other in sequence. The timing and sequencing are not arbitrary, as the example shows, for they affect information sets strongly. Moreover, the timing and sequencing must match the way orders and units of production come into the relevant markets.

MODELLING MARKET TIMING AND PRICES

Markets are inherently sequential because production and consumption are. Some things must be done first, so that they can supply other processes downstream, and production periods differ among goods and processes. If all markets are forced to meet according to an exogenous and common schedule the trading processes they support will be caught mid-stream of production in some cases and too late in others.

Order Processes

Assume that 'buy' and 'sell' orders stream into the market as a discrete stochastic process. If the market meets each day, the traders accumulated during the day will establish a clearing price and volume each day. If the orders are accumulated for a weekly market instead, the number of orders will be larger and will reflect information from the larger sample. How often should the market meet? Relative to the 'grand' equilibrium price of the Friday market, the daily trading would be at disequilibrium prices. On the other hand, the daily market lets traders trade sooner and act on more recent information. If Monday's information is stale by Friday, the case of daily auctions is stronger. But, for a given order process, the number of traders will be thinner in the daily than in the Friday market. The daily auctions will be more liquid (immediacy of trading) and the Friday market will offer more depth (lower price variance).

Following De Vany (1987), let traders arrive at the market according to a stochastic process. The sellers arrive according to a Poisson process at a rate λ_s and the buyers arrive at a rate λ_b. At time T, the expected number of buyers and sellers is $(\lambda_s + \lambda_b)T$. The distribution of buyer values is $G(x)$, and the distribution of seller values is $F(z)$. The market runs as a double-oral auction with discrete values. The price or price interval that clears the market at time T depends on the realisation of the stochastic arrival processes and the draws from the buyer and seller distributions. The clearing price or price interval will be a discrete random variable.

The cost of co-ordinating the market is the cost of bringing buyers and sellers with different desired order execution times together at one time. Sellers who arrive before the auction must hold as inventory those items they have already produced. Buyers must wait to place their orders. Both have a desire for immediacy in the processing of their orders that is related to their holding and waiting costs. We can summarise the demand for immediacy as the demand for liquidity *L(T)* which is lower the longer the interval *T* between markets. The delivery mechanism settles immediately contracts made at the auction. The auction meets at time *T*.

Liquidity and Price Variance

A market with more participants will draw a larger sample of values, and the auction will clear at a price that is nearer to the expected price than an auction with fewer participants. This means that the variance of price is related negatively to the number of traders, so that *V(T)* is inversely related to the time between successive markets. The optimal market timing balances the desire for liquidity for immediacy against variance in price:

$$\max_{T} U\,(V(T), L(T)) \tag{2}$$

Variance gives disutility and liquidity gives utility. So the optimal timing of the market trades off variance and liquidity. The timing will differ among commodities according to their attributes, such as durability, and the depth of the demand and supply processes, the arrival rates of their order processes, and the production period.

Market Frequencies

Imagine that each market has its preferred frequency; financial markets meet daily, machinery markets meet quarterly, meat and produce markets meet weekly, labour markets meet monthly and so on (how they might become synchronised is discussed later). What we have, then, is on ongoing process in which each market meets according to its own frequency. The frequencies are regular even though the order processes are not; they are stochastic and stationary. We can see the process if we express the market frequencies as sine and cosine waves, as shown in the Figure 13.1.

Consider the peak of a wave, or the portion that is above the origin as the time the market is open; this time is also the settlement date of the contracts made in that market. As may be seen, some markets coincide, but most do not. Frequency affects the price signals and their lags. Low frequency markets receive price signals from higher frequency markets.

Figure 13.1 Fourier analysis of market frequencies

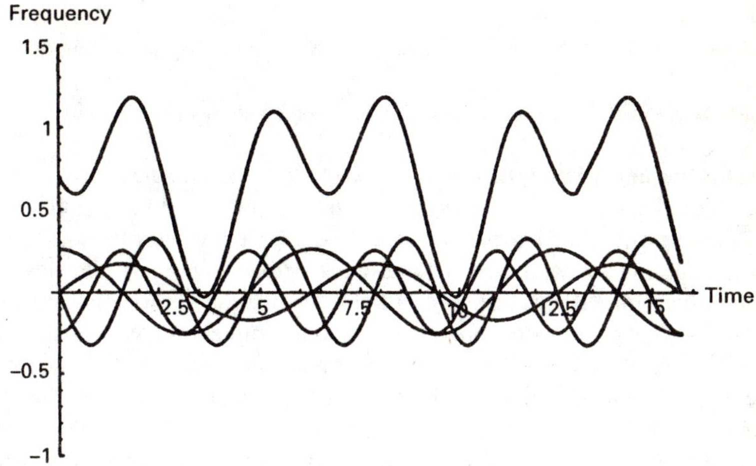

Higher frequency markets receive fewer signals, and the highest frequency markets get just one or two signals from other markets. Timing automatically makes the markets hierarchical, because it permits the commodities that trade in low frequency markets to condition on the high frequency markets. Index and futures markets are examples of low frequency markets, whose prices at settlement, condition on prices in high frequency markets.

Aggregate Activity

The weighted sum of the frequencies is the curve lying above the others (the other waves are the components that come from fitting sine and cosine functions to the upper wave). It shows the duration and frequency of the aggregate is positive so long as any market is open, and reaches a higher value when more markets are open. The longer and more irregular waves reflect aggregate economic activity of which the component markets are a part. Different weights of frequencies of the underlying components will induce different frequencies and amplitudes on aggregate market activity. The properties of the aggregate are emergent as there is no attempt to co-ordinate market frequencies in order to obtain some aggregate; rather, the aggregate reflects the local co-ordination among markets. And, the local co-ordination is driven by all the things that set frequencies for the commodities in question and by the degree to which their frequencies are interdependent.

EMERGENT PRICING AND RATIONAL EXPECTATIONS

Prices are made in the double oral auction according to the rules of the process. Buyers and sellers both make and take prices by making bids or offers and by accepting the best ones. Because the market is stochastic and timed at a set frequency, at some times there will be few buyers, fewer sellers, or both. The traders have different values and information. Some will have just come from a market whose price influences the value in this market and they will bid or offer accordingly. In order to protect themselves from adverse selection by more informed buyers, sellers will set reserve prices. The reserve will reflect the frequency of the market, for that determines how long sellers must wait for the next market, and on their past experience concerning prices. They may also condition the reservation price on other markets that inform them as to value in this market. Buyers will do this as well.

The Corridor Hypothesis

If the order processes are stationary over a period of time, buyers and sellers will set reserve prices that reflect the past value of the commodity, the prices in high frequency markets that have met since the last meeting for this market, and that reflect how long they must wait to try the next market instead of selling immediately. The result will be a bid–ask spread. A buyer would bid no more than his/her reservation bid on the grounds that it is better to wait for the next market; similarly for a seller.

These reserve prices form Axel's price corridor if they are stationary over time. Prices are confined to the corridor, though they are random within its limits. Episodes of high demand, for example, would lead some buyers to defer purchase to wait for the next market, and the carryover of demand will raise prices in the next market. Thus there is some persistence to unusual events and shocks.

The stability of reserve prices spreads information among products; the reserve prices take into account the prices of products in higher frequency markets. They are conditional reserve prices and they can work because the markets do not trade simultaneously. Prices become cointegrated and changes that are random, but stay within the spread, do not bring changes in reserve prices; they are noise.

Emergent Rational Expectations

Diverse agents, each following reserve price strategies, create a continuum of reserve prices. Noise from the random order process drives the system. The rules of the auction shape local interaction to make prices. Reserve prices emerge in each actor and have a one-or-more-period look-ahead quality based on past experience and the frequency of the market. Stable rules and reserve prices coupled with a noisy trading process makes prices random. Every price is a recent transaction; it does not mean that more transactions are available at that price. This process will make prices within the corridor without anyone having any knowledge of how it works. The price-making process is emergent, and no one needs to compute it or know how it is done for prices to be bounded within the corridor and random, but centred on the expected price.

Prices and reserve prices based on expectations formed out of experience are emergent computations if these properties aid the system in performing successfully. It is this feature that grounds the notion of emergence; the emergent information processing is used by the system to improve its performance. The observer makes the interpretive loop that information processing and improved performance are achieved. Prices are made and expectations are formed, but the agents do not know the model; prices are the emergent product of their collective actions.

SYNCHRONISING MARKETS

If markets have different frequencies that are adapted to the characteristics of the product, how are they co-ordinated? They must function together to overcome co-ordination failure, and this means that their frequencies must support one another. They become coherent through feedback.

Futures and Risk

The clearest example of co-ordination between markets with different frequencies is the futures market. Futures let producers hedge the product they have in production against price changes. If the futures price and the spot price are correlated, then a futures contract can lock in a price or reduce the variance. A futures contract lets a producer, in effect, set a reserve price several markets into the future, all the way to the settlement date. We have already argued that a reserve price is protection against adverse selection in a spot market. So the institution of a futures market permits a producer to lock in a reserve price for the contract term. But the contract must extend beyond the term of the spot market and into the future, so its settlement date must have a lower frequency than the spot market. Then spot prices can inform trading in the contract, and arbitrage will make both an element of price-making.

Only when the futures contract settles at a frequency that is lower than the frequency in the spot market can the futures contract permit hedging of price risk. The same point holds for price indices. One cannot simultaneously clear the futures and spot markets when the price of the futures contract is conditional on spot prices, and still hedge price risk. This is because once the general equilibrium auctioneer announces prices for all states and time periods, there is no price uncertainty left to hedge. Futures contracts or options do not have any value in allocating risk when they are traded simultaneously with spot and forward commodities. Similarly, forward contracts that are indexed on price indices cannot serve any role in hedging inflation risk in a general equilibrium model, where all markets clear simultaneously; there is no inflation risk to hedge, since the prices in all forward markets and the price index are known.

The solution is in clearing markets sequentially in a hierarchy of frequencies, spot markets having higher-frequency, and forward contracts lower-frequency, settlement periods. Hierarchical ordering of frequencies breaks symmetry by staggering markets in time and frequency. Symmetry is restored at settlement periods. All this means that part of the synchronisation problem is solved when the underlying markets evolve a hierarchy of settlement dates and market frequencies for spot, forward and conditional contracts.

Emergent Synchronisation

Looking back at Figure 13.1, it seems clear the market period waves follows a hierarchy of frequencies. High frequency markets can be hedged in low frequency markets, and this holds all the way up the hierarchy. For example, futures can hedge price risk in spot markets, index trading can hedge risk in futures, and derivative indices can be used to hedge risk in indices, and so on. Each level of the hierarchy clears at a lower frequency than the level below; price discovery feeds forward to higher levels and backwards from forward contracts; and higher levels to lower. The conditional nesting of contracts and prices drives the markets to coherence so that all the market frequencies support one another.

Competition among frequencies

One way for market frequencies to become coherent is for coherence to emerge from a competition among frequencies. Futures markets that have a frequency well adapted to the frequencies of the underlying commodity markets will receive good support, and this will strengthen these markets relative to others. Feedback from the futures market will then drive the frequencies of the spot markets and so on. A competition among frequencies, driven by feedback, will give markets coherence, much as the frequencies in a laser are driven to coherence by competition and feedback which 'slave' the frequencies to one another.

Constrained coherence

Unlike the laser, the market frequencies cannot all become equal without sacrificing efficiency and hierarchy. Each market is constrained from fully cohering to other frequencies in order to serve the order process that drives it. Their frequencies are tied to the production period. They must match the order processes that operate in each market, and so cannot become fully coherent. Liquidity constraints require that some markets operate on different frequencies from others. Some prices depend on signals from other markets, and they must work hierarchically in order to feed price information to the right level. These constraints limit coherence in market frequencies.

PRICE VELOCITY AND TRANSMISSION OF SHOCKS

Price shocks will propagate at a velocity that depends on market frequencies. Suppose there is a price shock in a market. In general equilibrium, all prices will reflect the shock; they are fully determined in the simultaneous auction for the state in which the shock was transmitted. Shocks are transmitted instantly in general equilibrium trading. When markets operate in a sequential hierarchy on different frequencies, price shocks unfold according to the periods of the markets. The transmission speed is tied to the frequencies. The high frequency markets will quickly reflect the shock. Then the high frequency prices will inform price-making in lower frequency markets, and these will inform still lower frequency markets. Prices will show a lagged response because they are made in markets with different frequencies. Price shocks will show persistence over time periods on the way up and slow decay after they are over.

Consider a simple example. The producer sells output at time t and goes to the input market to restock inputs at $t + 1$. S/he has reserve prices in both markets, a minimum acceptance price in the output market, and a maximum reserve price in the input market. His/her last sale in the output market was within the reserve price, but when s/he gets to the input market s/he finds the price is higher. S/he hasn't hedged, so must buy at that price or defer buying to the next meeting of the market. If s/he defers, this pushes back the time s/he can deliver output to a later market. S/he must consider that cost against the prospect that the input price will fall later when that market next meets. If s/he buys the input, when s/he next delivers to the output market s/he will condition his/her asking price on the higher input price. This will induce some correlation between the input and output markets. If the higher input price persists, then reserve prices in both markets will adjust until there are no arbitrage opportunities between inputs and outputs.

INFLATION

Inflation exposes one to adverse selection. This widens ask–bid spreads and breaks down the mutual information in prices. Liquidity suffers from the widening ask–bid spread and loss of mutual information. Trading volume declines and there is a tendency to move to higher frequencies in markets. In the transition to anticipated inflation, where reserve prices are adjusted at the right rate, market frequencies begin to change. They become less coherent, liquidity breaks down, and information is lost. During this transition, markets must re-establish coherence at higher frequencies and new indices must be developed to hedge and develop new methods of price discovery.

Consider two random prices p^i and p^j, with a joint probability mass function $\pi(p^i, p^j)$ and marginal mass functions $\pi(p^i)$ and $\pi(p^j)$. The mutual information in these distributions is

$$I(p^i, p^j) = \sum_{p^i \in P^i} \sum_{p^j \in P^j} \pi(p^i, p^j) \log \frac{\pi(p^i, p^j)}{\pi(p^i)\pi(p^j)} \qquad (3)$$

The mutual information is the reduction in uncertainty of one price when you know the price of the other.

Suppose two markets operate at different frequencies. The mutual information in their prices $I(p^i, p^j)$ depends on the timing and institutions in each market – the contracts, frequency of trading, market period, the information contents of price signals, sector specific shocks, and so on. As we showed, markets operate in hierarchical fashion and do not clear simultaneously. Hence the mutual information is incomplete and relative prices are not completely synchronised in the short run, and even in the long run.

Let $\gamma^i(t) = \dfrac{\log p^j(t)}{\log p^j(t-1)}$. If the mutual information in prices is $I(p^i, p^j)$ then

$$\gamma^i(t) = I(p^i, p^j)\gamma^j(t) \tag{4}$$

The variability of relative prices conditional on the price index $Z(t-1)$ at time $t-1$ is

$$\vartheta(\gamma^1(t) - \gamma^2(t)) = \vartheta((\gamma^1, \gamma^2, \gamma^Z)(t) - I(\gamma^2, \gamma^1, \gamma^Z)(t))\gamma^Z(t-1)) \tag{5}$$

This equation is not symmetrical because mutual information is not symmetrical. In general, $I(p^i, p^j) \neq I(p^j, p^i)$. Prices do not convey the same information about one another, nor does the price index convey the same information about relative prices.

CONCLUSIONS

A system of interrelated firms generates endogenous uncertainty and complex dynamics that cannot be co-ordinated within an Arrow–Debreu market framework. The problems that result from such a system and the way it breaks down the Arrow–Debreu mechanism give rise to virtually all the problems with which Axel has been concerned in his work.

We argue that markets must be organised hierarchically in order to overcome the co-ordination failures to which the interrelated production system is subject. Hierarchy is essential so that risk can be hedged and prices of the hedging instruments can be computed. Markets operate on frequencies that are tuned to the production periods and the stochastic order processes that deliver buyers and sellers to the market. Market frequencies trade off liquidity and price variance, and they evolve competitively to higher levels of coherence.

Price shocks have a finite velocity that is related to the frequencies and hierarchical structure of markets. The transition to inflation destroys the mutual information in prices, reduces market liquidity, and breaks the coherence in market frequencies.

Note

1. With the Friday input, Monday output market timing, the information set in (C_{12}, I_{14}) for the output market and (C_{14}, I_2) for the input market. In the Friday output, Monday input market timing, the lagged price information to the output market is (C_{14}, I_{12}) and (C_2, I_{14}) to the input market.

References

Chichilniski, G. (1992) 'Existence and Optimality of a General Equilibrium with Endogenous Uncertainty', Working Paper, Columbia University.

Leijonhufvud, A. (1981) *Information and Coordination: Essays in Macroeconomic Theory* (Oxford University Press).

Leijonhufvud, A. (1989) Information Costs and the Division of Labor, *International Social Science Journal*, XLI, pp. 165–76, May. (Oxford University Press).

Leijonhufvud, A. (1993) 'Towards a Not-Too-Rational Macroeconomics', *Southern Economic Journal*, vol. 60, 1–13, 1.

Milgrom, P. and Roberts, J. (1990) 'The Economics of Modern Manufacturing: Technology, Strategy, and Organization', *American Economic Review*, vol. 80, pp. 511–28.

De Vany, A. (1987) 'Institutions for Stochastic Markets', *Journal of Institutional and Theoretical Economics*, vol. 143, pp. 91–103, March.

14 Information and Co-ordination: A View from the South

Daniel Heymann*

This is a special occasion for those of us who had the chance of learning directly from Axel Leijonhufvud. Leijonhufvud is more than a standard teacher: he transmits an attitude towards economic research. Both in writing and in conversation he makes one well aware that economists must remain curious and prepared for constructive puzzlement. Leijonhufvud has made it clear that the search for precise arguments can, and should, be combined with an open-minded approach, and with respect for intellectual diversity. He insists that one should learn and use the tools of the craft, and that it is essential to keep in mind the practical applications of economic analysis; but he does not dwell on technical details, nor does he offer universal policy recipes. This, no doubt, may bother some people. But, for many of us, Leijonhufvud's work continues to provide a non-dogmatic guide for thought and an incentive to keep exploring the field.

Interest in macroeconomics is often motivated by particular experiences of large-scale economic disturbances. From the perspective of a Southern Cone analyst, many immediate concerns are related to the ability of economic systems to absorb large shocks, and to the way in which they adapt to frequent changes in policies and in their institutional setting. These have been recurrent themes in Leijonhufvud's contributions.

My aim here is to play some variations on such themes. Although I have borrowed freely from Leijonhufvuds's ideas, and some of the following notes are based on joint work, they do not pretend to represent his views. I trust that he does not object in principle to this exercise.

ECONOMIC CO-ORDINATION WITH RECOGNISABLE AGENTS

A quick glance from a window overlooking a commercial district is sufficient to show a striking variety of economic activities. Trucks load

*CEPAL, Buenos Aires office. Thanks are due to F. Navajas, E. Vesperoni, C.F. Bramuglia and C. Lavandera for useful comments. The author is responsible for any errors and opinions contained in this chapter.

and unload goods, shopkeepers arrange merchandise in their stores, people look around, enter shops and sometimes leave with packages. Even in this minute section of the economy, trades involve a large set of goods; making a full list would probably consume a considerable amount of paper. One also realises that much happens behind the scenes: merchants set prices (although in very high inflation, tags and menus are changed before one's very eyes); they review the flow of sales of particular goods; they place orders with various suppliers; and they arrange credit transactions with banks or other firms. And this is just the surface of a huge web of decisions and exchanges. An enormous number of people, each acting with some purpose, have participated to make possible the trades that we observe from our window (see Leijonhufvud (1989)).

This seems too trivial to mention, were it not that we are in the business of making sense of how people conduct their economic life, and that even the systems that support commonplace, everyday transactions do break down once in a while. Sometimes we find that people are walking idly, apparently without the desire or the means to get goods from the shops; in other instances, they carry money around, but are confronted with signs stating that the stores are 'closed for the lack of prices'. We have good reason to be concerned with the extreme cases of economic depression and hyperinflation, for their own sake, and for their informative value regarding the conditions under which economies may or may not perform reasonably well. But, as Leijonhufvud often indicates, the very fact that in 'normal times' the activities of vast numbers of agents do fit together, producing and distributing goods in a way that at first sight seems almost natural, ought to deserve attention.

Looking down from the window once more, one can recognise signs of the 'market process' at work. Customers go from one shop to the next, comparing prices and quantities. Shopkeepers presumably watch closely which goods sell well and allow good profit margins, and which do not. The messages they receive are transmitted to producers through their orders. Sometimes, one may find vivid evidence of adaptations to economy-wide impulses: a store that sold locally-produced textiles, for instance, now shows stacks of imported electronic goods after tariffs have been brought down. And, once more, at certain moments, the mechanism seems to go wrong. On occasion, one sees merchants standing by, waiting for customers who do not come; and sometimes people rush to buy without even trying to check prices two blocks away.

These 'microscopic' impressions drawn from an infinitesimal piece of the economic system may help to remind one that the objective of economic modelling is to represent concrete actions by hosts of people. At the same time, one is also struck by how intricate the picture soon becomes if one tries to describe in detail the decisions and interactions that take place in the street below. A model built on the basis of actual 'microeconomic' elements seems hopelessly complicated. Suppose that we are offered information regarding all the trade undertaken by all the agents in the economy during a certain period, specifying each transaction in every potentially relevant dimension: the data would identify the trading partners, the location of the exchange, the moment in which it happened, and would characterise precisely the goods and assets that changed hands. It is likely that we would not be willing to pay much for that mass of information or, at least, that we would not consider analysing each particular piece of data. Even assuming that we could store and process numerically the data for free, it seems probable that we would ask questions of a 'statistical' nature. For example, in high inflation, one may want to know what proportion of sales – perhaps classified into broad categories – are done in domestic money, or find out the average interval between price changes in various 'industries', but would not try to observe, or to interpret, each transaction individually.

So perhaps one chooses to start from the other end: take a set of aggregate figures, look for some patterns in their movements over time, and devise a model on the basis of those regularities, within an accounting framework that incorporates what one knows to be solid 'conservation principles': goods produced must be put to some use, purchases must be financed in someway, and so on. This leads to a mode of analysis that relies greatly on phenomenological 'constants': nominal prices or mark-ups, propensities to consume, and the like. However, one knows that those are not 'physical constants': they have the habit of moving without warning. And, once one tries to rationalise the arguments or the equations of the model, the 'micro' language creeps back again: one is not surprised to find that, in a more or less stable economy, investment is related negatively to interest rates, for example, because individual firms can be expected to act in that way.

The tension is apparent. A well-developed micro-based description of the economy as a whole does not seem to be at all feasible. But neither can one easily accept reliance on models composed of 'functions' that make no clear reference to the behaviour of agents. Moreover, in order to study the way in which individual activities are matched, there must be some representation of how people decide their trading plans and how they interact in the market.

Now all this can sound too abstract and far from the problems of an applied economist. However, in the quite commonplace work of doing macroeconomic analysis of specific economies, one is recurrently confronted with the choice of a basic 'approach'. Consider, for example, an economy that is just coming out of hyperinflation. One suspects that disinflation will have sizeable effects on the supply of goods and that, especially if the stabilisation programme uses an exchange rate anchor, it is likely that aggregate demand will shoot up; in past episodes that may (or may not) bear some resemblance to the present one, the inflation rate dropped suddenly, but did not go rapidly to a near-zero value, so that the currency appreciated substantially in real terms. This, clearly, is a language based on 'aggregate variables'; starting from such observations, one may build a small model (whether formalised or not is beside the point here) leading, for example, to the proposition that if fiscal and exchange-rate policies hold, there will be initially an economic expansion, followed by a slowdown as the inflation rate eventually converges to 'international' levels. But, will people necessarily react in the manner predicted by the aggregate functions one postulates? To what extent may firms revise their price-setting practices depending on the 'credibility' they assign to policy announcements, and how can one find out what their beliefs, in fact are? If one observes that people increase their spending at a fast rate, should one be worried that they may be misjudging their future income opportunities? And so on. These are not uncommon questions for applied macroeconomists. They are still posed in aggregate terms: they refer to broad trends and to the 'average' behaviour of whole groups of agents; but they force one to consider how people choose their actions, and how they look ahead. And, for this, one has to be guided by some sort of theory.

The literature has consecrated a particular way of modelling behaviour; it is applied to study matters that go from, say, the interaction of duopolists in a given market to the aggregate response to an economy-wide shock, or to the analysis of policy decisions. Each specific model defines a setting where agents, knowing that they want, make optimal choices given 'external' constraints. Individuals may lack data about the values of relevant variables, but they are assumed to process the information they do have available in an optimal way: their 'knowledge' of their economic environment, so to speak, is unrestricted. Barring the possibility of sunspot phenomena, the results of the model then appear to depend only on 'deep' parameters representing preferences, technologies (including those used for gathering data), 'institutions' that define the rules of interaction between individuals and, perhaps, the impulses caused by some outside agent such as the government.

This modelling strategy has produced a large mass of applicable work; whether this is explicit or not, many of the arguments one utilises in describing the evolution of particular economies trace back from, or can be reduced to, this line of analysis. After all, optimisation plus rational expectations is a conceptually simple way of depicting purposeful agents who are alert to changes in the environment. However, the common claim that those assumptions have a universal range of applicability and, even more, that they define the only sensible way of doing economic analysis seems, at the least, a gross exaggeration (see Leijonhufvud (1993)).

Models are built to be solved, and not every problem is tractable. So, the analyst must cut the problem down to size. The art, of course, is to simplify the environment that is modelled so as to be able to arrive at a solution and get interesting results. This is an art, indeed: there is no 'objective' way of making the cut. When we assume that agents optimise in the context of the given model, we are not only assuming that they use at least the same problem-solving abilities as the analyst (and, often, specialists have to struggle hard to make the model give an answer), but also that they share the 'frame' that is implicit in the model. It seems possible that some actual economic actors may find that a number of auxiliary hypotheses in a model are '*ad hoc*', and that these individuals carry out their business using other assumptions. Moreover, each model – and for good reasons – leaves under the *ceteris paribus* umbrella what may be the core of another one. Thus, again, the problems are 'uncoupled' in order to make them solvable, one at a time.

The implicit assumption of optimising models seems to be that agents act similarly: they 'focus' on one particular decision problem – for which optimisation is feasible – while 'holding constant' variables which will result from another calculation, and which may well interact with the first ones. When seen in this way, optimisation appears to be an heuristic procedure (rather than a universal device), more or less useful according to the circumstances.

In some instances we are confronted – both as analysts and as agents – with sudden changes in the economic environment. For example, we can have a disinflation package, a tax reform and a tariff reform all at once. These may well be the 'right' policies in the given situation. But, in any case, people must reconsider simultaneously their plans across a broad range of dimensions: pricing practices, portfolio choices and consumption and investment programmes are all thrown down on the table for revision, and each agent knows that others are also busily making up their minds on how to react. Sorting out the 'best' response in those conditions would probably require agents to perform large-scale computations, and to have at hand a full-scale model with which to form expectations. The argument stating that individuals have trained themselves to act 'as if' optimally loses its weight when the changes in the economic context are 'without precedent' (see Sargent (1993)). One may still choose to stick to the optimising framework of analysis, but it seems to need a leap of faith to disregard the possibility that the representation may be inaccurate. And, again, the issue is not mainly one of methodology: it bears on how one analyses real-life experiments.

In normal economic times, people can rely on routine procedures for most of their activities. Important decisions are spaced over time. The economy evolves but, on the whole, each state does not differ much from the previous ones. The 'continuity' of the environment allows agents to concentrate on solving the problems that look more relevant at each moment: they can change some aspects of their behaviour by 'taking a good look' at them. For the system as a whole, this continuity also makes co-ordination much easier: commonplace exchanges are carried out more or less as a matter of course. The quality of individual decision-making and the overall performance of the system are put to the test when the economy faces turbulent conditions or is subjected to a sudden change. Leijonhufvud has long insisted that episodes of this type should be given close consideration. We shall refer to them briefly in what follows:

EMERGING MARKETS AND VANISHING MARKETS

Perhaps the most elementary feature of an economy with a developed division of labour and decentralised decisions is that people meet and exchange. We know that the states the economy can attain depend on the set of markets in which agents participate. When an economy has 'incomplete' markets, some trades are restricted. A case with strong macroeconomic implications is that of liquidity constraints (see Clower (1965); Leijonhufvud (1973)). Market transactions also have a 'public good' element to the extent that they disseminate information to third parties.

But markets are not created from nothing. They have evolved in history. And, from an analytical perspective, it must be recognised that markets are 'made' by economic agents. Clower and Leijonhufvud have often stressed this point (see, for example, Clower and Leijonhufvud (1975)). If we could have a 'constructive' model, showing how markets arise and what form they take, we would probably get a much better understanding of issues such as pricing and the 'functions' of money. However, such a thing is not available. Instead, what we find are arguments starting from the other end, trying to explain why markets that 'should' be there are not actually observed.

The explanations are of several types. One group of arguments points to physical conditions that may prevent trade, as is the case of (literally interpreted) overlapping-generations models. Another set of reasons may be institutional: people will refrain from making contractual arrangements if there are no means of enforcing them, and will not consider making an exchange when they are not confident that the other party has the right to deliver what is offered as quid pro quo. The informational arguments, on their side, are based typically on the existence of variables which are observed by one agent and not by others (as in the 'market for lemons' models); this asymmetry can be viewed in a way like another type of physical constraint.

We can find in recent history striking episodes in which to study the creation and destruction of markets at the scale of whole economies. It would be tempting to consider at the same time the two (not disjoint) sets of cases of the Eastern countries in transition, and of those that have undergone high or hyperinflation. However, for obvious reasons of comparative advantage, the following comments will focus on the second group of cases, with a few, and tentative, references to the economies in transition.

Despite their specificities, unstable economies and those in transition seem to have one feature in common: in both, people face strong uncertainties about the performance of institutions that will govern the ownership and availability of resources. Modern literature has brought back the old theme that economic behaviour depends on the policy regime. But what defines a regime? On the one hand, institutions and policy patterns themselves can be considered as the result of an underlying process, and so on. The reasoning leads easily to a multiple regress (see Sargent (1982)); once more, the problem is not merely one for the analyst, but concerns agents in their daily business. On the other hand, written codes are not equivalent to concrete policy actions: people must learn from experience what to expect from governments. Given those expectations, moreover, agents have to develop behaviour patterns adapted to the policy framework. In some conditions, this adaptation may be rapid: for example, it is striking to see in the accounts of the European hyperinflations of the 1920s how closely some reactions resemble what was observed in similar Latin-American episodes, despite the different experiences people had. However, these are relatively simple responses – for example, getting rid of heavily taxed money – to strong and clear signals. It is likely that building up a system of market exchanges requires more difficult learning. Seen from this far away, the behaviour of Eastern economies seems to show it.

The hyperinflationary episodes provide vivid indications of how systems already organised on a market basis react to large disturbances. Economic analysis has had trouble in pointing out precisely 'what may be wrong' with high inflation. At the same time, people often wonder how economies still seem to function without collapsing under, say, double-digit monthly inflation. In fact, day-to-day trades, involving the use of domestic money, do show a noticeable resilience, even in the face of very high inflation. However, they can be disrupted in 'true' hyperinflation and, even in milder cases, closer look reveals serious damages to economic performance. To a large extent, these derive from the shortening of planning horizons and from the disappearance of markets.

In a certain way, inflation and instability are logically different phenomena. But they are associated in practice. This can be understood by considering how inflation originates: governments which can execute systematic policies do not rely on the inflation tax; this is typically an expedient to accommodate pressures from various sides. The high inflation scenario, then, is one where not only prices rise at high speed, but also the future moves of the authorities are very uncertain, and this is combined with a matching volatility in private behaviour.

Until not so long ago, the effects of inflation were analysed almost exclusively in terms of the substitution between money and real capital. This generated the prediction that higher inflation rates would be associated with more capital accumulation. In fact, the opposite seems to be the case. Recent models have started to explore the possibility that the inflation tax may have more subtle effects: for example, that inflation could aggravate adverse selection problems in credit markets, thus leading to less investment financing (see Azariadis and Smith (1993)), or bias the bargaining position of buyers and sellers in favour of the latter, which would reduce output (see Cassella and Feinstein (1990)). However, such models still do not account for the turbulence that characterises high inflations.

The shortening of decision horizons manifests itself even in the way people choose to refer to the inflation rate. The use of monthly figures – and, in hyperinflation, the demand for higher frequency indices – does not come from a reluctance to handle 'large' figures: after all, the same agents routinely deal with prices quoted in millions or billions of monetary units. The fact is that annual data are irrelevant: the 'memory' of the system does not extend that far back, and no one would venture to make a forecast of the price level a whole year ahead.

This makes sense, for two reasons. First, it is practically impossible to predict the cumulative effect of policies. Just trying to map the government's 'decision tree' from one short-run move to another quickly leads to an explosion of possibilities. Some months ahead, a major stabilisation programme might have been started, but also, high inflation might have degenerated into hyper. Second, the economy loses its buffers, so that the response to impulses is fast and strong. In particular, prices are revised at short intervals. And it is not as if nominal prices changed while remaining fixed in terms of some other 'standard': the reference to past price increases in high inflation, or to the exchange rate in hyperinflation, does not preclude wide movements in individual and aggregate prices relative to those variables. The pace of price decisions accelerates. Consequently, although inflation may show a degree of inertia, imparted by surviving backward-looking adjustment practices, individual prices react to high-frequency signals, as processed by each agent. If people realise that this will happen, they also know that they themselves must move at a similar speed, and that aggregate impulses will soon have an effect on the markets in which they participate. But predictions can only be made a few moves ahead, and this means that individual plans will be defined over short time spans.

The difference between stable and unstable economies seems not so much that in the latter the inflation rate is higher: one could perhaps say that the (relevant) rate of price growth is similar, but the time frame has shrunk. This is clearly revealed in the reluctance of agents to make commitments extending over relatively long periods of time. Inflation does stimulate the appearance of contractual instruments with sundry adjustment clauses. On the whole, however, high inflation economies end up with a poor menu of financial assets, most of which have a very short maturity. In fact, the recovery of credit and the emergence of longer-run contracts and of more sophisticated instruments is one of the characteristic features of economies in the stabilisation phase. From a certain perspective, this behaviour looks paradoxical: one might expect instability to provide incentives for the development of new contingent markets. But simple contracts do not provide good protection against the type of uncertainty that agents face (for example, people may not want to enter into backward-looking indexed contracts extending beyond the period for which they can be reasonably confident that the inflation rate will remain near the current levels), and considering in detail the possible shocks that might influence significantly the outcome of agreements does not appear to be feasible. When the range of contingencies which agents should contemplate grows large, they seem to take the simple option of quitting the game. A similar attitude prevails regarding physical investments: firms will only buy new equipment if they can expect quick and large returns.

This preference for flexibility (which is often associated with capital flight) reflects people's uncertainty, especially about outcomes further in the future. The scarcity of reliable information is not of a 'technological' nature. It is not that informational asymmetries in the usual sense have become more intense, or that agents have less access to data. People in high inflation situations handle much more 'macro' data for the purposes of their routine decisions than do their counterparts in stable economies. What is lacking is the capacity to interpret the signals that individuals receive in quick succession. When agents perceive that they cannot make sense of those signals, they refuse to trade. The length of the period over which information 'loses meaning' depends, of course, on the type of inflationary process. In comparatively mild forms of high inflation, people can still plan ahead for intervals of some months, although, even there, market decisions may incorporate sizeable differences of opinion regarding the future evolution of the economy. In hyperinflation, the time horizon shrinks to almost nothing: even selling goods over the counter becomes a sort of gambling activity.

One of the facts for which there is more or less ample empirical evidence is the positive association between inflation and the variability of relative prices, both over time and across locations (see Palerm (1990); Dabús (1993)). Although many prices are dollarised in hyperinflation, this association seems also to be valid in such episodes. What the evidence suggests is that 'local' markets become more and more segmented as the degree of aggregate price instability increases: in the language of the traditional metaphor, the activities in the different 'islands' that constitute the economy tend to lose coherence.

This can be rationalised as a consequence of the system's 'loss of memory' (see Tommasi (1992)). When prices in individual shops are revised very frequently, and embody either specific cost shocks or idiosyncratic expectations, consumers who observe that a certain price is low, relatively to what other firms are presumed to charge, cannot rely on this also being true in the future. The information they gather through search quickly becomes obsolete. Since investments in search depreciate at a fast rate, people will be less informed about prices: even if they choose to 'visit' more stores (and thus increase their expenditure in looking for information), they will not know precisely which firm sells cheap and which does not. So competition through prices becomes a less effective way of attracting customers. In turn, this implies that prices quoted by individual firms will be less homogeneous. Moreover, price instability allows high-cost firms to retain 'captive' customers, and it tends to increase average mark-ups.

An image that comes to mind is that of prices being kept in line by the activity of consumers, like particles on a magnet: the 'law of one price' would correspond to a state where all the particles point in the same direction. Inflation, in this picture, would act in the way of 'thermal noise', weakening the effect of the interactions that tend to hold together the orientations of the particles. As the noise grows stronger, long-range coherence first gets lost; after a certain point, the system goes to a state of complete 'macroscopic' disorder. The analogy of a phase transition (it may be pointed out, by the way) seems to correspond to the observation that in some instances the passage from high to hyperinflation takes place suddenly – prices accelerate in a discrete steps, many become dollarised, and reference to past inflation is erased as a few weeks elapse – as if there was a discontinuity between one type of inflationary regime and the other.

In any case, the main point here is that the co-ordination of economic activities across space and time does not take place automatically. Hyperinflation throws agents into a situation where their ability to plan is very much impaired. At the limit, they just refuse to exchange. Though in a much less dramatic form, 'better times' can also test people's capacity to make consistent decisions.

REAL BUSINESS CYCLES

Current 'real business cycle' models have shown that the equilibrium paths of aggregate quantities and prices need not be steady. The use of an explicit intertemporal framework, in addition, has served as a reminder that present outcomes depend on how agents perceive future conditions, and that decisions over time are linked together. What these models take for granted is that, no matter what shock the economy receives, agents will be able to estimate 'rationally' its present and future consequences, and will act correspondingly. There is another tradition which, by contrast, suggests that economies may have trouble making the 'traverse' between one growth (or non-growth) path and another, because there is no direct mechanism that reconciles intertemporal plans, and these may be based on inconsistent expectations (see, for example, Leijonhufvud (1968, ch. 4); (1981)). Such inconsistencies may result in 'real', but not equilibrium, business cycles. Here, too, the difference in analytical perspective is relevant for the way one interprets concrete data, especially in economies undergoing rapid change.

In very high inflation, people's reactions are mainly defensive: they place assets in liquid form and in a safe place; beware of selling hastily;

are cautious about making irreversible capital investments, and so on. The implicit rates of time discount are high, and agents' wealth perceptions are probably low. (Some, of course, are made worse off than others; in particular, the inflation tax hits mainly those with low incomes, who are the last to abandon the domestic money.) If a viable stabilisation programme can be put in place, the habits that people have developed during inflation will not disappear immediately. Nevertheless, the economy's performance will improve through a variety of effects: a supply-induced recovery, as firms cut precautionary mark-ups and are better able to 'mind their own business' and look for ways to reduce costs, a revival of credit, and a return of flight capital, among others. At the same time, in order for the stabilisation to be sustained, fiscal policies will have to be reformed, so that the configuration of taxes and public spending is likely to be modified substantially. All these are major changes, which can be expected to have strong effects on relative prices and the real incomes of different groups of agents.

Consider a stylised post-disinflation scenario, in an open economy which has just gone through a period of inflationary turbulence. In order to focus on the 'real' aspects of the case, assume that the transitional problems associated with bringing about the disinflation have already been solved, and that the government has put together a fiscal and monetary programme such that the public trusts that the authorities will no longer use the inflation tax to finance expenditure. Therefore, neither inflationary inertia nor policy 'credibility', nor 'expectations of expectations' about the aggregate price level (see Phelps (1983), 1992)), are relevant issues under these assumptions.

In such circumstances, one can probably expect aggregate spending to increase initially at a fast rate. A smaller perceived risk in keeping assets within the country is likely to induce capital inflows; the authorities can monetise those flows so as to satisfy a larger demand for real balances while, at the same time, real interest rates fall. Even if explicit taxes are raised, the removal of the inflation tax increases the purchasing power of a large group of consumers. Many households who had found themselves liquidity-constrained will now be able to buy goods on credit, since stabilisation removes strong disincentives for lending. Enterprises may start considering investment projects which they had shelved, waiting for less uncertain times. From the supply side, firms will probably try to increase sales, and will realise that market competition is to become more intense; they may count on receiving trade credit on a much larger scale than before. These are all ingredients for a quick economic recovery. A stabilisation will always have distributive consequences but, on average, agents can reasonably feel wealthier. They realise that their real incomes are increasing, and future receipts are discounted less heavily. Moreover, aggregate output can be expected to keep growing if investment recovers and businesses can remove the sundry inefficiencies in their operation that are likely to have developed during the inflation. Optimism about future prospects will make people more disposed to borrow in order to finance current spending, and foreign credit (or funds earlier sent abroad) will be available: the balance of trade may shift rapidly from a surplus (to accommodate capital flight) to a deficit. Assuming that the effects on supply are not too biased towards the production of non-traded goods, the increase in aggregate demand will tend to raise the prices of those goods relative to the prices of articles that can be bought or sold abroad; thus, some real appreciation can be expected. Qualitatively, all this fits with the changes that the economy has undergone. But, by how much has average wealth in fact risen?

Stabilisation will cause people throughout the economy to re-evaluate their estimates of 'permanent income'. Take the case, for example, of a producer of a certain non-traded good. S/he observes that demand has gone up, and is confident that s/he can raise the prices s/he charges and still make a larger volume of sales. If s/he expects such conditions to last, or perhaps even to improve further, s/he may confidently increase his/her own consumption spending immediately, and make investments in his/her business, counting on good returns. If our agent does not simply base his/her forecasts on the extrapolation of current trends, but applies a more sophisticated reasoning, s/he may realise that competitors are also going to increase their production capacity, and that s/he must try to anticipate the behaviour of aggregate demand, since it will have an effect on sales in the industry. The agent can see around him/her that real activity is moving up, and that people are spending more in a wide variety of goods; many are financing their purchases out of borrowing or (s/he guesses) out of foreign exchange previously hoarded. In the papers, s/he may read warnings by some analysts about the danger of an excessive real appreciation and about the emergence of a trade deficit. But s/he can also find replies from other people insisting that those concerns are out of proportion, and that large-scale capital inflows simply reflect a well-founded trust in the economy's growth prospects, since firms (including those producing internationally traded goods) will now be able to make substantial productivity gains by reorganising their plants and improving their commercial strategies.

These are somewhat ambiguous signals. The agent's future income clearly depends on the plans that other people are making at the same time, based on their own readings of the information they get in the markets in which they participate, and guided by their own guesses about where the system as a whole is headed. Our 'typical' agent has to imagine a future scenario where the behaviour of several variables is of critical importance. Some of these variables – such as the probable supply by competitors – are relatively 'close' to his/her range of personal experiences, but may still be hard to forecast. Other features of the future environment – such as whether the production of traded goods will be sufficient to sustain a growing domestic demand – depend on decisions that are distributed widely across the economy, and are quite removed from his/her areas of direct observation.

So, everyone makes a bet, in order to programme current and (we may assume) future actions. Now, these plans may or may not be compatible with one another. The non-traded-goods producer can well predict a growing market and high prices for his/her goods (maybe because s/he anticipates a large output effect in the traded goods sector due to the removal of '*X*-inefficiencies') while at the same time a 'typical' producer of traded goods is acting under the assumption that the real appreciation will be sharply reversed sooner or later. Both types cannot be right simultaneously; their beliefs are inconsistent with one another. But if they hold such beliefs, agents would in the aggregate overestimate their wealth. And current information may not be sufficient to make people change their perceptions: an observed fall in the savings rate, for example, can be interpreted either as a symptom of over-spending, or as a sign that agents are rightly discounting that their incomes will grow.

The previous example of inconsistent beliefs (see Heymann (1984)) is simply one possibility among others. But the question remains whether agents will be able to make correct inferences from the information they have at hand. Now the problem of forming expectations comes both from the fact that (as stressed in the real business cycle literature) the 'equilibrium' of the economy is a shifting one – because, otherwise, people may use past observations to make more-or-less accurate forecasts – and from lack of precise knowledge about the direction and, especially, the size of those shifts. It is not enough to realise, for example, that the economic recovery will probably be associated with a larger current-account deficit; agents (and analysts) must judge whether a deficit of a given amount indicates or not a state of excessive aggregate spending. After a big shock, the parameters whose value would be required to pass such judgements are deeply hidden.

In circumstances where 'fundamentals' are undergoing a non-recurrent change and people are engaged in the complicated task of guessing each other's plans and expectations (for example: What do I think they are thinking when they decide to spend more and do I think they are right in thinking so?), it is likely that beliefs will be heterogeneous. The heterogeneity itself implies that a good number of individual plans are not going to be fulfilled, and that many agents may end up in the future having balance sheets that, with hindsight, they will regret having chosen. It seems possible also that the expectations of agents have an aggregate bias. If, for example, people tend to overestimate their own future incomes, this would induce a cycle in spending, with an initial boom followed by a period of adjustment, which could be more or less traumatic according to the size of the debts that have been accumulated in the process, and to the response of lenders.

Now, of course, we know that in every economy people make mistakes of one type or another, and yet the performance of the system as a whole need not be disturbed significantly. What we are interested in finding out is whether individual errors will be corrected in good time and, if not, whether they add up in such a way as to produce observable 'macroeconomic' consequences. However, in some moments of rapid change, the observer is not in a much better position than the agents he is watching in action: there are no compelling reasons to believe that the economy will behave 'as if' the assumption of individual foresight was valid and, at the same time, it may well be that the optimists are right in this particular case, and the transition takes place smoothly. As in Leijonhufvud's version of J. M. Keynes' argument, intertemporal co-ordination appears to be neither automatic nor impossible. Clearly, this leaves the analyst in a somewhat uncomfortable situation where s/he is unable to offer a solid prediction, and must adapt his/her opinions on the basis of indicators that often show ambiguous evidence. But, after all, recognising the limitations of one's ability to predict seems to be a form of rationality.

RULES OF THE GAME AND CHESS PLAYING

The evolution of economic and, especially, monetary institutions, seems to undergo cycles. Often, the turns are motivated by a crisis of some type. After hyperinflation, for example, there is a general call to restrict the margins for 'discretionary' policy-making, while financial crashes or depressions typically produce the opposite reaction. The changes in mood are reflected in the language itself: what was once named 'policy flexibility' is now termed 'arbitrary activism', or what was referred before as 'rule-bound behaviour' is mentioned disparagingly as a 'straightjacket preventing desirable interventions'.

This tension is also present in analytical approaches to public policies. The relationship between policy-makers and their constituency can easily give rise to principal–agent problems, and (as a mass of recent literature shows) there are many instances where even games between benevolent policy authorities and well-informed private agents may have undesirable outcomes. At the same time, except in very simple cases, there is no analytical presumption that simple, non-contingent rules are the best alternative for policy design.

Perhaps the most definite conclusion that can be drawn from the analysis based on the time-inconsistency argument is that the old concern about the nature of institutions was well founded. The consequences of policies cannot be understood without reference to the institutions (or the lack of them) that govern the incentives of the authorities and provide a frame for private expectations. Therefore, it makes sense to investigate what kinds of policy decision rules can help to improve the economy's performance. And it is probably natural to approach this problem as if it was equivalent to the design of an 'optimal contract' to which the authorities would be committed.

But even in the case of transactions between private parties we usually observe either simple contracts or informal, 'handshake' agreements. Writing an optimal policy rule would at least require having knowledge about (present and future) 'social preferences' over the relevant dimensions, about the set of contingencies to which policies will have potentially to react to, and about the way in which, given the shocks and the nature of the rule, government actions would influence the variables of social concern. This is 'clockwork' decision-making (as Leijonhufvud has called it) on a grand scale: once the 'book of instructions' for policy-making has been printed, an automaton could pick the appropriate action on every date in every possible circumstance. Such a system for setting policies once-and-for-all clearly appears difficult to conceive. At the same time, while a complete, contingent clockwork rule would in principle predetermine fully the actions of the government, it would do so in such a fashion that simple mortals trying to find out what to expect from the authorities may have a hard time doing so. From the point of view of the 'typical' agent, the rule would look very much like an exercise in policy 'discretion' – and that, in turn, would change the problem itself.

Policy rules can be seen not only as devices to modify the government's incentives, but also as means of simplifying people's decisions: they help to establish 'what kind of game' is being played and, therefore, they make economic co-ordination easier to achieve. In order to serve this purpose, the government's behaviour in matters of everyday business must be understandable without much effort: it is precisely the lack of such guidelines that makes policy instability so harmful, as seen dramatically in hyperinflation. But simple rules cannot be applied for every instrument: policies would then be overdetermined. In any case, there has to be a choice on what to fix and what to leave free to adjust. Moreover, simple rules, like any other heuristic decision procedure, will not always produce good results: it is almost certain that they will break down occasionally but, almost by definition, the nature of the 'emergencies' in which that would happen cannot be stated precisely in advance. Also, it is difficult to imagine a policy rule which will be useful and, at the same time, free of ambiguity in its definition, so as not to demand to be 'interpreted' once in a while.

This is a fuzzy field of rules with unspecified escape clauses, of 'policy independence' within not exactly defined limits, and of policy statements which are always somehow vague. There seems to be an 'art' of institution design, as there is an art of managing institutions – even those as metal-hard as the gold standard (see Bordo (1993)). Policy-makers, like everyone else, are engaged in a learning process, and there is no way of pre-codifying the decisions they will face. The use of judgement seems unavoidable in running policies; what appears to be needed is a framework of rules to bind behaviour and guide expectations while allowing 'competent discretion' to be exercised.

The choice of basic monetary and fiscal institutions has a general logic. One reasonably may ask why some countries cannot apply more or less systematic budget procedures to determine fiscal policies, or why central banks sometimes act as mere providers of funds to undisciplined government treasuries – and the question leads to the tricky matter of identifying what are actually the degrees of freedom in the design of policy regimes. Nevertheless, the experience shows clearly that economies without elementary rules for policy-making are plagued with time-inconsistency problems (as private groups and government authorities find it expedient to engage in 'opportunistic' behaviour) and suffer from the impossibility for individuals to form accurate expectations. And, often, governments that operate in a setting where they are apparently unrestricted in their choices end up simply moving to the tune of the pressures applied on them by various interest groups.

Policy games thus have general patterns which make it desirable to set some types of rule in order to avoid perverse outcomes. However, history matters too. The people who administer policies and those that react to them have concrete experiences that determine their perceptions and influence their modes of behaviour. Monetary institutions are seen quite differently, for example, in a country with a tradition of stability and in one that has just come out of hyperinflation. Central bankers with a reputation for solid management can count (maybe not always, but often) on inducing 'co-operative' responses in the public towards their actions and announcements. Such confidence permits the exercise of discretion within perhaps implicit but more-or-less well understood limits. When that reputation is lacking, it is likely that the choice leans toward strict rules, for fear that anything else would open the door for a return to the old habits.

The attitude then tilts towards strengthening private expectations, against keeping flexibility to meet contingencies. This seems quite natural when the public will not easily put much faith in loose promises that monetary stability will be the main objective of the authorities, and adverse 'real' shocks appear as a vague and distant possibility while, at the same time, the troubles of policy mismanagement are vivid in people's minds. In the midst of a high inflation, governments have no effective power to run policies; on the way back, they will probably have to make explicit surrender of the use of (especially monetary) instruments. Adherence to a simple, rigid monetary rule helps to solve the crucial, immediate problem of re-establishing some degree of certainty about the value of the currency; the possible future costs of the choice, even if perceived as such, are likely to be heavily discounted.

But, at any rate, the establishment of a definite monetary rule does not eliminate the need for policy management. This is quite apparent in stabilisation processes: they do not seem to take place as the unfolding of a sequence of actions preconceived from the start, but as a succession of policy decisions adapted to the circumstances of each moment, even if they follow an overall pattern.

The analogy of chess playing is often used to describe the behaviour of policy-makers when executing a programme. The image fits the case, and other decision-making scenarios as well. Here too, we have agents with a definite purpose and a (possibly well conceived) view of how the game should be played, but no exact knowledge of where they will find themselves more than a few steps ahead, and no chance of devising an 'optimal' strategy for every conceivable state. Without a game plan, the opponent is almost certain to win, and some situations call for well-studied, standard responses. The players will use their technical skill, and will take advantage of the analogies they may recognise with other instances of the game they have in their memory. But at some point or another, they will have to 'invent' a move. And, sometimes, they will surprise the analyst.

SOME CONCLUDING REMARKS

Macroeconomics remains a lively topic. The behaviour of actual economies presents problems for consideration constantly; old ones may fade out of sight, but are likely to reappear later on, perhaps somewhere else or in a somewhat different guise. This is also a field where intellectual nihilism leads to nowhere, but where unquestioning complacency seems particularly hazardous: we have seen several once-well-established orthodoxies become theoretical relics. The all-encompassing macro theory seems to be out of sight; the applied economist should probably prepare him/herself (especially here in the South) for many rounds of substantive learning which will not converge to become certainties. But economic agents are in the same predicament. This rapid excursion into a variety of subjects has tried to suggest that the limits to people's ability to make informed choices do have consequences in concrete, practical settings. It is perhaps time, as Leijonhufvud has pointed out, for theory to consider alternative techniques to model economic decisions.

References

Azariadis, C. and B. Smith (1993) 'Private Information, Money and Growth', Working paper, Department of Economics, UCLA

Bordo, M. (1993) 'The Gold Standard, Bretton Woods and Other Monetary Regimes: An Historical Appraisal', Federal Reserve Bank of St. Louis Review, vol. 75, No. 2, March/April pp. 123–91.

Casella, A. and J. Feinstein (1990) 'Economic Exchange during Hyperinflation', *Journal of Political Economy*, vol. 98, pp. 1–27.

Clower, R. (1965) 'The Keynesian Counterrevolution: A Theoretical Appraisal', in F. Hahn and F. Brechling (eds), *The Theory of Interest Rates* (London: Macmillan).

Clower, R. and A. Leijonhufvud (1975) 'The Coordination of Economic Activities: A Keynesian Perspective', *American Economic Review*, vol. 65, pp. 182–8.

Dabús, C. (1993) 'Inflación y Precios Relativos: Estudio del Caso Argentino', Doctoral dissertation, Universidad Nacional del Sur.

Heymann, D. (1984) 'Precios Relativos, Riqueza y Producción', *Ensayos Económicos*, vol. 29, pp. 53–90.

Leijonhufvud, A. (1968) *On Keynesian Economics and the Economics of Keynes* (Oxford University Press).

Leijonhufvud, A. (1973) 'Effective Demand Failures', *Swedish Economic Journal*, March.

Leijonhufvud, A. (1981) 'The Wicksell Connection', in *Information and Coordination* (Oxford University Press).

Leijonhufvud, A. (1989) 'Information Costs and the Division of Labour', *International Social Science Journal*, vol. 91, pp. 165–76.

Leijonhufvud, A. (1993) 'Towards a Not-Too-Rational Macroeconomics', *Southern Economic Journal*, vol. 60, no. 1, pp. 1–13.

Palerm, A. (1990) 'Price Formation and Relative Price Variability in an Inflationary Environment', Doctoral dissertation, UCLA.

Phelps, E. (1983) 'The Trouble with Rational Expectations and the Problems of Inflation Stabilization', in R. Frydman and E. Phelps (eds), *Individual Forecasting and Aggregate Outcomes: Rational Expectations Considered* (Cambridge University Press).

Phelps, E. (1992) 'Expectations in Macroeconomics and the Rational Expectations Debate', in A. Vercelli and N. Dimitri (eds), *Macroeconomics. A Survey of Research Strategies* (Oxford University Press).

Sargent, T. (1982) 'The End of Four Big Inflations', in R. Hall (ed.) *Inflation: Causes and Effects* (Chicago University Press).

Sargent, T. (1993) *Bounded Rationality in Macroeconomics* (Oxford University Press).

Tommasi, M. (1992) 'The Welfare Effects of Inflation: The Consequences of Price Instability in Search Markets', Working Paper, Department of Economics, UCLA.

15 Complex Economic Systems, East of Sweden

Christof Rühl[1]

> ... extreme episodes can teach us better than anything else *how crucial* are the things we take for granted. (Leijonhufvud, 1993)

Happy Birthday!

Unlike most of his peers, Axel Leijonhufvud commands an immense array of interests in 'doing' macroeconomics. Few are singled out by mentioning his masterful use of the history of the subject, his proficiency in integrating what is deemed to be useful of the endeavours of fellow scientists from other disciplines, and his skilful elaborations on the sometimes bewildering social consequences of alternative economic policy options. His writings are a tremendous source of wisdom and ideas; so large, in effect, that acknowledgement is not always easy. While I may have learned more from Axel Leijonhufvud than from any other living economist, I have also felt the perils which arise so easily from misallocating the inputs from this impressive spring of knowledge.

Obvious links between his work and this latest reallocation can be established, however, in a number of ways. First, Axel Leijonhufvud, in contrast to many of his colleagues, immersed himself immediately in the intricacies posed by the transition from central planning but without following the beaten path; what is critical of this path in my chapter has been shaped and refined to a considerable extent by the discussions we had. Further, in one sense, the story told here is just another sequel on the more general theme of how to model individual behaviour and human interaction appropriately. In its undertones, the chapter is critical of the assumption of 'unbounded' rationality, so routinely employed in macroeconomics today: this topic Leijonhufvud explored in depth. Thirdly, the paper employs two of the tools he shaped, (and left for others), helpful in finding a way out of the haze left by too many of the 'too-rational'[2] macromodels of contemporary persuasion; that is, it attempts to take the history seriously, and the history of economic theory even more so. I am not a master in applying these tools. That I can draw from the writings of someone who is, illuminates best the connection between Leijonhufvud's work and this chapter.

INTRODUCTION

This chapter is motivated by the diverse results that the transition from centrally-planned to market-coordinated economic systems has generated so far. Growing out of the conviction that understanding the legacy of central planning is essential for the success of reform, it is an attempt to investigate the organisation of an archetypal planned economy. In doing so, it assumes the eclectic perspective of discussing that economy as 'one giant firm', with the aim of concentrating on the principal–agent problems created by subjecting economic activity to centralised pricing and resource allocation.

Obviously, such an approach ascribes more rationality to the system than the 'planning is what the planner does' strand of the literature, without, however, pretending in the first place. But by concentrating on what made the system work (and what slowed it down), it becomes better accessible to our own economic categories, without denying that the nature of the beast is different from what we encounter 'at home'.

Ultimately, this should provide better grounds to understand and judge those parts of the system's legacy that are easily overlooked but which matter, because they hinder economic reforms.

The next section describes the basic macroeconomic problem confronted by the worst-case scenario. It is a simple problem: output falls and refuses to stop, inflation is high and refuses to come down. This section discusses the most influential of the many early 'blueprints' for reform, by contrasting what should have been with a not-so-common-sense account of what in fact happened. The third section concentrates on reasons for the decline in output and argues that, if economies of scale are present, it is rational for a central planner to implement a highly concentrated industrial structure where the degree of vertical integration may exceed what is warranted in competitive economies. The fourth section digresses into the literature on industrial organisation to review how agency problems raised by the separation of ownership and control are resolved in market-based organisations, and what this means for the organisational structure of complex firms, characterised also by different degrees of ownership dispersion. The fifth section tries to make use of these ideas by applying some of them to the organisational structure of the centrally-planned archetype. The sixth section comes back to the initial problem, and discusses briefly the relevance of the exercise by means of an example, namely the emergence of inflation in Russia.

SCYLLA AND CHARYBDIS

In the beginning, the task seemed formidable yet manageable. Though no blueprints existed with which to engineer the switch from central planning to market co-ordination, economic literature described these economies in detail. Moreover, centrally-planned economies were all considered to be members of the same species and subject to the same legacy, with differences only in degree. They seemed to have enough in common to justify a unified approach.

Contemporary literature portrayed these economies as being subject to profound distortions – either in terms of the 'command economies' populating the textbooks on comparative economic systems, or as Kornai's 'economics of shortage'. The former delineated the bureaucratic consequences of attempting to compute an equilibrium price vector; the latter the economic consequences of fixed price regimes with quantity rationing. In both cases, the focus was on the limitations imposed on economic systems by an extreme degree of bureaucratic intervention, and the resulting inefficiencies. From this perspective, the degree of intervention appears to constrain a prototype – the 'free', unconstrained or 'natural' state of affairs. In theory, this prototype was easily identified. In practice, the natural state came to be associated with the market economies already existing, making possible an easy switch back to the original purpose, that of comparison.[3]

In principle, there is nothing wrong either with comparisons or the generalisations which make comparison possible. However, such analysis can cloud certain features of the object under consideration – in particular, those which have no counterpart in the benchmark of comparison. The initial blueprints for reform suffered from such deficiencies.[4] To a significant degree, they were based on the idea of freeing a constrained economic system – by letting things fall into their place 'naturally'. In some cases, this did not work out too well. Things simply fell apart instead.

The most influential of these blueprints, outlined by Lipton and Sachs (1990a), came to be implemented in countries as diverse as Poland and Russia, but, more importantly for our purposes, this scheme also set the context for most subsequent discussions. In fact, it can legitimately be claimed that many of the proposals that followed theirs were merely additions, directed at supplementing the economic core which they first presented.[5] This core comprises a three-step course of action, which takes as its starting point a prototypical planned economy. Step 1 is the liberalisation of prices, ideally in one strike (as well as the freeing of exchange rate controls, wherever possible). Step 2 involves 'corporatisation', by which is meant both the imposition of hard budget constrains on the economy as a whole through exposing the (still state-owned) enterprises to competition, and simultaneously turning them into joint stock companies. Step 3 is privatisation, again pursued as comprehensively and quickly as possible, and under these premises, *nolens volens* with an emphasis on the sale of existing companies ('top-down').

The plan is ingenious, simple, and of intuitive appeal to the economist. Let's trace it briefly. Step 1 came under the heading 'to end excess demand'. The liberalisation of prices – which, if anything, can lay claim to the infamous 'big bang' phrase – was meant not only to establish the equilibrium price vector, *conditio sine qua non* for ultimate market co-ordination. It was also supposed to eliminate the monetary overhang accumulated in a system with soft budget constraints, fixed nominal prices, absent capital markets, and a government with the power to cover nominal 'losses' simply by extending the money base. It was therefore expected to result in a single jump of the price level, after which nominal prices would start adjusting to their equilibrium values, thus mirroring the correct configuration of relative prices. A single jump, however, was also seen as a check to inflation.

The next step, effectively levelling the last vestiges of central control, comes under the euphemism of 'increasing competition'. The purpose, again, was multiple. Most important is the imposition of hard budget constraints on state-owned firms. This would not only guarantee macrostability (understood as the absence of inflation), it would also set one of the preconditions for privatisation by solving the evaluation problem. Putting firms 'out in the cold' by exposing them to competition would distinguish the viable from the non-viable from those in need of restructuring, making calculations of a firm's net worth possible (swift implementation also produced the benefit of weeding out non-competitive firms fast, while political support for the transition was still strong). In addition, the corporatisation of the old entities was to greatly facilitate the emergence of new market-related institutions – from appropriate accounting procedures to the emergence of capital and financial markets.

Against this background, the extent of privatisation (Step 3) could be used as a proxy for successful reform. Not only would it establish and cement the foundation of the new order, private property rights, but it was also seen as having the potential – contingent on the method chosen – to fill the coffers of the government, to allow for a proper tax code to be implemented, and thus in general to establish the 'power of the purse', holding obstruction by the old elite in check. 'Top-down' privatisation (outlined early on in Lipton and Sachs (1990b)) mattered in this context: time was of the essence, and governments at any rate could do little to foster Schumpeterian spirits, other than to instil confidence in the success of reforms by pursuing good policy without hesitation. Availability of worthwhile projects for privatisation was not a concern.

To appreciate the impeccable logic of the proposal reverse the order of the steps, and start with Step 3 by assuming that market allocation in a private property economy is the ultimate goal of the transition. Then backtrack: it is clear that each step establishes preconditions for each subsequent step, taking into account the constraints of central planning as the starting point.

The blueprint is simple, ingenious, and intuitively appealing. In its simple form it is a variant of the J-curve idea, and can be discussed in that framework. However, the translation into the Phillips curve framework (see Figure 15.1) yields more insight. Neglecting, for the moment, the dotted line, the economic adviser finds herself confronting an economy which, for reasons not at all related to efficiency, is stifled in a situation which in theory coincides with the origin: (almost) no inflation and (almost) no unemployment. The one-time jump in the price level initiating the process will not cause the rate of inflation to increase; but the weeding-out of unprofitable enterprises will cause output to decrease (and unemployment to increase) until 'bottom-up' and, here more importantly, 'top-down' privatisation has made some inroads. Subsequently, aggregate production should increase to overcome the transformational recession – above the old maximum in the J-curve framework, and in the Phillips curve framework by enough to home in on the normality implied by the concept of a natural rate.

The question of how long it would take to cross the 'valley of tears', as Dahrendorf (and Sachs) had called it, and how deep it would be, yields another characteristic of the plan. While we resist the temptation to quote early estimates of the output decline, one of the principal assumptions of the strategy demands mention: the depth of the recession will depend on the economies' initial endowments, whereas the length of the process will also be contingent on the speed with which liberalisation measures are implemented. The tension between policy objectives is captured in a simple formula: liberalisation propels the transition process forward, stabilisation is its safeguard.

Ingeniously, the proposal puts into place the requirements needed to continue the investigation from a theoretical viewpoint. By taking endowments and technology as given (combined with the obvious presumption that preferences can be taken as given at any point in time

Figure 15.1 'The J-curve' idea in the Phillips curve framework

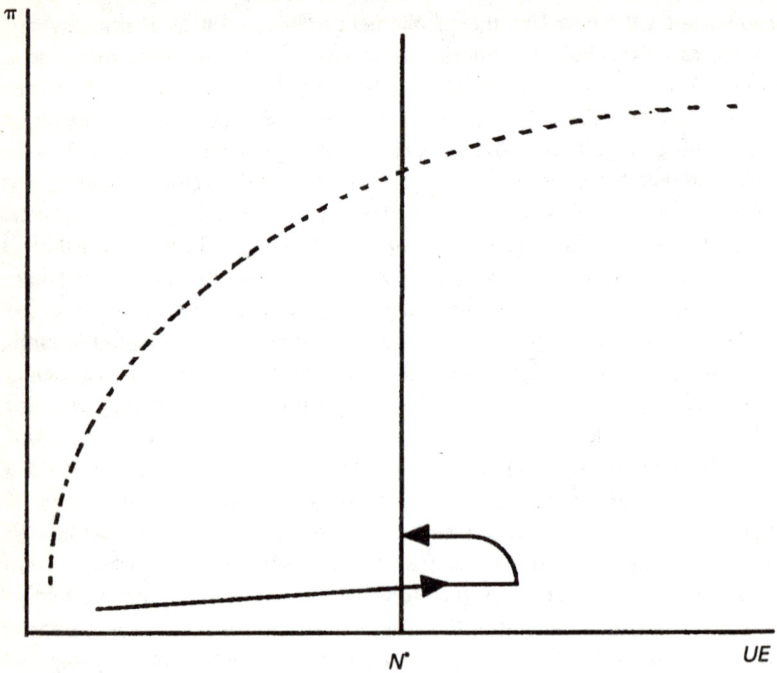

and, therefore, do not require special attention), it creates an environment in which the data set is adjusted to allow the intertemporal allocation of resources to be determined by the pursuit of individual preferences – in this case through establishing the missing private property rights and competitive markets. Not incidentally, this data set is identical with the information the economist requires to model this very pursuit.

On a deeper level, therefore, the proposal establishes the preconditions for rational economic man to act the way we suppose he should: it provides him with the information sets and also the constraints necessary to allow for intertemporal optimisation; and by having added the conditions for a competitive equilibrium, opportunity sets are defined. It thus establishes all the requirements of a choice theoretical framework from which Leijonhufvud's 'clockwork' of intertemporal optimisation can start ticking. There is no obvious reason why it should not. And providing the conditions to reap the clockwork's promises is, after all, a notion not without appeal to the economist.

In the meantime, we have observed mixed results in the transition from centrally-planned to market-coordinated systems. The examples we are concerned with here are the problematic cases, such as Russia, which cannot simply be explained in terms of having had the worst historic initial conditions in terms of wealth per capita (or the lowest standard of per capita consumption). Their problem has aptly been compared, metaphorically, to Odysseus struggling to steer his ship through the evils of Scylla and Charybdis (Leijonhufvud, 1994)). Charybdis, the monster relentlessly sucking in the ocean and everything on it three times a day, neatly corresponds to the continuous decline in output suffered in Eastern Europe. Scylla, devouring everything within reach whenever she has a chance, brings forth the image of erratic and high rates of inflation plaguing many of these economies. The problem, so the story goes, is aggravated because the ship – let's make it a Russian ship – has two navigators. One, a Western economist, has much knowledge and respect for Scylla and hence tries to avoid her at all costs, but has little experience (thus fear) of Charybdis. The other, from the East, understands well the evil consequences of crossing the path of Charybdis, but never having heard of Scylla, underestimates her destructive power. Figures 15.2a and 15.2b show the dimensions of the problem.[6]

The reasoning of the practitioner amounts to a familiar litany: History matters. Capital accumulation renders economic development path-dependent. The role of institutions and of their breakdown has been systematically underestimated. Interest groups block the process. But unless one is willing to argue that perfectly rational people choose to spend their time hovering around two evils, rather than trying to stay clear of both, we need an explanation for why rational agents could create a socially unwarranted outcome – or why behaviour should have differed from the standard macroeconomic assumptions. The key to the former lies in the past, that to the latter might well be in the assumptions.

Knowledge of the territory thus seems to involve two aspects. One is the legacy of the old system, and here especially of the industrial structure it left – where may be found reasons for 'path dependence'. Another would be the mind-sets of its inhabitants. We begin with the former.

CHARYBDIS, OR THE LEGACY OF CENTRAL PLANNING[7]

The (textbook) literature on the organisation of central planning provides two angles from which to categorise casual observations. The first, designed to clarify the economics behind the 'command economy'

Figure 15.2a Charybdis: Decline in industrial production in the Russian Federation (real terms, not seasonally adjusted), December 1990 = 100

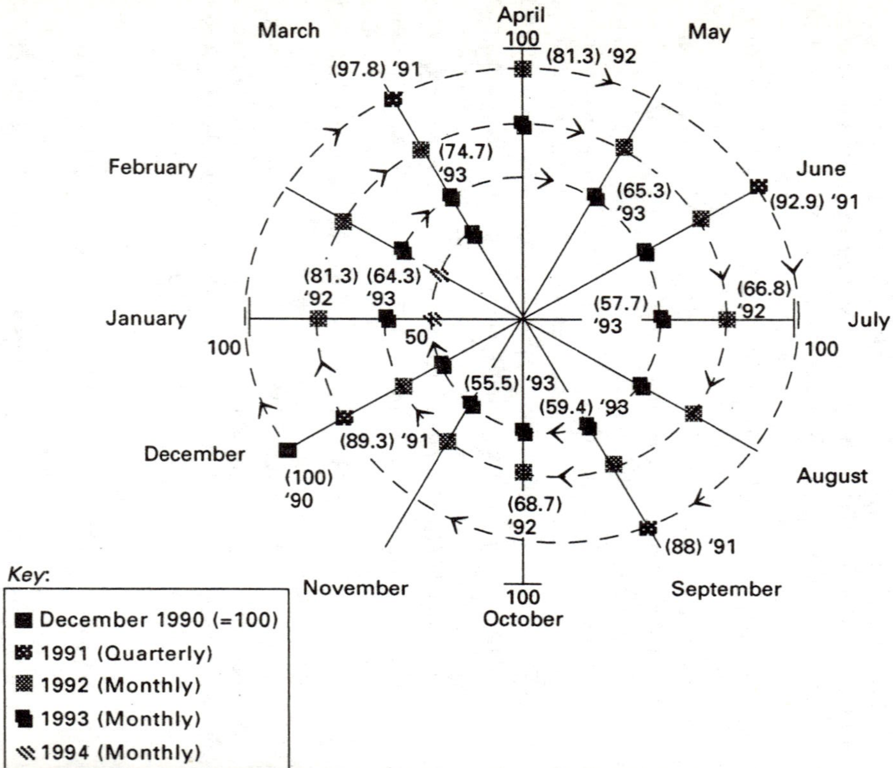

Month/Year	*Output Index*
J 1992	81.3
F	81.2
M	88.7
A	81.3
M	77.4
J	78.3
J	66.8
A	62
S	69.2
O	68.7
N	65.5
D	71.3
1993 J	64.3
F	66.7
M	74.7
A	68.9
M	65.3
J	65.9
J	57.7
A	53.3
S	59.4
O	58.1
N	55.5
D	59.8
1993 J	49.4
F	50.6
M	54.2

Figure 15.2b Scylla: CPI in the Russian Federation, December 1990 = 100

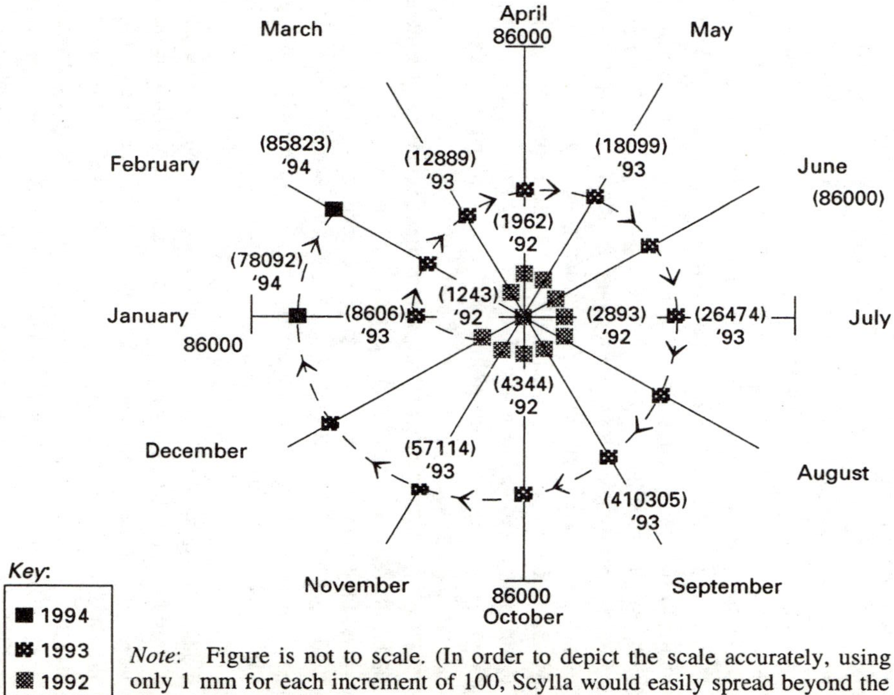

Month/Year	CPI
J 1992	
F	1243
M	1613
A	1962
M	2197
J	2606
J	2893
A	3135
S	3531
O	4344
N	5473
D	6841
1993 J	8606
F	10732
M	12889
A	15338
M	18099
J	21700
J	26474
A	33358
S	41030
O	49236
N	57114
D	64539
1993 J	78092
F	85823

Note: Figure is not to scale. (In order to depict the scale accurately, using only 1 mm for each increment of 100, Scylla would easily spread beyond the reader's desk.)

Source: *Goskomstat* (*Russian Economic Trends*).

paradigm, shows a pyramid which has as its floorplan an input/output table providing a mirror-image of the economy as a whole (Figure 15.3). Its foremost use is to compare degrees of central planning: from directing rough aggregates such as investment and consumption at the *B*-level (which may be achieved by means of tax and other incentives), to the direction of the 'material product' in minute detail, that is, the determination of the model economy's technical coefficients on the pyramid's ground floor (*A*).[8] As the figure suggests, the structure of such an economy is determined by the need to maintain an integrated, hierarchical bureaucracy, capable of issuing commands and monitoring obedience. The illustration also leads to an understanding of the most efficient configuration of production, when arranged by a central planner: Surely, it ought to mirror attempts to economise on transaction costs, and in particular, the costs of control and monitoring.

The second useful angle from which to classify our observations originates from analysing the 'economics of shortage'. The revelation that a fixed-price system, set in arbitrary fashion for all intents and purposes, would incur (and cover up) nominal 'losses' for the aggregate, makes for a good puzzle indeed. Though the mechanics of the process are easy to grasp, the rationale behind it is not. Beyond the inefficiencies implied by, for instance, the violation of proper accounting rules, the most perplexing elements of such a system lie in the subtle implications it has on the behaviour of firms. To so completely discount the importance of profits and losses (and to reduce money to Marx's 'circulation medium', negotiating only the link between commodities in the present) prompts us to seek rational reasons for such aberrations. But there is also the question of the consequences of 'letting firms loose' within the remnants of the financial system employed under central planning.

Stylised facts seem to support the first inference, namely that the structure of the capital stock was indeed adjusted to the needs of central planning, to minimise the costs of monitoring in a peculiar way (Intriligator and Rühl, (1994)).

- Industry and capital stock are highly concentrated. In many cases, bilateral monopolies and monopsonies exist at each stage of the production process for vertically integrated industries, and in general production is concentrated in relatively few facilities per output unit.
- Both firms and individual factories (several of which might comprise a single firm) are larger than would be justified under competitive conditions; their size cannot be explained by exploiting increasing returns to scale. In general, this includes the observation that vertical

Figure 15.3 Degrees of central planning and the 'command economy'

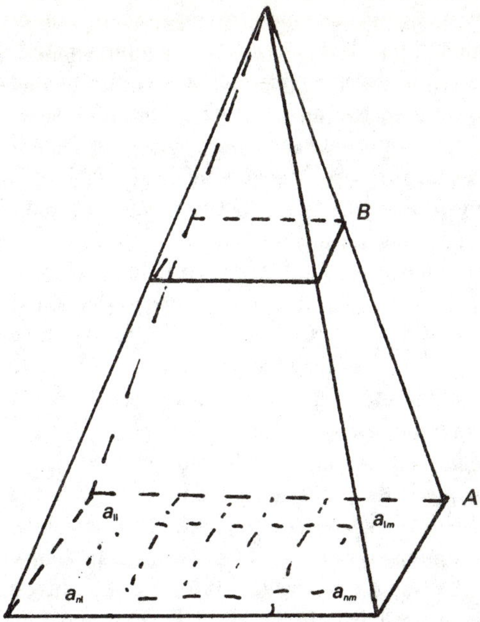

integration within firms was deeper than under competitive conditions, leading to the production of a relatively high proportion of primary and intermediate inputs at the firm level.

- The technology mix, mirroring the absence of competition, is characterised by a vintage structure which combines different degrees of efficiency. Consequently, parts of the capital stock, production technologies ('processes') as well as final goods ('products') cannot, at living wages, compete with established production possibility frontiers, domestic or abroad.
- Firms produced a peculiar joint product, in that industrial (and agricultural) centres of production traditionally have assumed responsibility for substantial elements of the social infrastructure and the tertiary sector. Day-care centres, schools, hospitals, housing and shops are operated typically by single enterprises.

Even before adding observations related to the organisation of the financial sector, we get a first, intuitive grasp on Charybdis. The story runs as follows. As a result of the desire to simplify the economy's input/output structure to keep implementation of the plan manageable and controllable, the number of columns and rows in the grid portraying the transactions between enterprises was minimised, by limiting the number of firms producing close substitutes, thus creating bilateral monopsonies, monopolies, and industrial concentration. By the same token, firm size and the extent of vertical integration on the firm level increased. Assume now that the remaining vertical chains between enterprises are broken, for example (to borrow Marx's phrase) as the result of technological obsolescence made obvious by 'corporatisation'. Disruption of the inter-enterprise links at single points will cascade throughout the system, creating bottlenecks external first to enterprises, and then to whole industries. Scaling down the operations of vertically integrated enterprises, on the other hand, whether caused by the obsolescence of single components, or by the 'unbundling' of one part of the enterprise as part of the overall strategy of restructuring, will create bottlenecks internal to the firms. These effects reverberate throughout the system in the same manner.

Technological obsolescence or indivisibilities, brought to light by corporatisation, trigger the process; concentration on the aggregate level coupled with the extent of vertical integration at the firm level, amplify it. The strategic location of enterprises, as well as the dependence of large parts of the population on the social services provided by employers, may then explain the emergence of powerful alliances across and within enterprises. All that is needed to extend the narrative also to cover inflation is reference to the institutional remnants of the old financial system, explaining how the chain of command could be reversed in order to create arrears, and to translate them into high powered money so smoothly. Maintaining the old pattern of interaction and institutional channels, while reversing the flow of demands running through them, seems to have been characteristic of the process. The infamous 'file number two', the inter-enterprise clearing account, is an obvious candidate to exemplify: under central planning it recorded enterprises' surpluses and losses, which were balanced semi-automatically by Gosbank, including the settlement of financial losses in the aggregate. This account provided for a natural way to accumulate arrears *post* central planning, without the need to invent new financial instruments (or patterns of interaction).[9] Scylla would then be the result of a bargaining game between industry and government, carried out via the extension of inter-enterprise IOUs. Attempts to use the threat of social unrest, and/or the strategic location of single enterprises, would force the government to forgive arrears by extending central bank credit. Whether understood as blackmail by 'red' enterprise directors, or the self-defence of a desperate populace, Charybdis's presence conjures up visions of Scylla. The struggle that ensued is inconceivable without the vertically integrated, and highly monopolised, industrial structure.[10]

This first glance at the conditions of Russia's transition makes for a plausible story, but it is not entirely satisfactory. Specifically, three questions are left unanswered:

1. The first, relating to the possibility of a massive breakdown, concerns the rationale for setting up the system the way it was. In what sense was it an efficient, rational organisation in practice, and in which sense was it not? Whose interests did it promote?
2. The second asks why the cascade of failures could not be prevented by the allocative power of the price mechanism. It proceeds from the assumption that the structural characteristics of a planned economy in principle present nothing out of the ordinary, except for a lack of information (for example, about alternative delivery channels), aggravated by incorrect pricing. Thus, it addresses the proposed link between the decline in industrial production and the onset of inflation, in a simple version, by doubting whether the decline in output was necessary at all, but also, in a more sophisticated version, by suggesting that the decline could have been prevented under all circumstances (even if the above description was half-way correct) merely by a functioning price mechanism.[11] The question has the potential to change causation: stop inflation, and by more precise signals and a more complete set of markets, the decline – which now has the distinct status of an especially vicious J-curve – can be stopped and eventually reversed.
3. The third question addresses the microeconomics beyond the proposed explanation for the onset of inflation. It asks how firms could find the power to steer the ship toward Scylla and to create an inflationary spiral, just to avoid hard budget constraints and the restructuring they entail. Understanding their motivation is rather different from explaining how they managed, given a government which, at least in word, was committed to reform. To turn the casual account into an economic argument requires that firms, to make rational choices consistent with the system's structure, must have been in a position of superior knowledge. The alternative, not easy, would be to denote reasons why firms' (or the government's) behaviour should deviate from the standard assumptions.

To clear matters up, however, a historical viewpoint will not suffice.

In an important article, Leijonhufvud (1985) addresses the 'Dahlman paradox', the observation that 'progress' in agriculture is characterised by a movement away from collective forms of land ownership towards single units (the family farm), while the reverse seems to be true for manufacturing since the Industrial Revolution. Collective organisation of large-scale production (the emergence of the factory) coincides with collective titles to capital goods (the joint stock company). Leijonhufvud traces the key to this paradox to a familiar but still somewhat unruly topic: increasing returns in manufacturing.

In this view, joint ownership is the economic solution to the problem of distributing the joint rent associated with complementarities in production. Following Adam Smith's dictum that the division of labour is limited by the extent of the market, increasing returns are realised by dividing production into individual tasks. Rearranging the organisation of the production process and the use of machinery in a way that standardises tasks increasingly, leads to a configuration which exploits all possible efficiency gains in the static context and also leads to the discovery of new ones in a dynamic setting.

Such gains in efficiency are associated with complementarities in production. A discussion of Smith's problem, or the transition from craft to mass-production systems, illuminates some Eastern European observations (Figures 15.4a and 15.4b). Let us assume, in the most simple case, that the production of a certain product (*X*) requires three distinct operations. Assume also that initially the labour of three craftsmen (*a*, *b* and *c*), each equipped with a complete set of tools (1, 2 and 3), will produce three units of *X*, independently from each other. Figure 15.4a shows the first possible rearrangement of production: pulling the craftsmen under one roof and sub-dividing the three operations among them, so that each is reduced to performing one task only. The claim is that the increased coordination resulting from reordering the process will increase productivity. The switch saves 'capital goods', since no set of tools need stand idle, as opposed to the initial state, in which each of the uncoordinated craftsman needed to own *all* the necessary tools. It also reduces the human capital requirements for production, as well as working capital, in inventory and overheads. The implied complementarities, however, create greater vulnerability of the process: the 'assembly line' stands idle as soon as one its 'workstations' is left unmanned.

The assembly line example indicates additional yields through the exploitation of time, an essential element in this analysis (Figure 15.4b). Assume that one of the tasks performed with the existing workstations can be done in half the time required for the other operations, for example,

Figure 15.4a Mass production: assembly line

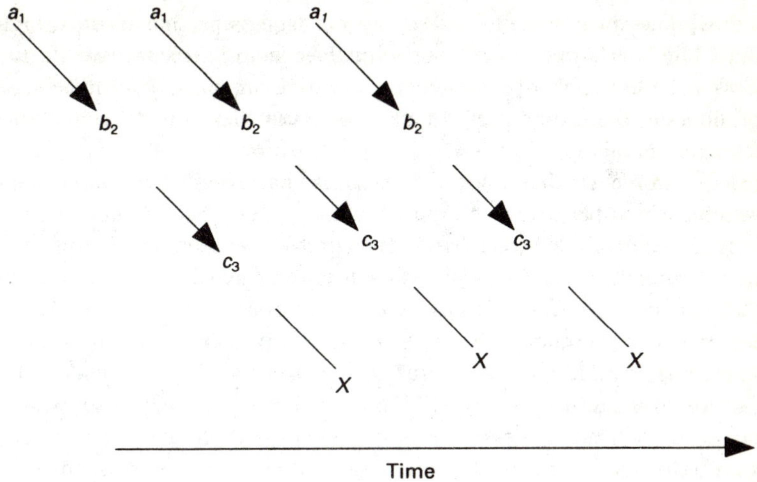

Figure 15.4b Mass production: parallel assembly lines
(*Source*: Leijonhufvud, 1985)

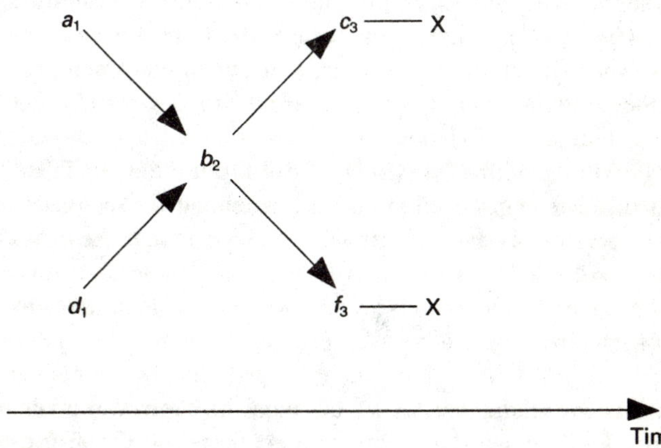

Step 2. If production can be increased, a parallel assembly line can be established, but inputs need not be doubled, because the configuration changes such as to make full use of Workstation 2 (which now serves both lines). Consistent with Smith's argument that the division of labour is

causally dependent on the 'extent of the market', further increases in production open further opportunities to optimise the configuration of already-known procedures and equipment, generating increasing returns to scale in a process which 'slips through modern production theory as a ghostly technological-change coefficient or as an equally ill-understood economies-of-scale property of the production function' (Leijonhufvud (1985) p. 209).

Note, however, that while for fixed activity levels a maximum exists beyond which performance cannot be improved, this need not be true as long as output is allowed to rise – though this source of scale economies will become less and less significant as long as the number of tasks remains constant. If, on the other hand, a process of discovery is allowed for, that is, if new machines are introduced to increase the number of serial stages and to support a further subdivision of the manual tasks, the possibilities become virtually unlimited.[12] The process discussed so far denominates an increase in the 'vertical' division of labour contingent on an increase in production. An increase in the 'horizontal' division of labour, dividing the production process spatially (and likely to increase the number of firms) is also conceivable, but subject to a minimum requirement of scale. Finally, it deserves mention that, while the vertical process is capital saving, an increase in the horizontal division of labour does not *per se* provide grounds for a conclusion to the contrary. This is in striking contrast to Marx's perception of technical change as associated with an increase in the organic composition of capital (the capital/labour ratio), as well as with his relative neglect of the horizontal division of labour.[13]

Leijonhufvud uses this approach to investigate how capitalist ownership of the means of production came to be superior to workers owning their own 'tools', and how the form this ownership too came to be 'shared ownership'. We shall follow him for a little along the way, if for a different purpose. We start with the second question and then ask what limits the size of firms and what allows for increasing the horizontal division of labour.

If, in the above example, allowance is made for each worker owning his own workstation, there arises a problem connected to the distribution of the joint rent. It is not clear how this rent should be divided among the machine owners, and marginal theories of distribution will not help to answer this question. In the simple example of a single assembly line, total output would be reduced to zero if any one of the workstations was withdrawn. Even sophisticated configurations would be affected substantially and in a way not captured by marginal productivity considerations (see

Figure 15.4b). Moreover, even if an agreement on the division of proceeds is reached among the owners, it will remain a highly unstable arrangement, because of the threat that any of the capital owners might 'hold up' production, thus reducing the earnings of all the others.

The solution is to institutionalise joint ownership. By owning shares in a firm instead of physical assets in a 'factory', the coordination of productive activities can be improved without the consequences of instability.

Since the process is driven by the fact that specialised machinery designed to support an increasing sub-division of labour is not easy to replace, we also come across a preliminary explanation of what limits the size of firms in a competitive environment. If a machine is easily replaceable, the threat of significantly reducing joint rent by withdrawing from the process loses strength. If replacements are found easily, their rental services face a thick market, similar to the market for unskilled labour. It will become worthwhile to establish a new firm: that is, to exploit a 'market gap' in situations where the market for the (intermediate) product resulting from a peculiar stage becomes thick enough so that firms on both sides of the gap are secured against the threat of hold-ups (Leijonhufvud (1985), p. 218).

Thus, the analysis extends to the horizontal division of labour as well. As processes on 'both sides' of markets become self-contained, a variety of firms facing thick markets for their inputs and outputs may spring into existence. What will determine their number and size? In a decentralised economy with voluntary contracting and free access to capital markets, there may be any number of reasons to invest one's capital in a new firm by switching away from 'old' equity. Among them are unsatisfactory allocation of the joint rent (dividends), costs associated with firm size outweighing the gains of specialisation (agency costs), and perhaps most importantly, the risk premium associated with marketing formerly specialised inputs, that is, the possibility of creating new markets. In general, the switch will be determined by individual calculations, comparing expected returns with the opportunity costs of not switching. For the moment, only two results of this process merit our attention: first, that the division of labour may be extended horizontally as well as vertically does not imply that the efficiency gains associated with static and dynamic parallel line economies will diminish in the aggregate, as activity levels increase. Secondly, the horizontal division of labour is bound to improve the prospect of 'discovering' new sources for increasing returns (new tasks), since it will partly be motivated by the risk premium attached to the creation of new products, and new markets.[14] It is therefore an important force beyond the process of discovery.

The analysis has important implications for the questions we have raised. To derive these propositions, we define the best possible case for a centrally planned economy, and incorporate our simple example of increasing returns. The economy has no private ownership of capital and therefore no markets for capital goods. In addition, a central authority sets all prices and decides what to produce. However, we assume that these decisions are implemented without any need for control or monitoring. The costs for implementing decisions (transmitting information) 'from above' can then be reduced to transaction costs in a one-to-one mapping; that is, they are an increasing function of the number of units to be addressed. Otherwise, information is assumed to be freely available.[15] We also shoulder a benevolent planner, concerned only with maximising the consumption and leisure choices communicated by his/her populace, which, given technology, is scattered across the economy initially in some random manner.

Under these circumstances, it seems that:

- The costs of planning are an increasing function of the number of firms. It is efficient to minimise the number of production facilities up to the point where the reduction in the costs for decision implementation, plus the gains from specialisation, equal the additional costs for the (spatial) transfer of all commodities and labour over the relevant time horizon.[16]
- The assumptions about information and monitoring made above in fact describe the (almost!) *unconditional* co-operation of economic agents with the central planner (except for spatial preferences and relocation and transfer costs). Under these assumptions, it will be possible for a centrally-planned economy to exploit static economies of scale as well as any idealised competitive system would, should the initial configuration of the production process not correspond to the optimal configuration. This holds true also for the dynamic context, provided that additional gains are restricted to an increase in parallel line production. In this case, the planner can calculate the maximum sustainable activity level *ex ante*, given the technical coefficients of production, and initial endowments.[17] The clockwork can be set in motion.
- However, if the issue is the discovery of new ways of specialisation, by enhancing the sub-division or the horizontal division of labour, the centrally-planned economy cannot claim to be systematically at the same level with the competitive economy, even in this best of all worlds. As long as the discovery of additional productivity gains

> involves costly trial and error processes, the results of which cannot be calculated, but are subject to both risk and genuine (Knightian) uncertainty, the likelihood of success may differ. Specifically, the path of both economies will diverge in a systematic manner, if it can be shown (a) that the probability distribution of anticipated success in uncertain trial and error processes is biased; or (b) if the costs of failure (which might be calculable) diverge between those borne by the aggregate, which is what the planner has to take into account, and those borne by individuals, making and implementing the decision to 'leave' the old structure to try something new, which is the level at which cost calculations matter in the competitive economy.

What this means is simply that under uncertainty as to the gains, and with risk as to the costs, the alleged uniformity of both systems' welfare performance need not hold, because consequences and expectations may differ between the individual and the aggregate. Individuals will calculate their opportunity costs in the case of failure, but the planner will estimate the costs for the aggregate; and individuals will weigh the expected gains of succeeding in uncertain trials against their opportunity costs, while the planner will weigh them against aggregate welfare losses. As a result, the rate of endogenous technical change may differ in competitive economies under plausible conditions, even if we neglect the planner's control and information costs.

These inferences more or less reiterate some of the conclusions derived in the so-called socialist calculation debate. In general terms, the debate showed that *in principle* a central planner could arrive at an equilibrium allocation of resources and the corresponding vector of relative prices just as the Walrasian system would, and with claim to the same optimality and stability properties – as long as s/he would not incur additional (as we would now call them) transaction costs; hardly a surprising result from a mid-1990s vantage point. In practice, however, this equal-efficiency result was derided and recognised as being unachievable. The reasons centred around the informational requirements of central planning and the feasibility of decision-making, that is, of computing the equilibrium price vector (at the time a hard constraint related to the limits of computational technology). Also, contingent on the underlying equilibrium concept, the essence of the debate was concerned with the efficient allocation of resources in a *static* framework; questions of dynamics were discussed, but not systematically, as the contemporary notion of equilibrium did not include the 1990s' intertemporal focus.

We have merely extended these assumptions somewhat, but depicted the same idealised case. The main difference, other than introducing a very simple example of increasing returns, was to partition explicitly the costs of the planning authority into a triad of (i) decision-making and implementing; (ii) control and monitoring; and (iii) information flows. By leaving aside (ii) and (iii), and by restricting (i) to a simple case of information transaction costs between sender (planner) and receiver (production units), that is, to decision implementation, the idealised picture furnishes its results before more worldly interferences are introduced.

Not surprisingly, it is rational, under these circumstances, for the benevolent planner to make use of the advantages offered by increasing the scale of operation (by weighing the operational costs and benefits, plus the utility of those involved). The economy will exhibit a degree of concentration determined by the blueprints of technology. The planner can, just as well as her Walrasian colleague, adjust an initially less than perfectly adapted configuration, with given endowments, to the existing output levels, and she can exploit the additional benefits offered by an increase in activity levels. Only when we take into account the fact that new discoveries might be made by costly trial and error processes which, although essentially unpredictable, might be initiated from 'above' or 'below', is the outcome of both systems likely to vary in a systematic manner. However, though we may be able to compare the merits of two hypothetical economies by developing hypotheses about motivation and markets, these are precisely the situations where Walrasian theory fails to deliver a description of the process.

The exercise, however, was not *l'art pour l'art*, because it suggests a preliminary answer to two of the questions raised by the inferences drawn from stylised facts, given at the beginning of this section. First, in this idealised world it is rational for a central planner to adjust the industrial structure, resulting in concentration, if economies of scale are present. *Ceteris paribus*, if a break-up occurs, with single production units trying to interfere with the allocation policies decided by the former centre, exactly the unstable situation suggested in the discussion of Figure 15.4b would result. Second, no vector of relative prices can cure that property.

On the other hand, while the planned economy operates as efficiently as any configuration under the standard assumptions of general equilibrium theory, it is less clear that this would happen if technological discoveries could be made by costly trial-and-error processes – because decisions to engage in them will be made, under uncertainty, by individuals, and communicated through decentralised markets in one system, and with an eye on aggregate welfare in the other.

The introduction of information and control needed to model the planners' co-ordination problem realistically, will not improve this best of all worlds. But it is necessary to show the true costs of planning, as well as to get a grasp on the system's ability to survive, and on what happens if it does break down.

Some more practical implications can also be drawn at this stage. The aggregate input-output structure representing the rational response to the planner's problem translates into a vertically integrated system, comprising bilateral monopsonies or monopolies. In addition, no mechanism limits firm size in potentially thick markets, and nothing at the firm level prevents 'oversized' plants. The degree to which initial output losses would be multiplied depends on the degree of industrial concentration. Had central control and co-ordination been part of this picture, and were then to be removed, we would expect the system to experience the 'hold-up' problem in the aggregate, causing aggregate output to decline. Once started, disruptions would tend to amplify rather than to settle down at a new level. Moreover, the existence of large units does not imply thick markets can be created simply by unbundling,[18] or that productivity gains can be realised by scaling back employment.

The example thus helps to explain the continuous decline in industrial production experienced in the state sector of all the former CMEA countries. But it also points to some of the remedies, and by so doing, indirectly provides explanations for the differentials in the transition to market economies observed across countries. It gives, for example, clear priority to 'bottom-up' over 'top-down' privatisation; countries that were more successful with respect to the former have indeed fared better than those that were not. It points to the importance of the privatisation of land, not as subject to complementarities as manufacturing, and easily divisible. The success of economies, in this respect from Poland to China, seems to confirm the idea. Finally, it serves as a warning against the inherent dangers of 'insider privatisation' as exemplified by the Russian privatisation programme because this might accelerate the destabilisation of the industrial sector by turning the existing state enterprises into units destabilising the system at large, as they attempt to survive by increasing their share of the 'material product', by exploiting their strategic position without being hindered by outside control.[19]

Time series data show that the decline of output and the increase of inflation occurred at about the same time. Employment fell with a lag. It could be understood in terms of agents trying to rescue their firms by using its strategic position to demand central subsidies, while the repercussions of artificially thin markets for intermediate goods already cascade through the economy. The old financial infrastructure transmits these demands. As delivery channels continue to be interrupted, the scramble for subsidies intensifies, since, in this case, it represents the only option to avoid rates of unemployment corresponding to an unchecked decline in output. However, this assertion must be substantiated by an explanation of why enterprises could use their strategic positions successfully to this end, that is, why they were, in fact, 'bailed out'. The account of the onset and perpetuation of inflation has to address the third of the questions raised previously: what could explain enterprises' informational advantage over the government?

CONTROL WITHOUT A CENTRAL PLAN

Our example of increasing returns in manufacturing leads to another problem. The proposed solution to the question of establishing a stable organisational form to appropriate the joint rent available from complementarities in production was, we recall, the institution of shared ownership. The joint stock company, however, raises the principal–agent problem. But the joint stock company is only a particularly obvious case in which this problem arises, because its ownership structure can be regarded as being extremely diffuse, leaving equity holders with little influence over the operation of their assets. In principle, the problem exists whenever control over assets becomes separated from the wealth effects derived from these assets or, in general terms, whenever residual claimants exist who bear the risk of ownership but do not have complete control over their assets.[20] Contracts become necessary to regulate these decisions, and their control and monitoring. This, however, is not costless.

The contract structure and other mechanisms alleviating the separation of 'ownership and control' is, of course, a venerable topic in the literature on industrial organisation. Although largely related to private firms and voluntary exchange, some of these contributions have proved to be helpful for the transition from central planning, mostly as far as the evaluation of alternative privatisation proposals is concerned.[21] But the discussion has also made the term 'control' more precise by yielding abstract accounts of complex decision processes; these accounts can be employed in situations in which the role of the residual claimant does not correspond to a principal who can easily be defined in categories of private ownership – in circumstances, that is, in which a control problem arises even though private ownership of the means of production is not involved.

This approach again suggests looking at the planned economy as one entity – though the analysis applied to it will come from the study of the (capitalist) firm. We too, to quote from the basic outline taken from Fama and Jensen (1983, p. 301), are concerned with the 'survival of organizations in which important decision agents do not bear a substantial share of the wealth effects of their decisions'. We thus begin by employing the common distinction between risk-bearing agents or residual claimants (principal), and decision agents (agent). On an abstract level, the framework investigates how the agency problem caused by the separation of risk-bearing and decision process is solved. It also contrasts two scenarios. The explanation is derived by sub-dividing the tasks of the decision process. The two scenarios refer to complex and non-complex organisations.

In the context of organisations which have to survive in the marketplace, residual claimants bear the risk of having a right to cash flows which reflect the difference between payments already contracted by other agents (outflows), and inflows with a stochastic component. If the claimants differ from those who decide the organisation's allocation of resources, they will refine mechanisms to defend their interest. Fama and Jensen argue that the features of organisations' decision processes, tending to be overlooked by economists' focus on the idealised entrepreneurial firm, include these mechanisms.

Within organisations, the decision process is allocated across agents in four steps. Decisions have first to be initiated, that is, proposals for resource utilisation and other tasks are generated; then they should be ratified, that is, the choice of what proposal(s) to implement has to be made; then follows implementation – the execution of the chosen proposal; and finally monitoring – the measurement of performance (including the implementation of performance based rewards).

Thus, the decision process can be divided into two somewhat broader categories, defined as follows: *decision management*, indicates the initiation and implementation of decisions; while ratification and monitoring is referred to as *decision control*. In addition, a distinction can be made between complex and non-complex organisations. 'Non-complex' refers to organisations where information, costly to transfer but relevant for the decision process, is concentrated with one or a few agents; complex organisations are characterised by specific information diffused among many members of the organisation. Based on these definitions, it can be argued that decision management will be separated from decision control if residual risk bearing is separated from the decision process, and the organisation is complex. On the other hand, in non-complex organisations, the combination of decision management and decision control will generally lead to residual claims that are restricted to those agents in charge of the (unified) decision process.

The argument is easy to follow in the case of non-complex organisations. Here, the benefits associated with risk-sharing must be weighed against the costs that would result from the agency problem if residual claims were not restricted to those in charge of both decision management and control. In small, non-complex enterprises, the agency costs associated with separating the decision process, as a rule, will be higher than the efficiency gains associated with unrestricted risk sharing. Proprietorships, small partnerships and closed corporations are the standard examples for these 'entrepreneurial' organisations.

Correspondingly, in complex organisations, where the relevant knowledge is diffused throughout the organisation, a separation of functions among internal agents will increase efficiency. However, to delegate parts of the decision process creates an agency problem: relief from this problem requires the separation of decision management and control. It also requires the adaptation of this separation at every layer of the organisation, thus tending to create an organisational hierarchy, where decision initiatives are ratified and their implementation is monitored (and performance evaluated) from higher to lower levels. It seems to be the nature of the beast that decision control within a hierarchy will strive also to make use of informal mutual monitoring systems, and will encourage their development. An organisation of that kind will have an apex, usually a board of directors, which is set up in such a way as to adjust the separation of decision-making and decision control to the organisation's highest level.

In this respect, the formal decision hierarchies (supported by informal mutual monitoring mechanisms) common to complex organisations neatly fit our description of central planning. However, the role and characteristics of the residual claimant ought to differ substantially in the 'one firm' planned economy from the corporate shareholder. In addition, the role of residual claimants affects the separation of the decision process at the apex of an organisation. Let us consider two examples.

First, the common stock company, where residual claims are widely diffused. Boards of directors will usually have a strong presence of outsiders, maintaining the control function (with insiders often reduced to supplying the necessary information). More often than not, these outsiders are either compensated appropriately or their service is considered to enhance their reputational capital, thus keeping possible conflicts of interest at bay.

Not so with large professional partnerships, where residual claimants come from inside delegated to a board. But the boards are made up entirely of residual claimants. Thus, while their common interests (as well as liability) make for strong mutual monitoring systems, the partners also know enough about the organisation's activities and observe the board's activities, so that there is little demand for outside expertise. Below the apex, however, the functional division of the decision process is maintained. Another interesting observation relating to situations where the residual claimants are not diffuse and reside on top of the organisation's hierarchy, springs from the fact that their shares are not alienable: while this implies the absence of other control mechanisms, such as hostile outside take-overs or the correction mechanism of changes in stock prices, it does not at all preclude the possibility of inside take-overs by 'dissident partners' (Fama and Jensen (1983), pp. 316–17).

For obvious reasons, the literature teaches us little about residual claimants in situations where voluntary exchange may be limited and capital markets are absent. But the main points are worth summarising:

- In complex organisations, decision management and decision control will be separated. In addition to a formal separation, informal mechanisms of mutual control are employed. Both mechanisms rely on a hierarchical organisational structure.
- The more residual claims exist outside the organisation, the more likely that the apex of the organisation (the board of directors) will also consist of outsiders. The more residual claims are held within the organisation, the more likely it is that the board consists of insiders.
- The more complex an organisation, the higher will be the costs arising from restricting residual claims to internal decision agents, defined as the foregone benefits from separating the decision process, plus the foregone efficiency gains associated with risk sharing. The less complex an organisation, the lower the gains from separating decision management and control (and the gains from risk sharing).
- The more residual claims are restricted to internal decision agents, the higher the foregone efficiency gains associated with the benefits of risk-sharing. However, in complex organisations, this can be overcome by allowing for the separation of decision management and decision control below the apex.

THE SOCIALIST ECONOMY AS ONE SINGLE FIRM

It was V. I. Lenin who made the point that the socialist economy had to be 'built like a single firm'. For him, this involved a hierarchical organisation subject to a strict control. The organisation's purpose was the welfare improvement of the revolution's beneficiaries: the working classes. But control was to be exercised by the vanguard of these recipients, the Communist Party. As laid out in the concept of 'democratic centralism', decisions were to be initiated and proposed 'bottom up', that is, on the lowest levels of the (Party) hierarchy. Reflecting the true will of the (relevant part of) the populace, they were passed up along the Party hierarchy, subject to re-evaluation and improvement along the way, and finally ratified by the central committee. Implementation was up to the masses, again, but controlled by the Party, in the very interest of these masses. Neglecting estrangement, performance was to be monitored by true representatives of those with a claim to welfare gains.

Thus we have a picture of 'residual claimants' who are widely dispersed and therefore cannot exercise control directly. But they have a medium through which they can do so. And in the end, this medium (the Party) is in charge of ratifying and monitoring its own decisions: Lenin, it seems, attempted to reap the benefits of risk sharing, as well as those from keeping the decision process separate. But on the other hand, who is to complain if the party is a vanguard of Walrasian proportions – best informed about what will increase everybody's utility most?

More recently, the idea of the centrally-planned economy being seen as one homogeneous production unit has been re-examined, from quite a different angle (Olson, 1993). Olson describes the system, first of all, as an efficient tax-collector, with the claimants located on top of the (still hierarchical) organisation. The picture he uses is the shift from marauding to stationary bandits – who have realised that it is not efficient to burn down the income-generating hamlet, if it can be taxed instead.

Phrased in the most general terms of the principal–agent problem, one could characterise his view of the planned economy as a most lavish form of monopoly – agents running the show, principals deprived of all takeover possibilities, forbidden to trade their shares, and coerced into buying the companies' product! On a more serious note, it seems that Olson describes the organisational structure of what was earlier denoted as a non-complex system. Residual claimants and decision agents are not separated; moreover, decision management and decision control are in the hands of the 'claimants'. Stationary bandits, indeed, will not easily trade the right to initiate and implement their own ideas.

Textbooks usually begin their analysis of the Soviet system with a description of two parallel structures. Both depict charts of a hierarchical organisation, with few connecting arrows indicating influence running from one to the other. The first chart reflects the organisation of the Communist Party; the second the 'government' or 'government agencies'. The arrows are one-sided, indicating influence by the party on governmental agencies (for example, Gregory and Stuart (1986)). However, the implementation of the plan is ascribed to the government structure, from Gosplan to the all-Union, and down to the regional branch ministries.

Lenin's picture is not very useful: the organisation of society seems to have worked much more along the traditional lines of 'top down' than as an efficient transmission mechanism for all those defined as claimants of the Revolution. From Olson's scheme, on the other hand, it remains unclear why Socialism should ever have collapsed.[22] Textbooks, finally, do not make it their first priority to translate observation into content. However, to consider the Soviet economy as one entity designed to organise production, or 'one single firm', has the advantage that it provides a sensible base for understanding the origin of the present troubles.

We may start by defining this organisation as a complex system; clearly, the knowledge required to run the economy was diffuse and specialised, representing specific information. The undertaking also requires an enormous need for monitoring and information-gathering, and thus for control. The tale of two parallel hierarchies is no fiction either.

Rather, it represents the separation between decision control (the Party) and decision management (the planning bureaucracy). At this point, one needs to be aware that 'government' includes what is often loosely referred to as the planning bureaucracy. The planners created feasibility studies and proposals of alternative ways to use existing capacity and regulate intertemporal resource allocation. The Central Committee, the highest organ of the Party, made decisions based on these proposals. Decisions were implemented by the planning bureaucracy, at all steps of the hierarchy, from readjusting Gosplan to the enterprise level. The Party, however, which always ratified decisions (whether on a central, regional or local level) also stood ready to evaluate performance. To this end, it employed the official structure of the organisation but also a network of informal 'mutual monitoring devices' [*sic*!].

But who was the residual claimant? Party bosses, Gosplan bureaucracy, the working class?

Let us first consider the so-called '*nomenklatura*'. It included Party officials as well as bureaucrats, the military, and members of the 'scientific–technical complex'. The *nomenklatura* was not restricted to Party officials, as opposed to membership. In our terms, this group comes closest to the residual claimants in large, professional partnerships. At the top tier of the organisation, we find the separation between decision control and decision management to be forsaken. Below the apex, however, we find it restored.

The second candidate for the residual claimant might be more difficult to accept. But gestures to that effect should lead us to believe that there is such a role for the population at large, at least since the idea of a 'peaceful competition of the systems' (the policy of attempting to catch up in material wealth and per-capita consumption with capitalist economies) became the official policy under Nikita Khrushchev. Furthermore, in many CMEA economies, the party felt itself to be under pressure to legitimise the system via an increase in *measurable* welfare, such as per capita consumption (including social services and the allegedly superior security they provided). This might be modelled as derived interest of those at the top of the organisation. It might even be modelled as some twist in their preference structure. But in the terms developed here, it seems obvious that some proportion of output increases went to those manning the workstations.

What defines the population as residual claimants, however, is not that they were fed more than subsistence. Rather, it is the hypothesis that this share was *not* part of the precontracted fixed income, but part of the difference between fixed incomes (in real terms) and aggregate production, where the latter had a stochastic component. Given the nature of the system, one must, of course, substitute 'goods available' for cash flows in our previous definition. But the common person is then elevated into the risk-bearing category, and the kind of unpredictability implied by this statement was, I believe, very much part of the system.[23]

Risk-bearing agents without property rights; and residual claimants who are decision agents but need not fear 'unfriendly' take-overs by the outside world, other than by their own 'dissident' kind – where does this leave us?

First, it implies that the system has more internal rationality than is commonly supposed. Olson's maximum tax collectors, for example, fit the description only with considerable extensions. As it stands, his system would have collapsed long ago because the organisational needs of the stationary bandit seem to be tailored to non-complex systems. Without separating the decision process further, the bandit is capable of running 'his' gang, but not a complex economy.

Secondly, it may lead to an understanding of how the system could disappear in the way it did. What happened at the very top of the organisational pyramid may, in fact, be described as a take-over by a (not so dissident?) partner. However, it also helps to explain an intuitive reaction many may have harboured as events unfolded: namely, that the organisation's survival capability was low. It was not an efficient organisation. The claimants in the *nomenklatura* could not have, all other things being equal, handled their business in the same way partners that of a professional partnership would. In a capitalist firm, the separation of the decision functions below the board of directors is the result of common interest and liability, and reflects specific knowledge; in our case, the same features over time came increasingly to be dominated by the need for decision control, ultimately because they were not subject to comparable, competitive correction mechanisms. The comparison is instructive.

Most important in accounting for low survival value is the fact that the benefits of risk sharing do not become operative if residual claimants cannot alienate their equity. With respect to the population, 'imposed risk-bearing' created all the disadvantages that non-complex organisations experience if the function of risk bearing is restricted to decision agents, but without the benefits of a unified decision process. Lenin's vision, in fact, turned risk-sharing claimants into subjects of decision control.

Also, a complex system cannot function without a whole set of formal and informal arrangements, of institutions and behavioural codes. Insisting on this point is perhaps reiterating the obvious, but a better understanding of the role of institutions and institutional change, including the structure of production and the arrangements of the financial system, would certainly have done no harm in considering the sequencing of the transition.[24] It also includes the observation that people were not capable (or not willing) to adjust from the constraints imposed by one system to those imposed by the next as fast as is required, which is hard to understand other than as a tendency to defend old modes of behaviour *vis-à-vis* the unknown.

Finally, the attempt to cast the centrally-planned economy in single-firm categories, should be helpful in explaining details of the problems encountered during the first years of transition. For example, it answers the outstanding question posed at the beginning, namely whether a rationale for the assumption exists that knowledge at the industry and enterprise level is superior to that of the different branches of government. Information about the conditions of production were better at the firm level and along the lines of the (now dismantled) planning bureaucracy than at the centre and along the (old *and* new) political hierarchy. The reasons are connected to the observation that the economy was a complex system, with information specific and disbursed, on the one hand; and to a lack of incentive compatibility, on the other – assuming that increasing aggregate wealth, to the benefit of both groups of claimants, was indeed the goal of its proponents.

Involuntary risk-sharing ought to increase risk adversity, and in part this may explain the rate at which socialist economies fell behind market-çoordinated economies, as far as technological change is concerned. However, we would expect such an economy to operate below its maximum at any point in time, by misallocating resources to the control of economic activity, and by foregoing incentive structures based on contingency markets.

On the real side of the economy, we observe the tendency toward concentration, previously discussed, even if we do not allow for 'over-concentration' (in terms of the technological requirements), caused by minimising the costs of a large organisational hierarchy. In addition, the longer the time horizon, the larger the relative losses from not making use of dynamic improvements in parallel series economies, and from not 'discovering' new combinations, including advancements in the horizontal division of labour. Discoveries of this kind are communicated as the results of costly experiments. If withdrawing resources is measured in terms of costs to the aggregate, by focusing on the disruptions caused to the existing structure, they are likely to be higher than if measured as the opportunity costs for individuals. Aggregate costs will also increase with the complexity of the system, especially if the system carries a large overhead of control activities, while individual opportunity costs have no reason to increase. The system can thus be expected to exhibit relatively low, and perhaps declining, growth rates.

To substantiate these speculations, we have to look at the informational consequences of central planning. Starting at the lower end of the hierarchy, the involuntary bearers of risk have only limited incentives to report their activities accurately. Under 'normal' circumstances, alienable equity is subject to the law of one price, excluding systematic attempts to lie about its value. Here, nothing is in place to prevent what amounts to a classical fallacy of composition.

Take involuntary risk bearer *A*, a member of, say, a group of people leaving the factory every morning to repair houses, and involuntary risk-bearer *and* decision-controller *B*, in charge of reporting the activities of a number of such crews up the hierarchy. As the general rule, we may infer from the theory of optimal risk bearing (shirking and free-riding not withstanding) that *A* – having contracted voluntarily for ownership claims in his enterprise – will increase output eagerly, report trouble and, in short, behave consistently with his firm's interest. Every other *A* in the system should behave in the same way, since all have decided that potential rewards exceed opportunity costs. In our context, they ought to see an increase in aggregate output as their task. But this cannot be true for *A* and his colleagues: they did not so decide. They face opportunity costs, but their gains as risk bearers are not computable – because their shares are not alienable. Leaving aside performance-related incentives, they will weigh the opportunity costs only against detection and punishment. They cannot, to any significant degree, calculate the implicit rewards of contributing to an aggregate called material product.

This is true *even if* they fathom their status as residual claimants and expect, at the end of the day, the aggregate to be distributed among all of them, and thus its level to have wealth effects. The reason is not only what is commonly known: namely that they lack any control over the process by which it is produced, what is produced, or the distribution they face. The reason is that they have no voice in whether to participate, and not even to exit. Hence they have no idea of how many others, if they had a voice, would choose to exit. Under these circumstances, 'risk bearing' reduces to uncertainty, and the required optimisation cannot be performed.[25] That shares are not alienable renders involuntary risk bearers unproductive and distinguishes this system from both the enthusiastic literature on labour-owned firms or cooperatives, and the neoclassical literature on shirking and moral hazard.

Person *B*, on the other hand, may be aware of *A*'s problem and also of his/her everyday solution. S/he is a residual claimant too, and should therefore be expected to take an interest in *A*'s performance. But s/he will act subject to the same limitations. Being caught in a position where future claims resulting from efficient performance on his/her part cannot be calculated, and where he cannot trade and re-contract, s/he will quickly lose interest in trying to figure out how to improve the common wealth. There is no invisible hand in this system and, counting out ideology, no visible handshake either. Worse, the system in the raw state described here has no inherent checks and balances. There is no reason why the behaviour just described should not prevail throughout the hierarchy.

The first result is malperformance in a static sense: the economy will at any point in time operate below the optimum achievable, given its productive capacity and also the available human capital. The second result is a tremendous loss of information. Knowledge almost freely available at the bottom of the 'pyramid' will not make its way to the top unharmed.[26] Next to direct, performance-related incentives 'from above', further tightening of effective control seems to be the only remedy, implying an increase in the relative weight of that part of the organisation in charge of monitoring, and thus creating additional inefficiencies, without changing the fundamental constellation.

However, the need for high investments in monitoring capacity has other consequences, despite alienating the owners of non-alienable stakes, and besides representing a rather unorthodox approach to economic growth as far as the efficient use of resources is concerned.[27] This consequence relates to the decision agents on top of the organisation's hierarchy. The gathering of information and the upkeep of a monitoring system is costly, but the need to maintain it creates an additional constraint on accumulation: changing the mode of operation will involve the relocation of those parts of the hierarchy guaranteeing continuous control – and monitoring. 'Keeping the shadow' therefore increases the true costs of changes in the structure of production, as well as risk (and the uncertainty) associated with adapting the configuration of the capital stock continuously and investing in 'discoveries'. The obvious consequences play themselves out in three parts, all impeding growth.

First, economic decisions under uncertainty are likely to become more difficult to be ratified: to the danger of physically interrupting the integrated structure of production, trading definite disruptions for uncertain benefits, is added the costs of readjusting the relevant mechanisms of control.

Secondly, the partners making up the highest office of the organisation will come from different strata of the organisation they command. Hence, they may pursue different interests. It is conceivable that proposals, to change the structure of production, say, are considered too costly by the representative of decision control; or it is possible that s/he wants to build an exquisite factory in an unruly province, which may not make sense to the representative of decision management. Differences rooted in different interests can be considered normal, and their solution efficiency-enhancing if interests and liability are shared. Not so in our organisation: its charter clearly states that the Party has the final say. This peculiarity, however, does not imply that failure of the enterprise (for example, a hostile takeover (read 'revolution') by those involuntary carriers of risk and misery at the bottom) would not result in joint liability. Nor does it mean that the top layer of planning, government or other bureaucracies should not be considered to be residual claimants in the sense previously defined. A high probability that, if conflict arises, economic decision initiation and the preferences of those responsible for decision-management will finish a distant second, if the ability to control is at stake, does not *per se* involve giving up our classification. What it means, however, is that in the absence of correction mechanisms as provided by the market, behaviour in the long run will be shaped by anticipating this very order.[28] Certain options will not be made available any more because of the perception that they are not open to discussion. Limiting what might be called the 'perceived' opportunity sets will thus slow down the rate of technical change, but also hamper adjustments of the capital stock required to optimise performance continuously. The behavioural implications of the system's incentive structure will create a long-run problem for growth, which is hard to grasp under behavioural standard assumptions.

The third aspect under which changing the mode of production is burdened with changing the mode of control is the need to explore the horizontal division of labour. Any such attempt involves (spatial) decentralisation. However, decentralisation also replicates that part of the existing organisation responsible for maintaining control, thus causing costs for the aggregate to appear higher than they would in the idealised case, and again making economic changes more of a venture than necessary. This will tend to diminish the survival value in the long run, especially if severe disruptions occur, or if outside competition ever becomes important.

As to the informational structure inherent in that system, the answer can now be given in an non-ambiguous way. The availability and the cost of reliable information will both be a function of organisational distance from the centre. Moreover, the mechanisms of monitoring and control, relying largely on involuntary residual claimants, together with the situation at the top where a systematic bias against attempts at 'discovery' may develop, give grounds for the conclusion that valuable information about economic activity will be concentrated on the lower levels of decision management. The decision control apparatus, by virtue of its selective approach, will have less knowledge related to the economy and to means of increasing aggregate wealth in the long run – but, and for our purposes even more importantly, it will also have less knowledge related to sudden *changes* in the economy, a fact the institution is bound to find out only after being challenged in this respect.

As to the question of how this organisation will react if one of its functions is eroded – or what in fact happened – we can be equally brief. Two aspects are of interest: the impact the system left on human behaviour and the institutional structures it left intact, both underestimated by many observers, but of importance as soon as the transition was seriously under way.

The perhaps-not-so-hostile take-over of one partner and his faction directed efforts towards improving aggregate welfare at the two columns on which the organisation rested. Addressing the first column consisted, essentially, in an endless number of campaigns or appeals – to show more initiative, improve the range of proposals, make an effort to speed up the transformation by showing more creativity, but without, toppling the existing structure. Parallel with appeals to initiative came measures to give them space. These involved loosening the process of decision ratification, and then those of decision monitoring; first at the political level, then in the economic arena. After our discussion, it could be judged to be a dangerous way. Too little, too late seems to be the more common perception. Either way, it was not successful.

Instead of entering into a detailed discussion of why the apparently widened opportunity sets did not translate immediately into the hoped-for increase in economic activity, I would like to point towards a phenomenon which is often overlooked, though it seems so clearly visible. This is the behaviour of the living objects of our studies. There is not much indication that economic behaviour at the beginning of reforms in Russia adapted to these changes in any consistent way. If the top level of the hierarchy was confused, the average person certainly was in no better shape – even if s/he realised that something noteworthy was going on at all. What one could observe was people repeating old patterns of interaction and clinging to them. As the transition dragged on, this tendency remained. This principle of *pattern repetition* made it difficult for some of the changes sweeping the country to generate the expected results – all the more so, since most of these patterns (and the institutions they used) were not recognisable for Western observers.

For an economist, accounting for the latest episodes in the economics of transition, this observation manifests itself as a difficult choice. Trained to explain the world as the efficient outcome of the interaction of self-interested and rational people, and used to discovering the hidden social optima behind some seemingly hopeless constellation, one is sometimes tempted to forget the ground rules. The trouble is that not casting the plot in terms of rational choice seems to leave few operational alternatives. On the other hand, Scylla (or the emergence of inflation in Russia) may well be the story of the consequences it has if economists stick to their guns too closely.

SCYLLA

In many respects, the break-up of a system as we have just sketched it was equivalent to assigning true residual claims to many of the former decision managers, by means of a mass privatisation programme guaranteeing preferential treatment to insiders on the one hand, while dissolving the old structure of decision control, by simply leaving nothing to report, on the other. However, in the process, decision management's task was also changed fundamentally, with the constraints 'from above' being replaced by those 'of the market'.

The replacement of one constraint by another should, in principle, not constitute a problem. In Russia it did, for it was during that shift that high inflation started to accompany the decline in output. In terms of the Lipton and Sachs programme, the second step went wrong: corporatisation, in the sense of imposing hard budget constraints, did not succeed. The Russian inflation, it is understood, is fuelled largely by the continued subsidisation of enterprises by means of central bank credit.

The situation is different from ordinary discussions in that it is not clear whether the Russian economy can be subjected to the standard Latin-

American/IMF approach to inflation (imposing ceilings on the budget deficit and the monthly rate, and providing a stabilisation fund for compliance); or whether it exhibits peculiarities which would turn this cure into a recipe for disaster. Both opinions originate in the observation that inflation was triggered by state-owned enterprises' refusal to accept hard constraints, and their ability to undercut this measure by means of accumulating inter-enterprise arrears, redeemed by the Central Bank, to finance production.

At the time, published opinion, the Gaidar reform team, and many in support of the three-step proposal, depicted the situation as being part of a more-or-less elaborate plot by the old elite to sabotage reform measures, supported by the central bank and not countered strongly enough by the new government. Consequently, they predicted disaster when the Gaidar team left and Chernomyrdin came. Though admitting that the industrial structure might have played a role in mounting this plot successfully, this camp essentially defends the view that decisive and traditional anti-inflation measures are called for. To stop inflation means cutting the deficit. To put an immediate stop on inflationary financing constitutes the most effective way for the government to signal credibility: realistically, a balanced budget is, given the temptation to monetise a deficit, the only safeguard for further development.

The alternative view regards the fact that firms resorted to arrears to abort the attempt at hard budget constraints almost as an act of self-defence, given the structure of industry. By using an institutional device to which most Western observers were not very attentive at the time ('file number two', the inter-enterprise clearing account semi-automatically 'refilled' by Gosbank), the rapid explosion of arrears serviced by central bank credits ushered in very high inflation – but it also turned the central bank into the first resort to prevent output (and employment from collapsing. Ever since, the development of arrears and nominal prices move in opposite directions, suggesting a bargaining game between governmental attempts to impose financial austerity, and firms' attempts to resist the hurtful consequences of restructuring (and bankruptcy). To stop inflation cannot mean to eliminate subsidisation in one move. To scale inflation down gradually shows both success of the government in recognising the situation, and gains in credibility (also by scaling down the monetisation of the deficit). Realistically, a balanced budget will bury the little development there is under the debris of the collapsing old sector.

There is no need to repeat the arguments in any more detail.[29] The analysis presented here lends support to the second account. It rationalises the monopolistic industrial structure, fits in with the institutional details of the process and, most importantly, supplies a reason for why the central government – a remnant of the part of the hierarchy concerned with control, as opposed to the central bank – could have been expected to be less well informed about the difficulties created by changing the constraints imposed on the economy overnight, than the former planning bureaucracy. Such a reason is needed to support the game sketched in Figure 15.5. The horizontal axis shows the cuts in aggregate subsidies (maintaining all subsidies coincides with the origin). The vertical axis measures the number of firms (as a share of all subsidised firms) which adapt to corporatisation on the micro-level, by restructuring, and by going out of business. The shape of the functions reflects the tension between credibility and hard budget constraints: attempts at reducing subsidies will be credible, and therefore introduce restructuring, only up to a point; beyond that point, all players expect industry-wide bottlenecks to cause output and employment declines too large to be sustained. But the expectations of firms (*F*) and the government (*G*) differ systematically, leading

Figure 15.5 Subsidy cuts and firms' restructuring: divergent expectations ((a–b) firms will try to finance production with arrears)

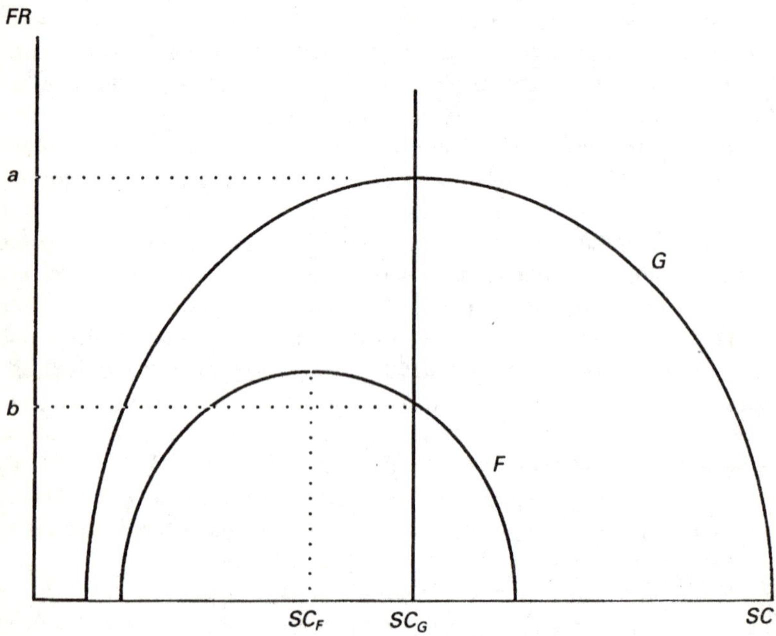

firms to resist the government's proposed cuts and to bridge the interim by issuing IOUs to each other. If firms' expectations are correct, they can force the government to renege. If the difference in expectations that has caused the government to systematically overestimate its capabilities can be explained, and hence forces it to readjust, the game has rational foundations, without any need for optimal solutions.[30] If not, a different behavioural explanation is required.

The analysis of the old system thus leads to a conclusion, supports certain policies, and in general, I hope, enhances our understanding of what has happened. It can give us an idea of how to capture behaviour under these circumstances. What it does not pretend to do, however, is to solve the problem of how to depict behaviour under extreme circumstances in a more general fashion.

CONCLUSION

Let me put the conclusion in a nutshell, by quoting two authors who phrased it better than I would. The first relates to the failure of socialism as we know it: 'An efficient social organisation, that is, one with a high probability of survival, never reduces error or experiment to zero' writes Demsetz (1988, p. 278). Markets serve as the playing field for errors and experiments, thus emerging as the ultimate training field for systemic survival in this age (up to a point, Karl Marx was right). Central planning, as long as it was exposed to outside competition, could not get round the conclusion that this basic training is not replaceable, however sophisticated the mechanisms of control.

The second relates to the failure of economic theory to realise the stakes more quickly: 'The rational expectant optimizers of today's standard theory do not need market interaction to teach them how to attain the efficient social outcome' writes Leijonhufvud (1993a, p. 12). The extremely simplistic rationality assumptions beyond the aggregate of mutually consistent decisions may in theory trigger the clockwork of intertemporal optimisation; in reality they did not. Here lies one of the reasons why a theoretical framework which, after all, is reducible to models of a central planner, failed to provide less than superficial advice to those scurrying away from the true legacy of central planning. Perhaps a little more trial-and-error in economic theory would have allowed for a little less trial-and-error in the transition facing a third of all people?

Notes

1. I wish to thank the editor of this volume, Kumaraswamy Velupillai, for both patience and helpful suggestions. I am indebted to Patricia Bartholomew for her help in compiling the data and editing the manuscript, and to Tatiana Nemeth for her relentless energy in discussing what, I know, are often strange topics. This chapter was finalised while the author was visiting the University of Western Ontario as a Bradley Fellow. I gratefully acknowledge the financial support of that Foundation. The usual caveat applies.
2. His term – cf. Leojonhufvd (1993).
3. And in rather extreme cases, proceeding to a scale depicting some measure of government intervention as a convenient yardstick for the degree to which socialism was *ante portas*. Kornai's original text, however (*The Economics of Shortage*) has to be excluded from this criticism.
4. In what follows, we concentrate on proposals made by the 'mainstream' of the (Western) economics profession.
5. For other timely discussion on the sequencing of reform, see, for example, Fischer and Gelb (1991) and Summers (1992).
6. One must always approach statistical sources in Eastern Europe with caution, for well-understood reasons. Yet, even if *half* of Russia's reported output decline is attributed to measurement failures, this still marks a substantially steeper fall than the one recorded for Europe during the Great Depression (see Kindleberger (1986), p. 279). The severity of the decline in industrial production seems to be confirmed, however, by statistics which indicate that the share of Russia's service sector exceeds that of the manufacturing sector for the first time (summer 1994), despite the fact that services are systematically underreported, being still to a large extent provided at the enterprise level (or by the new private sector, and not very well reported either).
7. Or 'The issue may be approached from either a theoretical or a historical angle' (Leijonhufvud (1985), p. 203).
8. Norway or Japan are often quoted examples for *B*-level countries, while life on the ground floor (*A*-level), of course, is reflected by economies where prices and quantities of literally millions of single goods were regulated by Gosplan (the textbook account stays at 23 million – see Gregory and Stuart, (1986)).
9. The existence of this account also helps to explain how a 'clearing house' could be established relatively easily in the summer of 1992. The popularity which 'Wechsel' more recently have acquired, on the other hand, reflects the need to finally come up with new financial instruments after 'file number two' was eliminated. Though tied to the old interfirm delivery channels at first, they have the potential to substitute actual payments due on a broader scale, for better or worse (for better, since, in theory, they reflect firm's assets and therefore may keep inter-enterprise debt from ballooning; for worse, if this feature is not enforced, that if Wechsel assume the role of additional means of payments, thus eroding any attempt at normalising the situation.)
10. To this should be added the arbitrary (politically motivated) spatial distribution of industries and firms throughout the former Soviet Union and the CMEA countries. The break-up of the empire thus had severe consequences

of the above nature for industries and firms in Russia, even before reform began in earnest. In the same way, the rouble zone as an attempt to hold the empire together while not abandoning the inter-republican re-finance practice amounted to the admission of multiple emission centres (the former subsidiaries of Gosbank in the republics) – which greatly accelerated the rate of rouble inflation, as it became rational for republics to maximise seignorage by outperforming each other.

11. Since no functioning price mechanism was in place, the difference may appear merely semantic. Yet it is not – the argument that the establishment of an equilibrium price vector is a sufficient prerequisite for curing the current malaise leads to policy conclusions different from an analysis which considers it necessary, but not sufficient. The latter questions the (global) stability properties of the old system, after central planning, independently of the established price vector.

12. We will not introduce new terminology and refer, in what follows, to the scenario in which the number of serial stages in the production process is unchanged, and the benefits of specialisation subject to the extent of the market, as 'static' (output levels are fixed) and 'dynamic' (output levels may vary) parallel-line economies, respectively. Extending the number of tasks, on the other hand, is referred to as the 'discovery' of new stages, or an improvement in the sub-division of labour. New products, as opposed to new processes, are discussed where the horizontal division of labour becomes an issue.

13. 'Akkumuliere, akkumuliere, das ist Moses und die Propheten!'. Marx had a very different understanding from Smith in this respect. The 'laws of motion' of his system relied on the efficiency gains realised by increasing the capital/labour ratio and imposed on the capitalist class by the competitive pressure to accumulate. Consequently, the rise of the organic composition was supposed to depict both: the rise of capitalism over feudalism as the result of a superior (more efficient) mode of production and its subsequent demise as the result of the 'law' of the tendency of the rate of profit to fall – both based on the very same unfailing trend of the capital/labour ratio to rise, caused by the individual capitalist's need to obey the rules of the game.

    It is indeed a fascinating question as to what extent Marx's early pupils might have got Smith's causality (from the extent of the market to increases in the division of labour to improvements in labour productivity), backwards by replacing them with the casual claim that, *pace* capitalism, increasing returns might (semi-automatically) follow from increasing the capital/labour ratio after the elimination of the profit motive – so that, as a tool of socialism, the latter would lead to increases in total factor productivity. It certainly would be of interest to learn whether, or to what extent, this 'ideological' component might have influenced early Soviet industrialisation. See Ehrlich (1967) for references.

    However, these and other plausible reasons for 'oversizing', such as the 'requirements' of Joseph Stalin's forced industrialisation programme will not be addressed here.

14. 'Capitalists unite' as long as the services of the machines they own leave them without alternative input sources and output customers. But having united, they can withdraw. Thus, in a competitive setting, the need to co-

ordinate activities will also tend to sub-divide specialisation across firms with prospects of a thick market, the prerequisite for sizeable firms (Alchian and Demsetz, 1972). Note that the question whether and to what degree increasing returns result in a non-competitive environment, and what the welfare implications are, is of no concern here. (We have already skirted it by assuming that an upper boundary exists for every output level, exhausting the gains from the vertical division of labour. The parallel series economies depicted here have asymptotically constant returns, if with a non-monotonic approach. They could, in other words, be modelled as a 'ghostly technical change coefficient' or as 'dents' in the production function, if one were to insist).

15. Information of the 'shoe-leather' variety, or as gains diminishing the transfer costs for commodities and people, are also conceivable, since information cost would again be depicted as an increasing function of the number of productive facilities.
16. 'Firm', in this picture, is a figure of speech, since state-owned enterprises are not surrounded by markets if they happen to produce primary or intermediary inputs.
17. What this statement implies is a restriction of the notion of technical improvement to the infamous 'book of blueprints' in neoclassical theory, or the choice of the 'best practice' technique in neo-Ricardian theory: the space for development is fixed at time zero in all its dimensions, because it is determined by the *choice* between *known* alternatives with respect to the location, alternatives, and the time path of production and consumption.
18. See Aghion, Blanchard and Burgess (1993) for a discussion of the problem of 'unbundling' diversified production units, organised functionally within large conglomerates.
19. It thus also provides grounds for reviewing the design of international assistance to strengthen outsiders' means of corporate control. See Frydman and Rapaczynski (1993) and Intriligator and Rühl (1994).
20. In general, residual claimants are those contracting for net cash flows which are subject to residual risk, as opposed to agents who face fixed or performance-related pay-offs. 'Residual risk', then, refers to the 'risk of the difference between stochastic inflows of resources and promised payments to agents'. We shall use this definition throughout (Fama and Jensen (1983), p. 302).
21. A prime example is Demsetz's (1988, 1993) hypothesis that alleviation of the control problem in market economies is related to a specific distribution of wealth in the economy as a whole. In situations in which the distribution of wealth is also peculiar – but almost egalitarian – Demsetz's productivity enhancing 'wealth effect' becomes relevant for the design of industrial privatisation, that is, for the discussion of insider *versus* outsider privatisation (Rühl (1994)).
22. This is assuming that it did collapse, while all that happened was, perhaps, that the last of the bandits (in the Soviet Union) made wrong decisions as how to increase the tax base. But I am willing to share the proposition that it would have imploded someday, provided gradual transformation (along Chinese lines, for example) was excluded – and as long as capitalist economies were allowed to be present.

23. The industrial organisation literature, for good reasons, will be vague on residual claimants without property rights. To address the planned economy as a 'single firm' results in terminology problems, if familiar language is not to disappear altogether. In the present case, residual claims are assumed to be distributed also to the powerless inhabitants of 'Gosplan Inc.' (insiders so to speak). 'Aggregate production' must then be substituted for 'inflows with a stochastic component'.
24. This point is drastically underscored if one compares the standard (Western) sequencing of reforms with the one blueprint published in the former Soviet Union (the 'Shatalin', or five-hundred-day plan). The latter almost reverses the sequence of reforms in an attempt to connect the path of the transition to the institutional initial conditions.
25. In reality, they will have an idea that many others would choose not to be claimants, if they could somehow exit. This may make the uncertainty less genuine, but does not change the substance of the conclusion. Rational expectations not withstanding, to *withdraw* from the sort of calculation will be their best bet.
26. A good example is the confusion sometimes encountered as to whether there was a tendency toward systematic 'over-' or 'under-reporting'. It depends: given that the overall goal on the part of the unwilling risk-sharers (that is, the population at large) can almost be described as keeping the individual level of productive activity as low as possible, it makes sense to understate at some occasions (when the new plan is discussed) and to overstate at others (when reports are due).
27. According to some estimates, the East German Secret Service employed almost one-tenth of the adult population in one form or other – spying on an opposition where every member prevented a huge proportion of potential output from being produced (and, with all these B'''s monitoring B''s monitoring B's, also representing a case of systematic under-reporting in the service sector).
28. This holds for firms in a monopolistic or oligopolistic environment, too – but it seems fair to presume that the competitive pressures caused by the 'competition of the systems' was far less than is conceivable in these cases.
29. The positions are discussed in more detail in Leijonhufvud (1993*b*), and Rühl and Serwin (1994).
30. The expectations of firms and the government will converge, as they learn, but even the agreeable optimum wil be characterised by the kind of subsidisation denied in the standard proposals. The need for subsidies is bound to decline as a new private sector emerges and the old one becomes more competitive. All this, of course, should not be taken to imply that subsidies are to be financed by seignorage.

References

Aghion, P., O. Blanchard and R. Burgess (1993) 'The Behavior of State Firms in Eastern Europe: Pre-Privatization', Working Paper, No. 12, European Bank for Reconstruction and Development.

Alchian, A. and H. Demsetz (1972) 'Production, Information Costs, and Economic Organization', *American Economic Review*, vol. 65, pp. 777–95.

Demsetz, H. (1988) 'The Control Function of Private Wealth', in H. Demsetz, *Ownership Control, and the Firm. The Organization of Economic Activity* (Oxford: Basil Blackwell).

Demsetz, H. 1993 'Enterprise Control, Wealth and Economic Development', in Demsetz, *The Economics of the Business Firm* (Cambridge: Cambridge University Press)..

Ehrlich, A. (1967) *The Soviet Industrialization Debate, 1924–28* (Cambridge, Mass.: Harvard University Press).

Fama, E. and M. Jensen (1983) 'Separation of Ownership and Control', *Journal of Law and Economics*, vol. 26, June.

Fischer, S. and A. Gelb (1991) 'The Process of Socialist Economic Transformation', *Journal of Economic Perspectives*, vol. 5, Fall.

Frydman, R. and A. Rapaczynski (1993) 'The Structure of Assistance for Russia', unpublished MS, CEU.

Gregory, P. and R. Stuart (1986) *Comparative Economic Systems* (Boston, Mass.: Houghton Mifflin).

Intriligator, M. and C. Rühl (1994) 'The Design of Industrial Privitization and the Consequences for International Assistance in Russia', UCLA, February.

Kindleberger, C. (1986) *The World in Depression, 1929–1939* (Berkeley, Calif.: University of California Press).

Leijonhufvud, A. (1985) 'Capitalism and the Factory System', in R. Langlois (ed.), *Economics as a Process* Cambridge University Press.

Leijonhufvud, A. (1993a) 'Towards a Not-Too-Rational Macroeconomics', *Southern Economic Journal*, vol. 60, April.

Leijonhufvud, A. (1993b) 'The Nature of the Depression in the Former Soviet Union', *New Left Review*, vol. 199, May-Jun, pp. 120-6.

Leijonhufvud, A. (1994) Commencement Address Delivered at the Central European University, Prague, August.

Lipton, D. and J. Sachs (1990a) 'Creating a Market Economy in Eastern Europe: The Case of Poland', Brookings Papers on Economic Activity, (Summer).

Lipton, D. and J. Sachs (1990b) 'Privatization in Eastern Europe: The Case of Poland', Brookings Papers on Economic Activity, Spring.

Olson, M. (1993) 'Why is Economic Performance Even Worse After Communism Failed?', unpublished MS, University of Maryland.

Rühl, C. (1994) 'Privatization and Corporate Control', Papers in Political Economy, No. 49, University of Western Ontario, London, Canada.

Rühl, C. and K. Serwin (1994) 'The A, B, C's of Industrial Restructuring in Russia', *Eastern European Series*, no. 10, Institute of Advanced Studies, Vienna, June.

Summers, L. (1992) 'The Next Decade in Central and Eastern Europe', in C. Clague and G. Rausser (eds), *The Emergence of Market Economies in Eastern Europe* (Oxford: Basil Blackwell).

Name Index

Note: This index covers persons both as authors and as subjects.

Subject Index